Soap, science, and flat-screen TVs

Soap, science, and flat-screen TVs

A history of liquid crystals

David Dunmur
Tim Sluckin

OXFORD
UNIVERSITY PRESS

OXFORD
UNIVERSITY PRESS

Great Clarendon Street, Oxford OX2 6DP

Oxford University Press is a department of the University of Oxford.
It furthers the University's objective of excellence in research, scholarship,
and education by publishing worldwide in

Oxford New York

Auckland Cape Town Dar es Salaam Hong Kong Karachi
Kuala Lumpur Madrid Melbourne Mexico City Nairobi
New Delhi Shanghai Taipei Toronto

With offices in

Argentina Austria Brazil Chile Czech Republic France Greece
Guatemala Hungary Italy Japan Poland Portugal Singapore
South Korea Switzerland Thailand Turkey Ukraine Vietnam

Oxford is a registered trade mark of Oxford University Press
in the UK and in certain other countries

Published in the United States
by Oxford University Press Inc., New York

British Library Cataloguing in Publication Data

Data available

Library of Congress Cataloging in Publication Data

Data available

Typeset by SPI Publishing Services, Pondicherry, India
Printed in Great Britain
on acid-free paper by
CPI Antony Rowe, Chippenham, Wiltshire

ISBN 978–0–19–954940–5

1 3 5 7 9 10 8 6 4 2

Contents

Preface

————◦◦◦◦◦————

LCD TV is a short-hand description that will be recognized by many, but understood by few. Modern methods of communication demand shorthand *txts* and acronyms, and these come close to constituting a new form of language. It is for others to decide whether such developments enhance or diminish the efficacy of communication. But, whatever the message, the medium through which this new language is transmitted and received is, more often than not, an LCD: a liquid crystal display.

This is the entry point for the general public to liquid crystal science, but for the authors it represents just one outcome of more than a century of scientific investigation into a remarkable form of matter. Our title of *Soap, science, and flat-screen TVs* is a whimsical description of the breadth of liquid crystal science, which indeed embraces soaps and LCDs, but also includes living cells and new materials.

Our inspiration, like that of many others, derives from the Greeks. It was Aristotle who divided up the natural world into earth, water, air, and fire. The first three of his elements survive as exemplars of our more rigorous description of states of matter as solid, liquid, and gas. This aspect of Aristotelian science has survived until the present day, while it was only a couple of hundred years ago that Aristotle's fourth element of fire as a separate substance was finally abandoned.

This categorization of the material world into gases, liquids, and solids lasted as an adequate representation of the universe until the beginning of the twentieth century. Then an age of enlightenment began to dawn, and it become clear that things were not quite so simple or clear cut. The three classical states of matter no longer sufficed.

Now, if you search the internet, you will find the fourth state of matter described as a plasma—a gas of positively charged atoms in a sea of electrons. Do not be deceived. This is simply the gaseous equivalent of a metal, and it is not very common. On earth, plasmas can be found in the glass envelopes of fluorescent lights and in the picture elements of some large-area television sets. Only in the heavens will plasma be commonly encountered, in the form of the northern (or southern) lights, or in stars, galactic trails, and other astronomical phenomena, but on our planet, there are *no* free-standing plasmas.

Liquid crystals, on the other hand, are indeed a terrestrial fourth state of matter. We shall present the scientific story of liquid crystals in this spirit. They take their place naturally in the scale of things, being something like solids and something like liquids, but different in essence from both. As such, liquid crystals contribute to many aspects of our world, but for the present generation of consumers, liquid crystals will be recognized as the materials of displays. Mobile phones, cameras, iPods, laptops, TVs, cars, as well as all manner of domestic and scientific equipment all employ a liquid crystal screen to display information or images. The ultimate manifestation of the liquid crystal display, or LCD as it has become known, is the large-area flat screen colour television. Today the LCD is ubiquitous and has had a similar impact on human life as the discovery of the wheel. Small wonder that we think liquid crystal displays are important.

Liquid crystals were missed by the Greeks because they do not readily proclaim their existence. They can be identified only by those with skill and experience, but they are abundant in nature. In particular, liquid crystals are to be found in living things. Animals and plants consist of cells, and cells function because they are surrounded by a liquid crystal membrane. It might not always have been so, but the formation of liquid crystal membranes was a crucial step in the evolution of cellular life. Our liquid crystal scientists made the connection between liquid crystals and life at a very early stage. Indeed some of them thought they had discovered the secret of life. It was not to be, but, as we will show, liquid crystals are nevertheless an important component of living organisms.

The science of liquid crystals began more than a century ago with a baffling observation of two melting points in a single pure substance by a German botanist (Friedrich Reinitzer). The botanist passed his results to a physicist (Otto Lehmann), who realized that he was studying something interesting, but misidentified the physical phenomenon as crystalline in origin. Despite his misidentification, Lehmann gave us a name that stuck, *liquid crystals*, and our subject was born. Later, a combative French crystallographer (Georges Friedel) realized that the materials were not crystals at all, but special liquids, albeit exhibiting some properties akin to those of crystals, and put the subject on a firmer footing.

The early story of liquid crystals involves botanists, chemists, physicists, mineralogists, and mathematicians, mainly from Europe, who pursued their interests through a turbulent period of history involving wars (the Great War and World War II) and revolutions (the birth of communism and the rise of fascism). The events of history took their toll on our scientists (Vsevolod Konstantinovich Frederiks published and perished in one of Stalin's gulags), but did nothing to stem their fervour or determination to promote their ideas.

In common with many other physical sciences, liquid crystals benefited from a post-war burgeoning in both fundamental science and applications. The former resulted in a Nobel Prize for one of the scientists involved (Pierre-Gilles de Gennes), while the latter led to the birth of the multi-billion dollar display industry, and fame and fortune for other players in the story of liquid crystals.

The origin of our interest in the history of liquid crystals goes back to an inaugural lecture given by one of the authors (TJS) in March 2000, celebrating his election to the Chair of Applied Mathematical Physics at the University of Southampton. The title of this lecture ('Fluids with attitude—the story of liquid crystals from oddity to technology') was an attempt to review the origins of the science of liquid crystals for a general university audience. A specific goal was to show how a multibillion dollar industry developed from Reinitzer's esoteric observations on compounds ostensibly extracted from carrots, but a secondary objective was to explore the personalities of the scientists involved.

A transcript of the original lecture has been published (with additions by DAD) in Portuguese, under the title *Fluidos fora da lei* (*Fluids outside the law*). We are grateful to the translator of that work, Dr Paulo Teixeira, for advice and encouragement in the preparation of this book. In 2005 the present authors, together with the German liquid crystal expert and science historian Professor Horst Stegemeyer, published a source book of the scientific history of liquid crystals entitled *Crystals that Flow*. That book was a collection of 46 classic papers on liquid crystal science spanning the period 1888 to 1980, a number of which had been translated into English by the authors for the first time. Accompanying the reprinted papers were five chapters of commentary by the authors to explain the background to the science and scientists, and set the work in context.

The present volume represents a natural evolution of our desire to communicate the excitement of liquid crystals to a wider audience. A volume such as this has required input from many sources. A principal source has been our earlier book. The many colleagues who provided advice and guidance concerning material for that volume have been recorded and acknowledged in the preface to that publication. We should also record our thanks to academic colleagues with whom we have interacted over the years, and who have helped to shape our understanding of liquid crystals.

In this book, the authors have cast the net wider, and it is a pleasure to thank additional colleagues who have helped us to access and understand more esoteric material. First and foremost amongst these must be our co-author of *Crystals that Flow*, Horst Stegemeyer. Indeed it was Horst's enthusiasm and knowledge that caused us to embark on this historical quest in the first place. Professor Stegemeyer and colleagues have established a permanent exhibition on liquid crystals at the Liebigs Museum, Giessen, under the auspices of the Deutsche Bunsen Gesellschaft. The material in this collection has been made fully available to us, and we are most grateful.

Another lasting tribute to the pioneers of liquid crystals is to be found at the Company Museum and Archives of Merck Darmstadt (Germany). Our especial thanks go to Dr Sabine Bernschneider-Reif, Head of Merck Corporate History, who made arrangements for a study visit. Not only did we have free access to the archives, we were able to interview former and present liquid crystal scientists from Merck who have played such an important role in the development of liquid crystal materials for displays and other applications.

Several chapters of the book owe much to primary research by Dr Peter Knoll in Karlsruhe. Dr Knoll's researches on the letters of the liquid crystal pioneer Otto Lehmann have been published privately in German, and will soon be available to a wider audience (Knoll and Kelker, 2010). We are grateful to Dr Knoll for giving us access to his unpublished work, explaining to us how he came across the material, and providing us with copies of early papers in German not available through the British library system. Finally he was good enough to read over our treatment of his primary material. We emphasize nevertheless that the responsibility for errors lies with us. Access to key German literature on the pre-war chemistry of liquid crystals was kindly provided by Professor Dr H Remane (Liepzig) and Professor Dr Holger Stark (Berlin).

We also thank in particular Professor Jacques Friedel in Paris for extensive correspondence on his grandfather Georges Friedel. He provided us with original material annotated by his grandfather, which gave insights into Friedel's views on the scientific controversies in liquid crystal research around 1930. Another primary source for material on the Russian physicist Vsevolod Konstantinovich Frederiks was the book in Russian by Anatoli Stepanovich Sonin and the late Viktor Yakovlevich Frenkel, published by Nauka (Moscow, 1995). We are grateful to Professor Sonin's son, Dr Andrei Sonin, for acting as intermediary in our correspondence with Professor Sonin, and also to Professor Igor Pavlovich Pinkevich (Kiev, Ukraine) for some assistance in translating relevant material from the book by Sonin and Frenkel. Finally, for material on the German–Czech–Brazilian physicist Hans Zocher, we thank Professor Luiz Evangelista for uncovering valuable sources, and in particular our thanks are due to Dr Alda Espinola (Brazil) for her recollections of Zocher and for directing us to relevant primary material.

Other important primary sources for the book were the Archives of the Royal Institution of Great Britain, and we are grateful to Professor Frank James, Head of Heritage and Collections, for his assistance. We also are grateful to the Special Collections Department of the New Bodleian Library, Oxford. The latter houses papers from the Society for the Protection of Science and Learning, which gave us the harrowing story of scientists under Nazi rule. We are grateful to the Council for Assisting Refugee Academics for permission to quote from some of the documents under their custodianship.

Many individual scientific friends and colleagues have given their advice and assistance on aspects of this book. Our thanks are due to Professor Lev Blinov and Valentina, the widow of Professor Victor Titov, for providing information and allowing the tale of Titov to be told. The understanding advice from Professor Wolfgang Weissflog and colleagues from the University of Halle on post-war liquid crystal research behind the Iron Curtain is appreciated. We are also most grateful to Wolfgang Helfrich for personal revelations on the invention of the twisted nematic display, the device that stimulated modern liquid crystal display technology. Our thanks are due to Dr John Lydon for his comments on our description of some biophysical aspects of liquid crystals, and for providing figures to illustrate his particular speciality of chromonic liquid crystals. We were alerted to new perspectives on liquid crystal membranes and disease by Dr Emil Poetsch, who kindly provided us with background information, for which we are also most grateful.

Many of the illustrations for the book have been taken from other sources, and we are grateful to all those for permission to reproduce figures and photographs. A separate list of acknowledgements for these illustrations is given at the end of the book.

Finally, our personal thanks are extended to Mike Caunt, talented graphic artist and son-in-law to DAD for reinvigorating the project at a critical stage, and for translating the incoherent descriptions of a scientist into pleasing images that hopefully help to tell the story of liquid crystals. DAD also wishes to thank his wife Juliet for reading and commenting on the entire manuscript, more than once! TJS wishes to thank his wife Celia, his children Ben and Rachel, and his mother Alice Sluckin OBE for their patience and encouragement in a second long literary programme. He is hopeful that by contrast with the previous literary ventures, this volume will be sufficiently transparent that they will be able to appreciate its contents.

We hope that this book will represent progress in bringing together material on the history of liquid crystals in one volume, and in a manner which is also accessible to readers outside the field. However, we do not claim here to have produced a definitive history of liquid crystal science. Indeed, this volume is not the first (and will not be the last) to attempt a history of liquid crystals. For other material we refer the interested reader to the publications listed at the end of the book.

Southampton,
December 2009.

Science and history: the two cultures

A note to readers

Our world has been transformed over the years by the progress of science. The primitive human of a hundred millennia ago interacted with raw nature. Only at the far ends of the earth, in the forests of New Guinea and South America, or in the deserts of Southern Africa and Central Australia, do a very few descendants cling to traditional ways of life. The rest of us, however, live a sheltered life, shielded from loneliness by our mobile phones, from disease by modern medicine, from predators by ancestors who killed them, from cold and heat by modern houses and electricity; in short by a modern culture made up from a myriad of technological advances.

Everywhere we look, we see the fruits of science and engineering. Until recently, a sensible maxim might have distorted Marx's mildly contemptuous remarks* about philosophers: the scientist has reinterpreted the world, but it is for the engineer to change it. The slogan seems oddly out of place now. Whatever the benefits to humankind of any particular scientific advance, there is always the threat of the unanticipated greater disastrous consequences. While we in the developed world currently enjoy unparalleled plenty, the spectre of pollution, the end of the age of oil, and anthropogenic climate change hang over us. If the scientists have made the revolution, somehow, always, the citizen and society must pick up the pieces.

It was the British novelist C.P. Snow, in his now famous Rede Lecture half a century ago, who coined the term 'the two cultures'.[1] One culture was that of the general intellectual class of the age. This was the culture of literature, of history, of music, of all things that entitle a person to invitations in the salons of the leisured classes. The second culture was that of the scientist and technologist. Snow was drawing attention to what he saw as the ever-enlarging intellectual gap between the two. No scientist, he implicitly pointed out, could be taken seriously

---

* Karl Marx, in his 11th thesis on Feuerbach (1888), famously stated that 'the philosophers have only interpreted the world; the point is to change it'.

in intellectual circles without, for example, familiarity with Shakespeare's plays, Goethe's writings, or Verdi's opera.

And yet, indeed it is the case that the *culture* of the world has changed as a result of science and technology. Where was the reciprocity, he asked, which would require those trained in the humanities to have a basic understanding of key scientific concepts? The literary class, he alleged, mocked science and technology as the province of those with high specific, but very limited general, intelligence. The very activity, they thought, would prevent the development of a general understanding of the world around us.

Snow chose as a metaphor for the lack of scientific culture within our ruling class ignorance of the Second Law of Thermodynamics. The second law is not just an arbitrary choice. Snow could have chosen Newton's Laws of Motion, or Maxwell's Equations of Electrodynamics, or Schrödinger's Equation in Quantum Mechanics, or Darwin's laws of Natural Selection. They are all key pillars upon which modern science stands, and without which modern technologies would struggle. If you are interested in modern science as a cultural construct alone, then a familiarity with any of these will serve as a qualification for membership of the elite.

But the second law is special. It was developed in the nineteenth century, at the interface of science and technology. The specific idea was to give a scientific basis for the steam engines which were powering the European Industrial Revolution, at that time in full flow. The everyday meaning of the word *energy* (ultimately derived from the ancient Greek word for work) was narrowed down and given a specific meaning by science. Work in this context does not mean toil or employment, it is a technical term applied to mechanical or indeed electrical action; it is the release of energy to do something. The First Law of Thermodynamics[2] was that this 'scientific' energy was *conserved*, i.e. if you lost it somewhere, it popped up somewhere else, and that one of the somewhere elses was in the form of *heat*. The Second Law is that you can't just turn heat into work.

There are many equivalent statements of the Second Law of Thermodynamics, and it was perhaps unfortunate that Snow chose this particular example for a scientific religious truth. Indeed it has been suggested by the popular science writer Peter Atkins that Snow himself may not have fully understood the Second Law.[3] Heat flows spontaneously from a hot body to a cold body. For example, if you put your cold wet sweater in front of the fire, it gets dry and warm. The fire does not heat up, and the water in the sweater does not turn into ice. A refrigerator works by reversing this spontaneous flow of heat, and transfers heat from cold objects inside the fridge, making them colder, to the hot surroundings, making them hotter. The second law says that this can only be achieved by doing work; the refrigerator requires energy to power it.

And more energy is required than the equivalent amount of heat transferred from the cold object to the hot surroundings. All this is not just experience, not just common sense, but scientific law.

Snow's novels were not always appreciated. His Rede lectures are sometimes read as part of a polemical exchange with his Cambridge colleague, the acclaimed literary critic F.R. Leavis. Distinguished as he was for the quality of his literary criticism, Leavis was not at all well-versed in scientific matters, a fact that Snow was eager to point out in the context of some withering remarks on the quality of his literary output. Snow is best known for a series of novels entitled *Strangers and Brothers*, a saga which follows a working-class boy through a scientific training to a civil service career in the corridors of power.

The working-class boy is, of course, a thinly veiled version of Snow himself. As the series progresses, Snow allows himself the luxury of reflecting on some of the problems of intellectual life. One of his novels, *The Affair* (1959),[4] dwells on the case of a brilliant young researcher who is accused, wrongly as it turns out, of scientific plagiarism. As an ambitious scientist, he is faced with a difficult dilemma. On the one hand his scientific career requires results, howsoever obtained, and on the other, the whole process requires that the results be obtained properly, for otherwise the process is simply meaningless. The storyline grips the reader precisely because of this perceived human conflict, which transcends the merely scientific.

Over the intervening 50 years, popular interest in scientists, science and the greater scientific endeavour has vastly increased. Illustrious biographers flock, often with great success, to work their art on the scientifically eminent, relating in some sense the history of science. Mostly, however, as a necessary function of the competences of both readership and writer, the account will omit any detail of the key technical contributions of its subject. Unfortunately, without these contributions the distinguished subject would merely have been the proverbial traveller on a Clapham omnibus. A history of science without science is like a book on theology without religion. There is no substance.

The authors of this book have spent their scientific careers, more than 30 years each, as journeymen scientists. We have been busily polishing postulates and testing hypotheses. We have spent long hours in the laboratory, in front of the computer, and perhaps most of all in quiet contemplation. In so doing, we have contributed, albeit in a very minor way, to the development of the current view of what has come to be known as *liquid crystal science*. In this book we shall develop the history of our art—or of our science, for in some sense it is the same thing. In so doing, we shall address many of the issues raised by Snow in 1959. Our narrative lies somewhere between history of science and popular science. Sometimes we shall stand in an area more accessible to the specialist. Where we do this, we have endeavoured to explain technical concepts in a manner that can be understood by a non-specialist.

For many scientists and engineers, the focus of their life's work never comes to public attention. It is our fortune, (whether good or bad is for others to say), that over the last 30 years one of the fruits of our collective activity has been the development of flat screen displays. Indeed, nowadays both public and private spaces seem to be dominated by these flat screens. They are no longer ugly and cumbersome like the old-fashioned TV with which the previous generation grew up. Rather, they are sleek and svelte, they hang on walls, and they hide themselves in a flash. Artists create whole exhibitions in fashionable spaces exploring the consequences of what might be called a *genre*. Through the screens, information and images are transmitted, efficiently, effortlessly, and instantaneously, from all corners of this planet. The liquid crystal display or LCD is seemingly everywhere, arriving apparently from nowhere in a short decade.

Liquid crystals, as we shall see, are much more than just flat screens, notwithstanding the fact that it is the flat screens that have drawn them to public attention. The tale we tell starts with a botanist investigating carrots. Then the story moves through a variety of themes with apparently little application, where the driving force is only that of scientific curiosity. It is almost half a century before there is even a single patent issued with the words 'liquid crystals' included. It is almost three-quarters of a century before work begins with a flat screen theme. It takes more than a century before liquid crystals begin to appear in complex devices such as TVs or computers. Meanwhile ideas based on liquid crystals are applied to soap, to living cells, to drug delivery, to armour used in warfare, and at the same time set a standard for ideas in mathematics, physics, and chemistry. Even if the next generation of flat screens no longer uses liquid crystals (which we doubt, but it could happen), there is a future for liquid crystal science.

We draw the attention of readers to the long time-lag between the scientific invention and the engineering application. Liquid crystals were not developed in order to create a flat-screen technology. Indeed when flat screens were first mooted, liquid crystals were not the obvious solution to the problem. Moreover, when liquid crystals were invented, the very idea of a screen (for cinematographic projections) was only just being developed.

Scientific funding agencies are keen on providing a link for the taxpayer between the money spent on scientific research and a technological payback to the citizen, but because historically the payback time is so long, it is right to urge caution for any very specific measures of return. For example, no bomb-making programme alone would have developed the horrific technology which destroyed Hiroshima and Nagasaki. On a happier note, no imaging programme alone would have developed X-ray imaging, never mind magnetic resonance imaging, and investigations of the chemical structure of compounds found in carrots are not an obvious place to start a programme on flat screens. But that is how it was.

There is some ambiguity in the task set to those who transmit our scientific culture outside the realm of academia. Some popular books about science merely transmit a sense of wonder at the range of phenomena with which the natural world can present us, but otherwise give the impression that there is little to choose between science and magic. Others ape the textbooks which we professionals use to train our students, but somehow go coy at the key points at which it is possible indeed to distinguish science from magic. There is a difference, however, between science and magic. The difference lies in the *process*, in the fact that concepts are confirmed by experiment and by precise mathematical reasoning. Thus, naturally, we are drawn toward an examination of the manner in which the new ideas have developed.

The investigation of the history of a scientific discipline is a profound intellectual task in itself. The discipline will have an internal dynamic, as small steps are gradually piled, one upon the other, to develop a bigger picture. For example, Galileo developed laws of motion which work, in heaven as on earth, for ever and ever, but he could not have done so unless the ancients had gazed endlessly at the sky and classified its working, unless Copernicus had reinterpreted Ptolemy's observations of planetary motions in terms of the earth's motion around the sun, unless Johannes Kepler had carefully shown that the observations were most consistent with elliptical planetary orbits around the sun with certain specific properties, unless...[5]

We must also consider the external social and political environment in which the scientist worked and, more specifically, the detailed way in which the scientist was influenced by this environment. Both Galileo and Archimedes, for example, were employed by political leaders in a military capacity because the projectiles launched under their tutelage did more damage than those launched by those less mathematically talented. Galileo's observations of four religiously incorrect moons of Jupiter famously led him into conflict with the Pope and the Catholic Church. In this case, the short-run effect was to enforce a retraction by Galileo, but the longer-run effect was more on the Catholic Church. For in science, at least sometimes, there really is a right and a wrong interpretation, and political authorities that end up on the 'wrong' side will end up hampering the efforts of their own societies to succeed.

And then finally there is the question of the individual talents and personalities of the scientists in question. J.D. Watson, for example, in his brilliant *Double Helix*,[6] tells the story of how his rivalry with competing teams acted as an engine which drove forward his scientific research. Some[7] have even construed his words as an admission that he improperly used data from a competing (woman) scientist. It appears that he persuaded himself into believing that Rosalind Franklin could not interpret the data herself, and further even persuaded himself that his inability to relate to her was in some sense her fault. There is perhaps an irony that it is this very story which has rehabilitated Franklin's scientific reputation, albeit unfortunately posthumously.[8]

Likewise, the movie treatment of mathematician John Nash, adapted from Sylvia Nasar's biography[9] *A Beautiful Mind*, concentrates on Nash's schizophrenia. The gripping theme of both book and movie is the contrast between Nash's scientific brilliance and his otherwise incompetent manner of dealing with ordinary human situations.

Popular media treatments of scientific subjects, even in otherwise quite serious publications, sometimes find themselves unable or unwilling to deal with deeper themes of intellectual or social history. As a result, they tend to concentrate on these personal issues. The scientist under study is odd, or brilliant, or is gay, or struggles against some family problem, or has an unpleasant rival with some personality disorder. In some sense these factors then 'explain' the progress made by particular individuals. 'Keep it personal' runs the motto, lest the audience lose concentration.

In this book, the reader will find elements of all these caricatures. We do not have a fixed ideological view about how to present science to the non-expert, or about which type of historical explanation is valid and which is not. We are, after all, neither professional historians nor professional writers, but professional scientists. Knowing our material, we have drawn eclectically from all the traditions mentioned above.

In particular, notwithstanding our evident scepticism about the utility of explanations which concentrate too much on the personal, and our snobbish addiction to serious themes, in truth we are just as addicted to gossip as anyone else. The reader will find a good deal of personal detail in our story, detail of how individuals came to their discoveries, what drove them, how they behaved when they were famous, and so on. For us, at least, although it has explained little about the global development of the subject, it has provided a personal link with those who founded our subject and whose names decorate the phenomena with which we work every day.

Before we pass to the real story, it is worth giving suggestions to our readers on how to read this book. A history of liquid crystals without any science would be impossible to tell, so we need to provide some explanation of various concepts, techniques, and experiments. To equip all our readers with a basic tool-kit of scientific understanding, we have included a series of technical boxes. Equations and mathematical rigour have been omitted, and no doubt our explanations will deserve the criticism of our expert readers and colleagues. Our only plea is that we have tried!

For the non-expert, science is difficult, more difficult than scientists usually recognize. Someone trained in science has an innate understanding of the natural world. Atoms, molecules, the nature of light, the meaning of magnetism and electricity, the force of gravity, the existence of the solar system, and the motion of the stars are all familiar, but not for the non-scientist. We try throughout the book to provide the reader with enough of the science that they can follow the narrative. In particular, the whole of the first chapter concerns science in the pre-liquid-crystal age. It explains, using a historical introduction, the world view of the physicist and chemist in the late nineteenth century just at the time that liquid crystals were first discovered.

Inevitably our explanations cannot always succeed. We ask readers who at first find our explanations lacking to be tolerant. Just carry on with the story, which we hope will grip you in the same way as it gripped us. You can come back and fill in the scientific details later. Finally, for the impatient, or for those (like us) whose senior moments overwhelm them more often than they would like, there is a glossary of scientific terms at the back of the book.

For us the story of liquid crystals, in all its aspects, is so compelling and important that we want everyone to understand sufficient of the science to share our excitement and wonderment.

1
It's all Greek to me: an introduction

⎯⎯⎯⎯⎯◦◦◦⎯⎯⎯⎯⎯

> If you cannot—in the long run—tell everyone what you have been
> doing, your doing has been worthless.
>
> Attributed to Erwin Schrödinger, theoretical physicist
> who shared the 1933 Nobel Prize for physics with Paul Dirac

This is the story of liquid crystals, a form of matter that is neither a liquid nor a solid. The scientific essentials are that liquid crystals are a new addition to our concept of the natural world, which traditionally is divided into three states of matter described as solids, liquids, and gases. The discovery of liquid crystals starts with an improbable investigation into extracts of carrots, and a persistent investigator who wouldn't take no for an answer. It was not long before attention focused on the remarkable optical properties of these strange materials, and there began a heated scientific war of words, which lasted for more than two decades. Scientific battles are not like other human conflicts: there is no blood and the only things likely to be broken are reputations. However, with the passage of time most of the protagonists emerged with some distinction. Above all, they were scientists, and eventually they reluctantly accepted the force of experimental evidence. What few of these pioneers could have foretold was the huge importance of their work for humankind.

As the story unfolds we move from speculation and controversy to acceptance and scholarly endeavour. But an unpredictable mix of political and scientific events opened the eyes of a few to the possibilities of liquid crystals for displays. The enormity of the change of 'mind-set' is hard to appreciate. All the accepted views of modern technologies had to be turned upside down—and this took some doing. Those who embraced the new ideas most readily were the emergent economies of Japan and the Far East. They were the winners, and they successfully exploited the science to bring the modern revolution in displays. Where would we now be without our laptop computers, made possible only by liquid crystal displays. There is a vast array of information and communication devices that all have a liquid crystal screen somewhere. Liquid crystal displays provide the interface between humans and machines. The screens convey information and increasingly allow an operator to act on that information, to initiate a new enquiry, respond to a question or even correct a malfunction.

But that is not all. The remarkable properties and behaviour of liquid crystals were seized upon as the answer to the mysteries of life. Not surprisingly, some allowed their enthusiasm to overwhelm scientific caution. Liquid crystals are not alive, but even now in the most recent of biological research we are beginning to realize that liquid crystals are indeed essential for life, and their remarkable properties may help in repairing human tissue and curing disease.

Many consequences of science, familiar to us now, were anticipated in the annals of fiction. The emergence of science fiction was itself a landmark in the development of human perception. No longer was science in the same category as magic or belief. By the nineteenth century there was sufficient confidence in 'science', or as it was then known natural philosophy, as 'truth', or at least partial truth, that fictional authors could start pushing the boundaries of reality. A new genre of literature emerged, and one of its early exponents was the American writer Edgar Allan Poe (1809–1849). These days, liquid crystal scientists like to speculate that Poe, through some remarkable insight, had imagined the liquid crystals of the future.[1]

In Poe's novel* *The Narrative of Arthur Gordon Pym of Nantucket*,[2] the hero voyages to a strange island where all is not as it seems. Happening upon a stream on the island, the eponymous hero investigates:

> *I am at a loss to give a distinct idea of the nature of this liquid, and cannot do so without many words. Although it flowed with rapidity in all declivities where common water would do so, yet never, except when falling in a cascade, had it the customary appearance of limpidity [ed. transparency]. It was, nevertheless, in point of fact, as perfectly limpid as any limestone water in existence, the difference being only in appearance. At first sight, and especially in cases where little declivity was found, it bore resemblance, as regards consistency, to a thick infusion of gum Arabic in common water. But this was only the least remarkable of its extraordinary qualities. It was not colourless, nor was it of any one uniform colour—presenting to the eye, as it flowed, every possible shade of purple, like the hues of a changeable silk.*

---

* The literary works of Edgar Allan Poe did not make a great impact during his lifetime. However, the iconic science fiction author Jules Verne (1828–1905) was sufficiently inspired by Poe's work to write a sequel to *The Narrative of Arthur Gordon Pym of Nantucket* entitled *Le Sphinx des Glaces*, translated as *An Antarctic Mystery*, one of many accounts of exciting journeys written by Verne.

Poe did not attach a name to his magic fluid, but its properties are very similar to a certain type of liquid crystal. Liquid crystal scientists have recognized the description,[3] particularly of the optical properties, as being those of a cholesteric liquid crystal, i.e. derived from cholesterol (see Figure 1.1). What Poe's hero is describing is of course entirely fictional, but the changing hues of the liquid and the effect of flow can be reproduced by a form of liquid crystal (cholesteric, or chiral nematic to give it its scientific description), but at the time of publication of Poe's work of fiction in 1838, the discovery and characterization of cholesteric liquid crystals were still more than 50 years in the future.

In describing the imaginary fluid, Poe makes reference to its unusual flow properties, as well as the dramatic optical properties, both signatures of real liquid crystals. In a further passage, the structure of the fluid is also described as consisting of veins, and again thread-like structures are often seen in certain types of liquid crystal. However, we can be certain that there was no scientific foresight in the novel of Poe.

Before we embark on the voyage of discovery traced by liquid crystals, it may be helpful to provide some background to the journey.

Figure 1.1 A droplet of cholesteric liquid crystal exhibiting the hues of a changeable silk. (See colour plate 1.)

States of matter

The distinction between solid, liquid, and gas is evident even to the non-scientist, and the earliest theories of matter in ancient Greek times emphasize the physical form of matter (under the guise of earth, water, and air) over its chemical basis.

Chemistry, by contrast, only slowly developed in the seventeenth and eighteenth centuries out of the medieval art of alchemy. Atomic ideas had been around since Democritus and Leucippus in the fifth century BC, indeed *atom* comes to us from the Greek word for indivisible, but originally the justification was more philosophical than based on evidence.[4] It seemed possible to break up some materials into simpler constituents, and the basic materials, i.e. those that could not be broken down further, came to be known as *elements*. The key experimental advance was made by the English chemist John Dalton (1766–1844). He established that chemical compounds came from constituent materials in definite proportions. His seminal work, *A New System of Chemical Philosophy*,[5] was published in 1808. This gave strong, but not yet decisive, evidence that matter consisted of basic indivisible particles which Dalton labelled as atoms, or compound atoms,* now known as molecules. How big the atoms were, or what other properties they possessed, was in 1808 anyone's guess.

Similarly, by the end of the nineteenth century a basic understanding of mechanics had already developed. Materials came in three basic families: solids, liquids, and gases. Liquids and gases flow, and change their shape to fill a container. The Scottish theoretical physicist James Clerk Maxwell (1831–1879) developed in 1866 basic laws which connected the motion of atoms (or molecules) in a gas to its pressure, temperature, and volume.[6] This work, leading to the so-called *kinetic theory of gases*, was the culmination of studies by a number of mathematicians and physicists going back more than a century.[7] Maxwell, of course, stands as a colossus amongst theoretical physicists,[8] comparing probably with only Archimedes, Newton, and Einstein. His work over the period 1855–1856 explaining the relationship between light, electricity, and magnetism led to a true change in perception of the natural world. But if the behaviour of gases was becoming clear, what was going on in a liquid was less obvious.

The modern theory of liquids is essentially due to the work of Dutch theoretical physicist Johannes Diderik van der Waals (1837–1923), described in his doctoral thesis in 1873. He supposed that liquids were simply dense gases and the molecules must be attracting each other at long distances, but repelling each other at close ranges because of their finite size. The long-range attractions, then introduced in *ad hoc*

---

* Combinations of atoms in a particular proportion, which we now refer to as molecules.

fashion, but whose fundamental origin is now understood, have come to be known as van der Waals forces. The thesis was entitled *On the Continuity of the Gas and Liquid State*,[9] and so impressed his contemporaries that, according to legend, the great Maxwell himself learned Dutch in order to understand it. Although the ideas were coherent, however, there were no numbers, as distinct from estimates, as to atomic sizes or masses.

Van der Waals' work rested on experiments carried out by the Irish chemist Thomas Andrews (1813–1885).[10] His 1869 Bakerian Lecture,[†] to the Royal Society of London, entitled *On the Gaseous State of Matter*,[11] established the idea of a *critical temperature*. Below this temperature a distinction could be made between gas and liquid. Here, as the pressure is increased by decreasing the volume, so the density increases as well, with a sudden jump from gas to liquid at a particular pressure known as the *saturation vapour pressure*. At temperatures above the critical temperature, however, the density just increases gradually as the pressure increases, and there is no difference between a liquid and a gas.

Later on it turned out that the idea of a critical temperature could be generalized. Sometimes different liquids mix easily to form a uniform mixture (alcohol and water, for example). Sometimes, like oil and water, they do not. The heavier liquid settles in a more or less pure state at the bottom of the container, and the lighter floats to the top. In fact it is more complicated than that. Some pairs of liquids mix at high temperatures, while at lower temperatures they do not, with a critical temperature in between. Usually (although not always) the separation occurs at low temperatures and the mixing at higher temperatures.

Whereas fluids can change shape to fit the container, the shape of a solid is largely fixed. A solid can sometimes be deformed, usually within limits, and on release of the deforming force the solid will return to its original shape. In the nineteenth century, it was said that solids possess *displacement* (nowadays we would say shear) elasticity, whereas liquids do not. We will find that liquid crystals also possess a rather special form of elasticity, different from solids, which firmly distinguishes liquid crystals from liquids, which cannot be elastically deformed.

Many solids are crystals, but what exactly is a crystal? Originally the word comes from the Greek word for ice and is related to the word for 'to shine'. Crystals shimmered and shone. Single crystals appear to possess a definite shape, often appear to be transparent and to reflect light at surfaces. The surfaces are important. They reflect light like a mirror because they are flat. The crystal appears to exist only with certain forms of surface because if you shatter the crystal and look at the

---

[†] The Bakerian Lecture is the premier named lecture of the Royal Society. In fact, this particular 'lecture' was never actually delivered, but only published, as pointed out by Rowlinson (2003).

fragments under the microscope, they appear to be smaller versions of the original with similar flat surfaces. So important are these surfaces in crystallography that they acquire a name: crystal *facets*. The faceting of crystals is exploited to great effect in fabricating jewellery.

The current picture of crystallinity in terms of a periodic structure of atoms, i.e. a three-dimensional network, is due to the French crystallographer René-Just Haüy (1743–1822). Although this picture had gained wide acceptance by the end of the nineteenth century, it did not yet rest on firm foundations and the evidence was all indirect. Indeed a vigorous debate was in full swing between the Austrian theoretical physicists Ludwig Boltzmann (1844–1906)[12] (in the materialist camp) and Ernst Mach (1838–1916) (in the idealist camp).[13] Mach,[14] remembered for relating speed to sound as Mach 1, 2, etc., was an idealist and contended that atoms were simply a convenient way of thinking, a metaphor, but to ascribe to them a material existence would be an error. Boltzmann, however, the materialist, asserted that the atoms were every bit as real as his pen or his blackboard. Unfortunately the microscopes of the day were unable to settle the dispute: they could not distinguish single atoms.

Some materials seemed to confound the classical categorization into solid, liquid or gas. There are many substances that defy a simple description as solid, liquid or gas: mud, toothpaste, and cream are examples of, well, what? Some such preparations can be identified as 'colloids', and the distinguished Scottish chemist Thomas Graham (1805–1869) is generally reckoned to be the founding father of colloid science. A colloid is a material in which microscopic particles of one state of matter (solid, liquid or gas) are suspended in another.* The microscopic particles are large enough to be seen under a microscope, but small enough that the naked eye only sees goo. It is to Graham that we owe many of the important terms in colloid science, including the appellation itself, *colloid*, from the Greek word meaning glue. This story is mainly about liquid crystals, but we shall find that again and again the colloid story runs along a parallel track converging and interweaving with liquid crystal tales.

Scientific revolutions[15]

The Ancient Greeks first contemplated what we now think of as physics and chemistry under the term 'philosophy'.[16] There was then a long intellectual hiatus while the western world concentrated on other matters

---

* A common example of a colloid is vinaigrette: one liquid suspended as small droplets in another liquid, with which it is immiscible, known as an emulsion. A fog or aerosol is a dispersion of a liquid in a gas, while a sol is a dispersion of finely divided solid in a liquid. All combinations are possible.

of human importance (mainly praying and killing). The rebirth of interest in intellectual matters in Europe is usually known as the *Renaisssance* and dates back to the fourteenth century. The start of the scientific revolution is often dated to the English philosopher and statesman Sir Francis Bacon (1561–1626).[17] Unlike his near contemporary Galileo Galilei (1564–1642), who laid the foundations for the modern understanding of mechanics, Bacon is not credited for the brilliance of any breakthrough he might have made. Rather, it is to him that we owe the explicit insight that in order to make progress in understanding the world, you have to observe it, not just think about it, as was the traditional view of philosophers. You have to look, and think, and look again, and think again, and still keep on looking. For our quest, looking was the key to success.

What does 'looking' actually mean? The *ordinary person* looks through his or her eyes and 'sees' something happen. The *scientist*, on the other hand, uses whatever is available to probe his object of study in order to find out what is inaccessible to the normal senses. At the outset of our story, the only 'scientific instrument' available was the optical microscope, but as we shall see the understanding of the strange new state of matter we now know as liquid crystals increased with every new addition to the scientific toolbox. Still, the optical microscope was the key to unravelling the mysteries of liquid crystals, and even today is an important instrument for study. No liquid crystal laboratory worth its description is without a microscope. We need to know something about what the microscope can reveal, and that means understanding something about light itself.

Light

The Romans possessed glass lenses. Medieval painters record the existence of spectacles at least from the mid-fourteenth century. Spectacles only correct sight that is known to be deficient (even if one cannot be sure exactly why). A single lens, a magnifying glass, can enlarge an object or concentrate the rays of the sun. When one combines lenses in a tube, depending on the length of the tube and the strength of the lenses, one constructs either a telescope or a microscope. Now it was possible to see that which was previously invisible.

Galileo constructed both types of scopes. He is best known for his work with telescopes, during the course of which he discovered moons revolving around Jupiter. However, the title 'Father of microscopy' is usually assigned to the Dutchman Anthonie van Leeuwenhoek (1632–1723).[18] To him we owe the discovery of protozoa (single cell organisms), bacteria, spermatozoa, and muscular fibres, all objects whose small scale makes them inaccessible to the naked eye. The microscope is a practical device, and one does not need a sophisticated understanding of optics to use it. It just works. Whenever what it tells you is checkable with the naked eye, the naked eye verifies its correctness. There is no reason to doubt its veracity.

Other scientists were curious about the nature of the light itself. Coloured light was of course familiar to the ancients: fire was usually red or orange, the sun yellowish, the moon a bright white. The difference between coloured light and coloured objects was less clear. Painters had rules of thumb about combining primary paint colours (red, blue, and yellow) on the palette in different proportions to make any desired hue. It was Sir Isaac Newton (1642–1727) who used a prism around 1665 to split sunlight into constituent colours. Paint colours and light colours are different: colours on the palette combine to make black, but different coloured lights combine to make white. Newton's prism separated out white light to make a spectrum, in just the same order as the colours in a rainbow.

What sort of substance was the light? The Dutch mathematician, physicist and astronomer Christian Huygens (1629–1695) put forward the hypothesis that light was a wave, and developed a whole mathematical theory to justify this point of view. Newton, by contrast, was convinced that light consisted of particles, which he called *corpuscles*.

There followed a vigorous debate, which was eventually resolved, but only temporarily, in favour of the wave theory. Somewhere along the way, it also became clear that light travels at a finite speed, rather than infinitely fast, as had previously been believed. The finite speed is, of course, very, very, very, *very* fast: about 300,000 kilometres per second. In fact, the speed of light depends on the medium through which it is travelling, and a consequence of this is the process we call *refraction*. Viewing the bottom of a swimming pool from the side, the water appears to be less deep than it actually is. This is because the light wave travels more slowly in water than in air, and changes direction, is *refracted*, at the interface of the water and air. It appears to an observer on the air side as if the light has travelled from a lower depth than it actually has.

The wave theory prevailed for two main reasons. First, it can certainly explain refraction, but so also could the corpuscular theory. More significantly the wave theory provides a simple explanation of the phenomenon known as *diffraction*. The key feature of waves is that they can *interfere* with each other. If you have two sources of light, then there may be places where the heights of the waves from the two sources add up (and the light would appear bright) and places where they subtract (and so the light would appear dark). Corpuscles do not do this.* They always add up.

---

* Modern physics teaches us that both the corpuscular theory and the wave theory of light are acceptable, and both theories provide satisfactory explanations for optical phenomena. The corpuscles are now known as photons, and the two descriptions of light as streams of photons or travelling waves are just one example of the *wave-particle duality* that underlies the current understanding of fundamental particles.

TECHNICAL BOX 1.1 The properties of a light wave

Frequency and wavelength

We represent light as a wave, and the wave is the amplitude of its associated electric field. There is also an associated magnetic field at right angles to the electric field—hence the description electromagnetic radiation. In space (i.e. a vacuum) or in a material, the light wave travels with a characteristic speed, so that the crests of the waves move forward, like the ripples on a pond or the waves at sea. There is a theoretical maximum to the speed of light, which is the speed (or more correctly, velocity) of a light wave in vacuo: 299,792,458 metres per second.

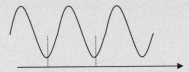

The distance between two equivalent points on the wave is the wavelength (λ) and the number of full oscillations per second is the frequency (ν). The colour of the light is determined by its frequency. The wavelength and frequency are connected by the speed of light (c): $c = \nu \times \lambda$

Optical refraction

Refraction is the process whereby light changes direction when it passes from one material to another. The material quantity that determines how fast the light travels is the refractive index.

The **refractive index (n)** is defined as the ratio of the speed of light in vacuum divided by the speed of light in the medium. In a vacuum n equals 1; the greater the density of the material, the larger the refractive index, and the more the light slows down. The frequency (i.e. colour) stays the same, but the wavelength decreases.

light wave in a vacuum   light wave of same frequency in a dense medium

The English polymath Thomas Young (1773–1829)[19] sent light of a single colour from a single source through two extremely thin parallel slits and thence onto a far screen. The result was a pattern of light and dark stripes, just as predicted from the interference of light waves from the two slits, a phenomenon now known as diffraction. This experiment also established the wavelength of the light as very small, about half a micron, where a micron is one ten-thousandth of a centimetre.

TECHNICAL BOX 1.2 Optical interference

If two light waves with the same wavelength are shifted with respect to one another, then they can 'interfere'. The resultant wave intensity is the sum of the two overlapping waves. If the waves have shifted by exactly half a wavelength, then there is destructive interference and the net intensity is zero—this appears as black. If the waves are shifted by exactly a whole number of wavelengths, then there is constructive interference and this reinforces the intensity and appears bright. If the light contains many frequencies (i.e. white light), then depending on the refractive indices of the medium and the path-length of the light in the medium, interference may only occur for particular wavelengths, hence the appearance of 'interference' colours, for example in thin oil films on water or sometimes in soap bubbles.

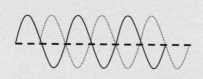

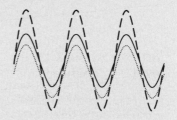

Destructive interference - dark   Constructive interference - bright

In the figures, the dotted line represents the resultant intensity of the two interfering light waves.

When light from an object passes though some material (water or a crystal or glass) on the way to your eye, its apparent position is changed. We see an image of the object, but not where the object actually is: this is the refraction effect. The Danish scientist Rasmus Bartholin (1625–1698) discovered in 1669 that refraction is more complicated in Iceland spar. This is a particular crystal type of the mineral calcite, otherwise known as calcium carbonate, familiar in another form as simple limestone. If light from an object passes through Iceland spar, there are *two* images, rather than the single image that normally appears. When a light beam enters a crystal of this material, it is split into two refracted beams. The material shows an effect we describe as *double refraction* and the crystal is *optically anisotropic*.

So how could a *single* incident light beam give rise to *two* refracted beams? The clue to this came in an observation by the French physicist Étienne Malus (1775–1812). In 1808 Malus was studying how the sun reflected from the surface of an Iceland spar crystal. To his surprise the intensity of the reflection seemed to vary with the angle of reflection. Malus deduced that a beam of light contained a quality perpendicular to the direction of travel, which he called polarization.

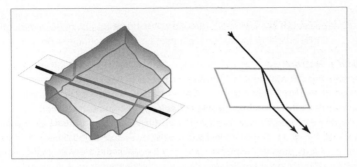

Figure 1.2 A crystal of calcite (Iceland spar) placed on a line shows two images.

TECHNICAL BOX 1.3 Polarized light, double refraction, and birefringence

A light wave has two directions associated with it. These define a plane which contains the direction in which it is travelling and the direction of the electric field associated with the light wave. This is known as plane-polarized light, e.g. the light wave shown below is plane-polarized in the plane of the paper.

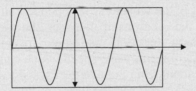

For many crystals the speed of light depends on the direction that the ray travels through the sample, but it is not just the direction that is important, it is also the polarization of the light that determines the time of passage of a light wave through certain samples. We have seen the dramatic consequence of this with double refraction in calcite crystals. For unpolarized light incident at an angle on a calcite crystal, two images are formed because two rays of different preferred polarization pass through the crystal with different speeds. Because of the laws of refraction, the rays propagate at different angles and so emerge as different images on exiting the crystal, i.e. the crystal exhibits *double refraction*: it appears to have two refractive indices. If the incident light is polarized, then it is possible to switch between one ray and the other by changing the plane of polarization of the incident light. In addition, by changing the orientation of the crystal with respect to the light, the double refraction can be maximized or caused to disappear altogether.

Thus whether or not you see double refraction depends on how the crystal is illuminated. On the other hand, regardless of the illumination of the crystal, it appears to have two refractive indices. The technical description is *birefringent* (i.e. possesses birefringence), the *bi* denoting two refractive indices. Some crystals, and even a few liquid crystals, have three refractive indices. In general

the passage of light in a particular direction with a defined polarization through a
crystal or liquid crystal depends on a combination of the refractive indices.

Optical anisotropy

Anisotropic means different in different directions. We use the term to describe
the properties of materials that have different values when they are measured
along different directions. The concept of anisotropy can be understood by
considering a piece of wood. The wood has different strengths along the grain
and across the grain. For the most part anisotropy in properties is restricted to
crystalline solids, and this was the origin of the controversy over liquid crystals,
since they also are anisotropic in many of their properties. It Is anisotropy in
optical properties that is central to the story of liquid crystals.

Optical anisotropy means that the refractive index, and hence the speed of
light, is different in different directions. The two images in Figure 1.2 occur
because the transmitted light rays with different polarizations have travelled along
two different directions in the crystal at different speeds. Because of refraction it
therefore appears that the light originates from two different sources, although in
reality the light rays come from the same source.

Malus developed his serendipitous observation of the sun's reflected
waves into a whole mathematical system. He was able to show that the
velocity of the light in the Iceland spar depended on its polarization, or
more precisely on the relationship between the direction of the light
beam, the polarization direction, and some special directions in the
crystal. It is this difference in wave speed which causes two differently
polarized waves incident from outside the crystal to be refracted at different angles. Nowadays we use the terms *ordinary* and *extraordinary*
for the two different refracted beams. Malus's work was well-received,
so much so that in 1810 he received a prize for his work from the French
*Académie des Sciences.** His tragically early death in 1812 was a result
of complications from the plague, which he had caught while serving as
an engineer with Napoleon's army in the Middle East.

Polarized light turns out to be important in sciences ranging from
astronomy through physics to biology. The polarization direction of light
from outer space is rotated by galactic magnetic fields. The polarization
of light from the sky betrays compass directions, even on a cloudy day.

---

* The polarized waves rapidly attracted attention from other scientists. As early as
1811, Malus's close colleague, the Catalan François Aragon (1786–1853), showed that
crystals of quartz rotate the plane of polarization of a polarized light beam; such materials were said to be optically active. Depending on the type of quartz, sometimes the
rotation went one way (*dextrorotation*, to the right) and sometimes the other (*levorota-
tion*, to the left). In 1815, their colleague Jean-Baptiste Biot (1774–1862) showed
experimentally that even liquid solutions containing certain dissolved compounds, often
of biological origin, were optically active; it was not necessary to have a crystal.

Humans and other mammals are only slightly sensitive to polarization. All *we* see is brightness and colour—bees, ants, and squid are sensitive to polarized light, which gives extra information about the environment.

An important technological advance in polarized light was made in 1828 by the Scottish physicist William Nicol (1770–1851). Using a special kind of prism made from two single crystals of Iceland spar glued together, he invented a device which separated 'unpolarized' light into two perpendicularly polarized rays. One of the rays was then redirected away from the original beam direction, while the other continued on the beam direction. This was the first polarizer, which produced a beam of light linearly polarized in a particular specified direction. The device soon came to be known as a Nicol prism. It was widely used in nineteenth century scientific experiments, so widely indeed that the name was shortened simply to a 'nicol'. Nowadays many experiments still require specially polarized light, but there are other ways of generating polarized light, most commonly by using a Polaroid sheet, which transmits just a single polarization of light.

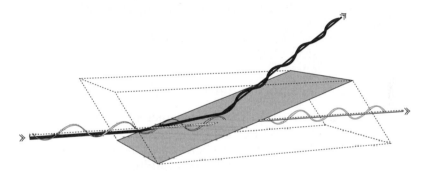

Figure 1.3 The Nicol prism, showing transmission of the extraordinary ray and diversion of the ordinary ray.

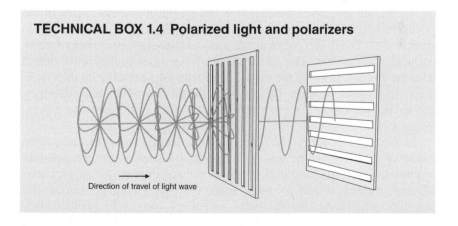

TECHNICAL BOX 1.4 Polarized light and polarizers

Direction of travel of light wave

For ordinary light in an optically isotropic material, or in a vacuum, there is no restriction on the direction in space for the associated electric field oscillation, except that it is at right angles to the direction of travel. This is unpolarized light. It is possible to select a single polarization for the light by using an optical polarizer. This only allows light of a particular polarization to pass through. If a second polarizer is placed in the light path, having its admittance direction at right angles to the first, then no light is transmitted by the second polarizer. These are referred to as 'crossed polarizers', and this arrangement is the defining feature of a polarizing microscope. In the condition of crossed polarizers, in the absence of a suitable sample, the field of view in the microscope is black, i.e. no light is transmitted by the second polarizer.

If an optically isotropic sample, such as a liquid or some crystals such as common salt, is placed between the crossed polarizers, then the field of view remains black. If, however, an optically anisotropic sample (refractive index varies with direction and polarization of the light) is placed between the polarizers, then depending on the orientation of the sample with respect to the axes of the polarizers, the field of view will be illuminated. Sometimes colours can be seen due to interference or black brushes corresponding to destructive interference. The colours of the viewed image map the variations of refractive index with position in the sample.

Liquid crystals—carrots to flat-screen TVs

Now we can begin the story. It will commence in late nineteenth century Prague with a botanist fascinated by carrots. His studies focused on extracts of carrots: compounds of cholesterol that will strike a chord with the health-conscious of today. However, his observations on the properties of these new materials defied explanation and he was forced to seek help. Help eventually came from a German physicist with a microscope looking for a problem.

Several decades later, and after many scientific battles, the topic of liquid crystals became established in the world of science. Most of the scientists involved miraculously survived the Great War, but a new generation then faced the political tensions and social upheavals of communism and fascism. The scientific story becomes inextricably linked with the human story of the times, with tales of persecution, imprisonment, and untimely death. Despite this, liquid crystal science advances, and by the middle of the twentieth century things are ready for a great leap forward.

In fact, the progress of research and development to flat-screen liquid crystal televisions was less a series of leaps, more a maze-like journey through the jungle of research and commerce. As applications loomed, money reared its ugly head. Not surprisingly, commercial conflicts developed. Applications scientists fought to protect their inventions at the same time as attempting to persuade industry to take up their ideas. Some were successful, others fell by the wayside. Nearly 100 years after experiments with carrot juice, a liquid crystal television emerged and today they are to be found all over the globe. But this is not the end: future applications for liquid crystals continue to emerge in medicine, optical communications, and new materials. This is the story we wish to tell.

2
Crystals that flow: fact or fiction

$\Longrightarrow\!\circ\!\circ\!\circ\!\Longleftarrow$

There are certainly such things as soft crystals, I am happy to concede flowing crystals, but liquid crystals, never!

G. Tammann's remarks to the 1905 Annual Meeting of the German Chemical Society, University of Karlsruhe

The double-melting liquid

The foundation of liquid crystal science is traditionally set in the year 1888, with the work of Friedrich Reinitzer (1857–1927; Figure 2.1). Reinitzer is commonly termed a botanist, although in modern terms he would perhaps be thought of more as a biochemist. He was 30 years old at the time and assistant to Professor Weiss at the Institute of Plant Physiology at the German University of Prague.

Nineteenth century Prague was the capital of the province of Bohemia in the Austro-Hungarian Empire, although in earlier centuries Bohemia had enjoyed periods of independence. What is generally known as the Charles University in Prague was founded in 1347 by the Holy Roman Emperor Charles IV. It is (or rather, as we shall see, was) the oldest German-speaking university in Europe, predating the foundations of the universities of Vienna in 1365 and Heidelberg in 1386.

The late nineteenth century was a time of great political ferment in Bohemia, as political pressure for Czech home rule within the Austrian Empire gathered in strength. In 1882 the Charles University was split into independent German and Czech sections, each following studies in their own language. A majority of students chose the Czech section, in keeping with their political aspirations, but in many ways it was the German section, in which Reinitzer worked, which continued the unbroken traditions which were by now more than 500 years old. Academic staff members included the physicist and philosopher Ernst Mach, who was professor of physics from 1867 to 1895, and, most famously, the young Albert Einstein, who spent 17 fruitful months in Prague in 1911–1912.

Figure 2.1 Friedrich Richard Kornelius Reinitzer (1857–1927).

If Prague is an unsurprising place to start a technological revolution, the subject matter which starts this revolution is strikingly unexpected. For Reinitzer was obsessed by carrots. More precisely, his experiments involved extracting cholesterol from carrots in order to determine its chemical formula, which at that time was unknown. He thought (wrongly as it turned out) that cholesterol was chemically related to carotene (the red pigment) and thus to chlorophyll. Cholesterol had been detected in plants and in the cells of many animals, and Reinitzer was keen to find out if the cholesterol from carrots was the same as from other sources or whether there were a number of closely related compounds.

Reinitzer examined various compounds formed from cholesterol by the action of other simple chemicals. He first studied the melting behaviour of his compounds, since a number of previous workers had observed some dramatic colour effects on cooling cholesterol compounds from just above the solidification temperature. He himself found the same phenomenon in cholesteryl benzoate, formed from cholesterol and benzoic acid.

The flashes of colours observed near the solidification of cholesteryl benzoate are not its most peculiar feature. Reinitzer found, to his amazement, that this compound does not melt like other compounds. Normal pure substances, in Reinitzer's experience, indeed in most of our experience, form crystals when they are cold and when warmed they melt into a liquid at a precise and repeatable temperature. Cool the liquid down and it freezes, reforming the crystal at the same temperature at which it had previously melted. The transition between the two states is known as either the melting point or the freezing point, depending on whether the normal state is solid (in the former case) or liquid (in the latter).

Cholesteryl benzoate was different. It appeared to have *two* melting points. At 145.5°C the solid melts into a cloudy liquid. Heat up the

Figure 2.2 Liquid crystal sample in a test-tube, warmed from room temperature. Samples go left to right. The sample starts cloudy then develops a region in which it is clear, with the interface between the two regions advancing until the whole sample is clear. On cooling, the process is reversed. (See colour plate 2.)

cloudy liquid to 178.5°C and the cloudy liquid goes clear (see Figure 2.2). Cool down the clear liquid and the phenomenon appears to be reversible. Near both transition points the system exhibits some dramatic colours. What is going on? Unsure of his ground, and out of his depth in what is now clearly a physics, rather than a chemistry, problem (and remember Reinitzer was not even a chemist; he was merely applying standard chemical methods to what he saw essentially as a biological problem), Reinitzer sought help.

History has not recorded how exactly Reinitzer was able to identify a suitable collaborator. All we know is that somehow, by asking around and reading the scientific literature, he found his man. On 14 March 1888, he wrote to Otto Lehmann,[1] at that time an Extraordinary Professor (roughly equivalent to an Associate Professor in the USA), then the assistant of Professor Wüllner at the Polytechnical School of Aachen, close to the Dutch border in Germany. Lehmann's key skill was as a crystallographer.

We shall return to Dr Lehmann at greater length later because he plays a central role in our story. For the moment, let us follow the correspondence between Reinitzer and Lehmann. Reinitzer's first letter to Lehmann was 16 pages long and handwritten in Gothic characters. The colour phenomenon in cholesteryl benzoate is of interest to the modern observer. When he cooled cholesteryl benzoate below its second melting point at 178.5°C (later called by Lehmann and others the *clearing point*), Reinitzer observed that

> ...*violet and blue colours appear, which rapidly vanish with the*
> *sample exhibiting a milk-like turbidity, but still fluid. On further*
> *cooling the violet and blue colours reappear, but very soon the sam-*
> *ple solidifies forming a white crystalline mass.*

Reinitzer observed the appearance of colours twice! At that time the mere existence of the double melting and the colours was sufficient to excite interest. In fact, nowadays we are also able to understand why in one material two sets of colours were seen, and in others only one. Indeed it is a tribute to the exactness of Reinitzer's experimental method that he observed and recorded rather subtle phenomena whose significance could not have been understood at the time. The explanation itself is complicated and involves concepts that are of extremely recent origin.

Following Reinitzer's initial enquiry there was an exchange of letters with Lehmann, and presumably of samples as well, throughout March and April of 1888.[2] Lehmann examined the intermediate cloudy fluid and reported that he had seen tiny crystals, or crystallites. When the exchange of letters ended on 24 April, although definitive answers to the nature of the cloudy phase had not been elicited, Reinitzer felt that he had enough to publish. His results were presented, with fulsome credits to Lehmann, and also to his Viennese colleague von Zepharovich, at a meeting of the Vienna Chemical Society on 3 May 1888.[3] The important point here is that these first observations of liquid crystals (although not yet recognized as such) were a serendipitous by-product of an apparently unrelated piece of research. Neither for the first nor for the last time, Nature had sprung a surprise on an unprepared investigator.

After accidentally stumbling into ground-breaking territory, Reinitzer more or less disappears from this narrative. He was promoted to professor in Prague and then to a professorship in Graz in 1895, where he later took the position of Rector. His one further contribution to the story came some 20 years later, in a rather unedifying exchange with Lehmann in the pages of *Annalen der Physik* concerning scientific priority.

The scientific puzzle is now taken up by the 33-year-old Lehmann. Reinitzer had stumbled onto an inexplicable observation, but Lehmann realized that he had come across a new phenomenon. Furthermore, Lehmann, unlike Reinitzer, was in a position to launch a systematic research programme to investigate it.

Herr Professor Dr Otto Lehmann

Otto Lehmann (see Figure 2.3) was born in 1855 in Konstanz, close to the Swiss border. His father, Franz Xavier Lehmann, was a mathematics high school teacher. More important for his son's future career was his interest in microscopy, which led him to develop a laboratory at home. In this laboratory he examined carefully the spiral forms on snail shells

Figure 2.3 Otto Lehmann (1855–1922).

and traced around the outside of leaves, seeking always a connection with mathematical formulae.

Otto Lehmann's childhood was peripatetic, for his father's post in the Baden-Würtemberg school system meant that the family was constantly being moved around the state. Successively the family moved from Konstanz to Freiburg, from Freiburg to Offenbach, and finally from Offenbach to Rastatt, near Karlsruhe. Otto was an only child, and the frequent relocations must have interrupted his social interactions with his peers. Instead he created a social life for himself in his father's laboratory. By the age of 17 he was already using his father's microscope to carry out quite sophisticated studies of growing crystals (in particular snowflakes).

These studies prepared him for a life in science in a number of ways. He became adept at self-teaching, requiring only himself and a book to learn new material. He became a careful experimenter, keeping detailed notes and fastidious records. Historians of science must track progress through records that are available. If political history is written by the victors, then scientific history is written by those who keep the most complete notebooks, and Otto Lehmann's notebooks were up with the best. He began to understand the importance of first-hand observation in the development of a scientific picture. Later in life, he faced opponents who treated his observations with scepticism because they did not fit a previously conceived world view. He was hard on such opponents, and would strongly defend his opinions against accepted scientific wisdom. Otto had developed a pride and confidence in his own work, but this later led him easily to take offence when—as naturally occurs as part of the scientific dialectic—his work was challenged. Of all his qualities, perhaps only Lehmann's pride was not always scientifically fruitful.

At 17, in 1872, he was already off to university in *Straßburg*. No longer Strasbourg, the city was then newly reintegrated into the German Reich following the Franco-Prussian war of 1870–1871. Its university was recruiting famous academic names from all over Germany to head up their programmes, and the young Lehmann benefited from contact with the distinguished professors.*

By 1876 we find him receiving a doctorate for a thesis in physical chemistry supervised by crystallographer Paul von Groth (1843–1927). His research involved studies of crystals of different isomeric compounds.[†] The principal tool in these studies was the so-called 'crystallisation microscope', which, following his youthful experience, he designed and built himself. The important feature of this microscope, which would make it ideal for studying liquid crystals, was that it was equipped with polarizers. Thus the sample could be examined by illumination with polarized light. Crystals had particular optical properties that were only apparent if illuminated by polarized light and examined through a polarizer.

The doctoral qualification that Lehmann achieved also included more general studies of other sciences, philosophy, history, as well as Latin and French. The French, examined orally, would prove of more than just cultural interest. More than 30 years later, as a famous man, he would find himself addressing colleagues in the heart of French academia in their own language.

Following his doctorate, Lehmann then held junior academic posts in Freiburg, *Mülhausen* (Mulhouse, in Alsace, and hence at that time in Germany) and Aachen (where Reinitzer found him). The postdoctoral years were spent building up expertise in crystallography. The principal weapon in his scientific arsenal was experimental microscopy, for which, as we have seen, given his home background, Lehmann was well-prepared. It was Lehmann's jealously guarded and increasingly prestigious microscope, not yet available off the shelf, which had attracted Reinitzer's attention. With Reinitzer's peculiar double-melting liquid, a problem in search of a scientist had met a scientist in search of a problem.

In fact the letters from Reinitzer had not come at the best time. For Lehmann had just been appointed as an extraordinary professor in Dresden beginning in October 1888. However, after a very brief sojourn in Dresden, on 1 April 1889, Lehmann received a call to a full

---

* Here Otto was taught by the distinguished petrographer Harry Rosenbusch (1836–1914), by the chemist and (later) Nobel Prize-winner Adolf von Baeyer (1835–1917), and by the crystallographer, Paul von Groth (1843–1927). He himself named the physicist August Kundt (1839–94)—best known to generation of schoolboys for the eponymously named tube, which measured the wavelength of sound waves—as his major influence.

† Isomeric crystals consist of molecules with the same chemical *formula* (i.e. ratios of different elements), but not the same chemical *structure* (i.e. three-dimensional molecular shape). Such molecules are said to be *isomeric* with respect to each other.

professorship of physics at the Technical High School in Karlsruhe. This was a prestigious post, for he was the successor to Heinrich Hertz (1857–1894).[‡] Back in Karlsruhe, where he was to spend the rest of his life, Lehmann now had time for Reinitzer's double-melting materials. He launched a systematic study, first of cholesteryl benzoate and then of related compounds which exhibited the double-melting phenomenon. With his microscope, he was not only able to make observations in polarized light, but also, and this was the original feature of his micro-scope, was able to make observations while samples were at a control-led temperature. His microscope possessed a sample holder that could be heated or cooled, an accessory that has become known as a *hot stage*.

The intermediate cloudy phase clearly sustained flow, but other fea-tures, particularly the appearance under a microscope, convinced Lehmann that his materials were also crystalline. By the end of August 1889 he had his own article on the mysterious flowing crystals ready for submission to the *Zeitschrift für Physikalische Chemie* (Journal of Physical Chemistry).[4] The tone of this article, whose first paragraphs will be of some interest to readers, not only isolated the key problem, but also give some idea of the nature of Lehmann's personality.

On flowing Crystals
by O. Lehmann
(With Plate III and 3 wood-engravings.) *

Flowing crystals! Is that not a contradiction in terms? Our image of a crystal is of a rigid well-ordered system of molecules. The reader of the title of this article might well pose the following question: 'How does such a system reach a state of motion, which, were it in a fluid, we would recognize as flow?' For flow involves external and internal states of motion, and indeed the very explanation of flow is usually in terms of repeated translations and rotations of swarms of molecules which are both thermally disordered and in rapid motion.

If a crystal really were a rigid molecular aggregate, a flowing crys-tal in flow would indeed be as unlikely as flowing brickwork. However, if subject to sufficiently strong forces, even brickwork can

---

[‡] Hertz is now a familiar name. He is the Hertz in the '50 Hertz' describing the fre-quency of the alternating current that powers our houses. In Karlsruhe, Professor Hertz had experimentally verified James Clerk Maxwell's prediction of the existence of elec-tromagnetic waves.

* The 'wood-engravings' indicated that special efforts had been made to reproduce pictures in the journal, and there was also a plate, containing two photographs, at the end of the journal issue to which readers' attention was drawn.

be set into sliding motion. In a certain sense, the resulting motion corresponds to a stream of fluid mass in which the joints between the individual bricks open. The bricks then run out of control, moving over and rolling around each other in a disorderly manner, rather like single granules in a turbulent mass of sand.

Lehmann was certain that the cloudy liquid possessed simultaneously liquid and crystal attributes, and believed truly to have discovered 'crystals that flow'. Much of the rest of the article is concerned with advocating the coexistence of liquidity and crystallinity in the same material, and is not without flashes of rhetoric.

He must clearly have expected to meet with significant opposition, but as to the observations themselves, even if others were sceptical, he was sure:

These observations indeed contain many contradictions. For, on the one hand a liquid cannot melt on increasing temperature and also at the same time exhibit polarization colours between crossed nicols [polarizers]. On the other hand, a crystalline substance cannot be completely liquid…Despite all these contradictions, in my own investigations I have really been able to confirm Reinitzer's results. The impossible here really seems to become possible, but as to an explanation I was at first totally helpless.

He found that the cloudiness of the intermediate fluid occurred when what we would now call nucleating droplets merge, and that sometimes the individual droplets exhibited a black cross when viewed between crossed polarizers, which he refers to as nicols. The cloudiness itself was the macroscopic manifestation of 'large star-like radial aggregates of needles'.

Lehmann's observations were quick to attract the attention of colleagues. As early as 1890, the organic chemist Ludwig Gattermann* (1860–1920) wrote:[5]

It was with great interest that I read your article on flowing crystals in Zeitschrift für Physikalische Chemie. For some time I have had several substances here which also exhibit the same properties. To begin with I thought I was considering mixtures of several materials, but the properties remained unchanged after several crystallization cycles. Following your article I am now clear as to what is going on.

---

* At that time Gattermann was a mere *Dozent* (Lecturer in UK currency, or Assistant Professor in the USA) at the University of Heidelberg, before moving to a full professorship in Freiburg-im-Breisgau in 1900. He was later to become famous for his ironically labelled eponymous *Gatterman's Cookbook*, a comprehensive textbook known more formally as the *Practice of Organic Chemistry*.

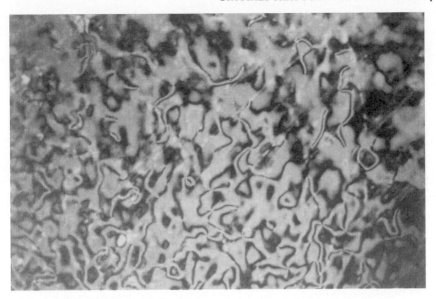

Figure 2.4 The schlieren texture. This is an optical pattern seen in a polarizing microscope, characteristic of Lehmann's liquid crystals. (See colour plate 4.)

Gattermann had made a compound whose properties at first seemed peculiar. After some puzzlement, and correspondence with Lehmann, he realized that his new material—para-azoxyanisole (PAA)—shared the properties of Lehmann's flowing crystals. Gattermann seems to have been the first to use the term *liquid crystals* to describe the strange new materials. For many years, even until after the World War II, PAA was to be the standard material on which to study liquid crystal properties.

The examination of PAA under the polarizing microscope is a testimony to Gattermann's scientific imagination. His bulk samples often exhibited peculiar streaks, *Schliere* (stains), he called them. Later it turned out that many (though not all) of the liquid crystals showed this pattern, which they called a *texture*. An example of what has come to be known as the *schlieren texture* (even in English) is shown in Figure 2.4.

Liquid crystal droplets also exhibit dramatic and striking optical patterns, and can sometimes amalgamate. When they do so, the patterns change rapidly. Gattermann defined this phenomenon as *copulation*. It was not to be the last time that comparison would be made between the physical properties of the liquid crystals and biological process and function. As time passed, as we shall see presently, the comparison would be made with a deeper purpose than that of mere rhetorical flourish.

In the years following, Lehmann made exhaustive studies of the optical properties of what were being called flowing crystals.[6] Because

Figure 2.5 Liquid crystal droplet showing the characteristic dark cross when viewed through crossed polarizers.

the essence of their unusual optical behaviour seemed to occur in the droplets, he made a virtue out of necessity. Often he deliberately prepared fluid mixtures from which the flowing crystal phase would settle out in droplet form.

Lehmann found other materials which exhibited, as in cholesteryl benzoate, two melting points. Some materials even exhibited *three* melting points. He found a phase which he sometimes called *Fliessende Kristalle* (flowing crystals) or sometimes called *Schleimig flüssige Kristalle* (slimy liquid crystals). There was another phase with different properties which he named *Kristalline Flüssigkeit* (crystalline fluids) or *Tropfbar flüssige Kristalle* (liquid crystals which form drops). If these two phases existed in the same material the latter was always the higher temperature phase. The latter was cloudy, but the former was clear, although very viscous.

Lehmann's slimy liquid crystals were obviously solid-like, if only because of their reluctance to flow. The drop-like variety also showed one physical property that had hitherto been uniquely associated with crystallinity, that of *birefringence*, which explained the peculiar dark crosses seen through the polarizing microscope in droplets (Figure 2.5).

Lehmann continued to insist on his interpretation of his microscope observations as representing materials combining all the properties of fluidity and crystallinity, while freely admitting his ignorance of the precise molecular explanation. By 1900, he was prepared to subsume all the new phenomena under the more general classification of *Flüssige Kristalle* (liquid crystals). The amount of material that he had collected was multiplying to an encyclopaedic degree. By 1904 Lehmann had published 10 papers on his new substances. It was time to pause for breath, take stock, sort the wheat from the chaff, and summarize his findings for posterity. This he did in a generously-sized 260 page tome, including no less than 483 illustrations drawn from his microscopic observations, which was published by Wilhelm Engelmann in Leipzig. Never one at a loss for words, Lehmann

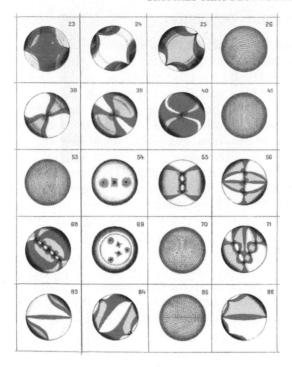

Figure 2.6 Coloured pictures of liquid crystal droplets. Reproduced from Lehmann's 1904 book. These are a few from an enormous collection of similar images. (See colour plate 3.)

entitled it *Flüssige Kristalle sowie Plastizität von Kristalen im Allgemeinen, molekulare Umlagerungen und Aggregatzutandänderungen* (liquid crystals as well as crystal plasticity in general, structural changes and changes in the state of aggregation), or just plain *Flüssige Kristalle* for short.[7]

Lehmann had long experience of preparing sketches and photographs from microscopic observation. However, the pictures of liquid crystal droplets which emerged when he developed the (black and white) images from his camera did not do justice to what he saw as he gazed through the eyepiece of the latest in his series of beloved microscopes. The blacks and whites, and even greys, failed to communicate the brilliance of the visions nature was presenting him. He was not the last to find liquid crystal textures addictive.

Nowadays we can faithfully reproduce nature's colours electronically. Lehmann was forced to resort to a more labour-intensive solution. His research assistant was set to work laboriously colouring in the camera's black-and-white reproduction, so as to simulate the real thing. It must have taken months, but the result is dramatic. Some of the images are reproduced in Figure 2.6.

Supporters and opponents

Notwithstanding the lack of proof of its essential nature, crystallinity certainly seemed incompatible with fluidity. Lehmann's assertion that he had observed liquid crystals was not, as we have seen, made trivially. The liquidity of his materials was plain for all to see. The crystallinity was less obvious, but here too the grounds were strong. He had two principal reasons for believing that he had crystals. One was simply the evidence of his own eyes. You could *see* the crystals in the microscope.

The other reason was more subtle. The liquid crystals looked dramatic, but only under the polarizing microscope, that is, between crossed polarizers. This meant that somehow birefringence was playing an important role. The cloudy double-melting phase appeared (at least to Lehmann) to break up into many regions when looked at as a layer under the microscope. Each region itself seemed to be transparent but anisotropic and birefringent. The cloudiness visible to the naked eye was the result of the light constantly changing direction as it was refracted at the boundaries between different regions, each of which had a different special direction. The cloudiness was therefore a secondary, rather than a primary, characteristic, that is, it depended on how you looked at the sample. Finally, Lehmann knew that birefringence only occurred in crystals because this was the only circumstance in which it had been observed, and this was the only circumstance in which mathematical theory seemed to allow it. Lehmann had followed Sherlock Holmes's adage of eliminating the impossible and had arrived at what he regarded as merely the improbable. Liquid crystals it had to be.

All pioneers in strange territories acquire acolytes willing to seek their fortunes in the new country. Gattermann can partly be regarded as such, even though most of his time was spent exploring elsewhere. A more enthusiastic acolyte was the young Rudolf Schenck (1870–1965) of the University of Halle. The ancient city of Halle, birthplace of the composer George Frideric Handel (1685–1759), 250 km to the south west of Berlin in Saxony–Anhalt, has played, and indeed continues to play, an important role in the history of the liquid crystal phases.

In fact Schenck's interest in liquid crystals was somewhat serendipitous. He was a student in Halle, his home town, and his interests were turning from organic chemistry to physical chemistry. He was trying to carry out a difficult experiment involving heating up gases. The work was relatively unsuccessful, but then came to an abrupt and dramatic end when a cleaning lady in the department accidentally smashed his apparatus. Following this incident he retired grumpily to the library. There, whiling his time away reading the latest periodicals, he came across Gattermann's recent publication on PAA. This stimulated him to study liquid crystals.[8] He continued this work at the University of Marburg, where he obtained his Habilitation* degree in 1897 and was then appointed to a lectureship. He made contact with Lehmann and although they did not publish together, they influenced each other's work and Schenck joined the small but growing group of liquid crystal scientists.

Others were less keen. Perhaps this is not surprising, given the circumstances and intellectual climate of the day. Lehmann's point of view elicited more than a little scepticism from some of his scientific peers concerning the very existence of the liquid crystal phenomenon. They were worried by the apparent explicit contradiction between the existence of a crystal lattice and liquid crystals.

For them, Lehmann's unlikely explanation was impossible. They sought more conventional explanations. This usually involved some sort of colloidal mixture that combined solid and liquid components. In such a case properties intermediate between those of a solid and those of a liquid would be less surprising.

The first such suggestion seems to have been made in 1894 by Georg Quincke (1834–1924), professor of physics in Berlin and later in Heidelberg.[9] It was clear therefore, averred Quincke and the Russian theoretical physicist Georg Wulff (1863–1925),[†] that liquid crystals must really be colloidal, suspensions of small crystallites inside a liquid. An example of such a colloid is white paint (at least while it is still in the tin!), which consists of crystallites of titanium dioxide suspended in a polymer resin. According to Quincke and Wulff, the existence of a solid component would explain the birefringence. The opaqueness or strong light scattering, called turbidity, in colloids results from the interaction of light waves with individual colloidal particles, which are just the right size to reflect the light in all directions. Interestingly, the theory which showed this explicitly was produced by Gustav Mie (1869–1957), a former assistant of Lehmann's, in 1908, but the general phenomenon had been known for a long time and had been investigated by, amongst others, Michael Faraday.

The most prominent and most vigorous and persistent sceptic was the physical chemist Gustav Tammann, later distinguished as a pioneer of modern metallurgy (Figure 2.7). Tammann was a Baltic German, born in Jamburg (since 1922 Kingisepp), near St Petersburg, in 1861. All his early career in chemistry was at the by no means undistinguished

---

* The Habilitation degree is the requirement in Germany to be appointed to a Professorship.
† Later to become famous for explaining why macroscopic solids take on their characteristic crystalline shapes.

Figure 2.7 Gustav Heinrich Johann Apollon Tammann (1861–1938). The physical chemist Tammann vociferously and aggressively opposed Lehmann's idea of liquid crystals.

University of Dorpat (now Tartu, in Estonia), which despite then being part of Russia, used German as its principal language. After 1893 the medium of instruction switched to Russian and the increasingly alienated faculty (many from Germany itself) looked for other positions. Tammann himself was invited to apply for the professorship of physical chemistry at Göttingen in 1903, where he remained for the rest of his life, dying in 1938.

Those not familiar with German culture should be aware of the premier position held by Göttingen in the German-speaking world—comparable to Oxford and Cambridge in England—so the move from Dorpat to Göttingen should be seen as real recognition of Tammann's perceived promise and talent. Indeed, in the late 1890s and early 1900s he was gaining an impressive reputation as *the* up and coming man in thermodynamics. In particular his specialities were the study of the effects of heat and pressure on materials, and the phase behaviour and mixing properties of complex mixtures. With its peculiar phases, the behaviour of liquid crystals was a natural theoretical problem for him to tackle. Peculiar phases required peculiar talents, and Tammann was certainly not a man who lacked self-confidence.

Tammann vociferously propounded the view that the underlying cause for the anomalous 'liquid crystal' behaviour would be found when the purity of these substances was carefully examined. His first contribution to the debate in 1897 only elliptically referred to liquid crystals,[10] but in an article submitted to *Annalen der Physik* on 27 December 1900, sarcastically entitled 'On the so-called liquid crystals',[11] he suggested an alternative viewpoint. He made an analogy with the behaviour of water–phenol mixtures. The simple organic compound phenol, commonly known as carbolic acid, in former times was the cleaning agent that gave rise to the characteristic smell in hospitals and other institutions. Phenol is only partially miscible with water, depending on the relative amounts of phenol and water. Below a certain *lower critical temperature*, mixing does not occur and the resultant mixture of water and phenol appears cloudy.

Perhaps, suggested Tammann, the so-called liquid crystals were really a mixture of some sort. These mixtures were also exhibiting a lower critical temperature, he proposed, but of a type not previously observed. Tammann compounded his legitimate scientific disagreement by suggesting that a volume change observed by Schenck in Marburg at this critical temperature, the clearing point, was probably the result of poor measurement, and that the sensitivity of the clearing point to impurity concentration was good evidence that the whole phenomenon was impurity driven. In a mixture, no abrupt change in volume is observed when it separates. Incidentally, he remarked, since the liquid crystal patterns observed in the microscope were not easily disrupted by poking the sample, almost certainly everything was occurring at the surface anyway. The tactless suggestion of error and careless experimentation transformed a disinterested scientific discussion into an unpleasant personal confrontation that lasted many years, and which still echoes through the ages.

Tammann's underlying objections were really twofold. On the one hand it was known that colloids with particles whose sizes were of the order of magnitude of the wavelength of light strongly scatter light. In fact Tammann preferred the idea that liquid crystals were *emulsions*, in which droplets of one liquid are suspended in another. Emulsions are often thought of as a particular form of colloids. Familiar emulsions include milk (oil droplets in water) and vinaigrette (oil droplets in vinegar). Both systems are cloudy, and in this way superficially resemble the turbid liquid crystal state. It was the strong physical resemblance of liquid crystals to such systems which convinced Tammann, on the basis of what we now know to be inadequate evidence, that this was also the case here. His other objection was due to his strong attachment to the as-yet-unproven lattice theory of solids. Liquid crystals, he believed, really were a contradiction in terms.

Tammann's published article enraged Lehmann. On 25 February 1901, he fired off a lengthy reply.[12] In principle Tammann could be right, he admitted (and he had himself seen some examples of such phenomena in other materials), if only there had been no other observations. But there *were* other observations! Indeed (Lehmann too was no slouch with heavy irony) in a lengthy recent article he had explained how he had been led to the idea of a liquid crystal, but this article seemed to have escaped Tammann's notice, for it was not mentioned in Tammann's paper.

Lehmann's points of rebuttal were somewhat technical. He noted the existence of a surface tension between the cloudy and clear phases, which is inconsistent with the picture of an impurity-induced phase. He pointed to the fact that one could observe clear extinction directions (regions of destructive optical interference) in his polarizing microscope (and which, as we have seen, should only occur in real honest-to-goodness crystals). As to Tammann's comment that the effects must really be confined to the sample surface, if that were really the case, then the colours exhibited by the liquid crystal samples under the polarizing microscope would always be the same, but actually they depend on sample thickness. Then, if Tammann could be rude about Lehmann's experimental technique, he could return the compliment. Tammann's attempts to purify the sample, he asserted, had actually had exactly the opposite effect! Tammann's (but not his own or Schenck's!) experiments had been carried out on impure samples. No wonder Tammann had observed impurity effects! But if pure materials are used, in this case, not much is changed. As a final warning shot, he also escaped from Tammann's criticisms of experiments on cholesteryl benzoate by pointing out that anyway this compound is not strictly a 'liquid' crystal, but a 'flowing' crystal.

Tammann was unconvinced, but remained silent for a year. On 3 March 1902, however, he was back, submitting yet another paper to *Annalen der Physik*, with the same title as before.[13] Given Lehmann's rather robust response to his previous missive, Tammann shifted his ground somewhat. Look, he said (implicitly, one has to read between the lines), last time I only skated the surface with my objections. Let me put them with full force. Lehmann's liquid crystals do not 'shake' like real solids (technically, they possess no shear elasticity, so they cannot possess an underlying lattice). The (re)definition of cholesteryl benzoate as a flowing crystal was *ad hoc* (i.e. invented just for the purpose of explaining these experiments) *and* otherwise implausible. Liquid crystals, pointed out Tammann, were *all* milky and equally milky, but solid crystals, on the other hand, were perfectly clear, so where was the commonality? And then, if liquid crystals really were mixtures or emulsions, then distillation (i.e. boiling them off in a special apparatus in such a way as to eliminate the impurities) would lead to a shift in the clearing point, and his former student Rotarski had shown experimentally that this was indeed the case. Tammann concluded by re-emphasizing that all so-called liquid crystal phenomena were typical of emulsion behaviour.

Lehmann was by now incandescent. His reply,[14] submitted on 31 May 1902, was 15 pages long, with again a numbered list of points. By now, however, battle lines were drawn, and many of Lehmann's explanations only made sense within the liquid crystal paradigm. An example of this is his assertion that liquid and solid crystals differ because of their lack of shear elasticity. The substantive point made by Lehmann was, as we have already seen, that the cloudy phase is not really cloudy at all, contrary to Tammann's assertion. Under a microscope, magnified 700 times, it becomes clear, just like a solid crystal. Each droplet is *anisotropic*, and the cloudiness is because the whole system is an aggregate of differently ordered droplets. Lehmann's irony degenerated into sarcasm as he remarked that Tammann was persuaded by the new and untested lattice theory of solids, but unconvinced by the old well-tried and well-tested crystal optics. As for the remaining disagreements, they reduced to doubting the opposing experimentalist's competence. Little wonder that Lehmann was offended.

In fact Tammann's doubts were widely shared in the chemical community. Probably only Tammann—an expert in thermodynamics—was motivated and articulate enough to express them in public. However, the distinguished but elderly Russo–German organic chemist Friedrich Konrad Beilstein (1838–1906) from St Petersburg exhibited this scepticism in a 1905 letter to his Halle colleague Jacob Volhard (1834–1910).* Beilstein noted that (Georg) Quincke had shown him some work on ice and glacier formation that proved conclusively that liquid crystals were an optical illusion. Not only that, but 'Lehmann was a man who knew neither physics nor chemistry...!' Volhard was a friend of Beilstein's, but he had also supervised the Ph.D. work of Rudolf Schenck, who was contemporaneously playing a major role in the opposing camp; Schenck's reaction to this clash of loyalties is not known.

Karlsruhe 1905

It was in this climate that in late May 1905 the German physical chemistry community gathered together in Karlsruhe for the annual meeting of their society, the *Deutsche Bunsen Gesellschaft* (The German Bunsen Society).[†] The proceedings, published in the *Zeitschrift für Elektrochemie*, give us a grandstand view of the whole event. The meeting started on the evening of Thursday June 1—Ascension Day according to the proceedings—with a greeting party in the *Stadtgarten*, and finished with an excursion to the Old Castle in Baden–Baden on Sunday 4 June. In the event of bad weather, the walk round the castle was to be replaced by a visit to Baden–Baden itself. We are not told, however, whether (or how many of) the 166 participants actually made it to the *alter Schloss*.

---

* We are grateful to Professor Horst Stegemeyer for providing us a with a copy of Beilstein's letter to Volhard.
 † A good deal of detail about this meeting is provided by Knoll and Kelker (1988). We have augmented this from the outline of the meeting programme in pp 299–300 of *Zeitschrift für Elektrochemie* **11** (1905).

The meeting was a general physical chemistry meeting, with an eclectic choice of talks. Friday saw no fewer than 17 papers, including a presentation on a geological thermometer by the society's president, the Dutch physical chemist Prof. Dr Jacobus Henricus van't Hoff (1852–1911) from Berlin, another on the teaching of physical chemistry in schools, and yet another on the physical chemistry of wine.

The Saturday morning session was set to start, as the programme rigorously instructed, at *precisely* 8.30 a.m. The session was honoured to have as chairman the society's president. Van't Hoff had been professor of chemistry in Amsterdam, but bureaucratic pressures (and too high a teaching load!) had led him to accept a research-only post in Berlin in 1896. Many were envious then, but the level of his research was unquestioned. He had been the winner of the very first Nobel Prize in chemistry in 1901. His eminence meant that his authority was accepted by all. It is not impossible that the organizers had anticipated that the exercise of that authority would be required.

The session concluded, eight seminars later, with a presentation by *Privatdozent* R. Schenck from Marburg, entitled 'The nature of crystalline fluids and liquid crystals'. Lunch was then due to be taken in the *Stadtgarten* at 1.30 p.m. The proceedings also fail to inform us how many delegates lasted until Schenck's presentation, read presumably at around 1 p.m. Could some less committed delegates perhaps only have persisted until the seminar by Prof. Dr Cohen of Utrecht, on the explosive properties of antimony (with demonstration!), before tiring of the morning session?

Those few delegates—and this included Tammann—who were patient enough to wait until Schenck's paper witnessed an event in scientific history. Schenck's point of view is best presented in his book, *Kristallinische Flüssigkeiten und flüssige Kristalle* (Crystalline fluids and liquid crystals),[15] which appeared almost contemporaneously with the meeting. Originally intended as a thermodynamic appendix to Lehmann's 1904 book, it finally appeared on its own a year later, but with a blessing from Lehmann (and published by the same publisher). Perhaps the delay turned out well, for the Schenck book is shorter, less pretentious, more focused on possible objections, and altogether easier to read than Lehmann's effort. But the Karlsruhe paper provided a theatrical dimension to what turned out to be surgical unpicking of the Tammann point of view.

If previous exchanges between Tammann and Lehmann had been noted for their heat rather than their light, Schenck's presentation dragged the experimental status of the field into new pastures.[16] By 1905, the book reports, citing each of them in a list in the first chapter of the book, there had been 41 publications on or referring to liquid crystalline phenomena. By 1905 all of 21 liquid crystal compounds had been identified (there are now, in 2010, approaching 100,000!).

Schenck's paper made a number of points, all of which pointed strongly against the idea of liquid crystalline properties depending on the lack of material purity. One was to do with the dependence of the turbidity on the birefringent drops. Then there were the jumps observed in both density and viscosity at the onset of the cloudy phase. If the liquid crystal were an emulsion, there would still be a clearing point, but no discontinuities, and (as Tammann had chosen to cite this example, although of course Schenck politely did not point this out) here were some explicit results on water–phenol mixtures to stress the point. Furthermore, Schenck had actually carried out experiments on the *exact same* experimental samples used by Tammann, generously provided by Herr Professor Dr Tammann himself. Apparently Tammann had been careless in preparing his materials. Indeed they were not pure, but if the materials *are* purified, it makes no difference.

The light absorption in crystalline liquids is temperature-independent, which would not be the case in an emulsion because the droplet number would change with temperature. Emulsions can be made to separate if a high voltage is applied, but experiments by the German physical chemist Georg Bredig (1868–1944) and his student Schukowsky had failed to cause phase separation of the presumed emulsion.[17] They failed at the usual 12 volts and they failed at a higher value of 70 volts. They even failed at the *enormous* value of 48,000 volts! There was no demixing, and if there was no demixing, it must be because there was no mixture.

Not only had Bredig been unable to separate out the 'mixture' into its components, but similar experiments by Coehn, working with Tammann in Göttingen, had likewise led to a negative result.[18] In an attempt to rescue the Tammann hypothesis, Coehn had suggested that perhaps some mixtures could not be separated in this way, a suggestion treated by Bredig with laughable contempt. For Bredig, for Schenck, and for Lehmann the circumstantial evidence against the mixture idea was overwhelming.

Finally, noted Schenck, the Hungarian physicist Baron Loran von Eötvös (1848–1919) had predicted in 1886 that in pure liquids the surface tensions varied in a regular way with temperature. This would not work in a mixture, as was suggested by Tammann for liquid crystals. In fact the surface tension of liquid crystals varied with temperature in the manner expected for pure liquids, so there was no *a priori* reason to suppose that the liquid crystals were impure.

As soon as the lecture was over, up popped Tammann with a long, involved, and impressively tactless restatement of his by now well-known position, thinly disguised as a question:

> *I am familiar with liquid crystals from personal experience. I want to discuss the central question: Are these materials anisotropic or isotropic? Equivalently, has birefringence in these cloudy liquids been conclusively established?*
>
> *First: the materials in question have been observed in parallel films between crossed Nicols. Professor Lehmann has described a situation in which the image is divided into segments. When the sample is rotated, their brightness changes. This seems to demonstrate birefringence, and I would not object to calling these liquids crystals. But then one observes that the segmentation and the change in brightness are not properties of the liquid itself. Rather there seems to be an anisotropic dust adsorbed onto the glass plates surrounding the sample. If the liquid is shaken violently the picture does not change. The phenomenon is thus not a property of the liquid. It can be disrupted by interfering with the adsorbed dust. Then no segmentation can be observed. I have made such observations on many occasions, and Professor Lehmann has essentially described the same thing. I conclude from this that the liquid is not itself birefringent.*

On and on and on he thundered (or perhaps droned, for the high emotion of the meeting sadly has to be inferred from the words on the page). We omit some of the rhetoric:

> *The anisotropy relates only to optical properties. In all other contexts there is complete isotropy, and this even applies to growth phenomena. All liquid crystals are optically turbid media. They are thus emulsions, and contain at least two components. However, a complete analysis of this phenomenon has not yet been successfully carried out. In my own view, distillation of the liquid crystals offers the best prospect for the resolution of the problem, but this is an extremely difficult task.*

Unsurprisingly (and correctly) he was focusing on the importance of identification of 'liquid crystals' for the lattice theory of solids.

> *We must also take into account that Dr Schenck has found discontinuous changes in certain properties as a function of temperature. However, a discontinuous change in any given property is difficult to establish beyond doubt. Finally I would like to come to the so-called flowing crystals. There are certainly such things as soft crystals. I am happy to concede flowing crystals, but not liquid crystals. The existence of liquid crystals is a key question when considering the lattice theory of solids. I would thus assign enormous theoretical importance to the question of the existence of liquid crystals!*

So the experiments were wrong and the interpretation was wrong. The lattice theory of solids trumped everything. According to Tammann, the key observation was the turbidity. Although the birefringence at this stage was not yet understood, in the fullness of time, it would sort itself out. Lehmann, of course, took a different view.

A reply was clearly necessary, but at this point session chair van't Hoff curiously and abruptly closed the discussion. Whether eight successive papers had caused delegates' stomachs to rumble so loudly as to impede further fruitful interaction, or whether simply the chair merely wished to prevent bloodshed, is not recorded. We read that the 'discussion ran out of time...and that consequently a reply to Professor Tammann's points was impossible.' In lieu of a reply on the spot to what was clearly a contentious issue which raised important scientific issues, van't Hoff proposed the setting up of a Commission of Experts. This could examine liquid crystal problems further at its leisure. It would, of course, include, among others, Tammann and Lehmann. Meanwhile they should all reconvene in the afternoon for a demonstration in Lehmann's own laboratory.

The aftermath of the meeting was no less unhappy than the meeting itself.[19] It was not just a scientific mistake for Tammann to take on Lehmann on his home turf. Lehmann felt humiliated and insulted by what he saw as an unanswered public attack on his scientific competence and integrity. There was a bitter exchange of letters. Turning down an offer of a private meeting to sort things out, Lehmann wrote to Tammann on 12 June:

> The treatment in the meeting must have convinced all the students that my work is completely unreliable...I must assume that you are aware that a private discussion just between us cannot repair the offence to my honour in the meeting. The people in front of whom this injustice has been perpetrated will necessarily discover that things are not like this at all...

Tammann claimed, in a seemingly disingenuous manner, that he was merely involved in a disinterested search for truth, claiming in a letter of 14 June that he 'valued (Lehmann's) work very highly'. Retreating somewhat (in a letter of 18 June), he claimed social rather than professional offence:

> Concerning the course of the meeting, I obviously made an error, since I could not see well in the auditorium. So I concluded that when van't Hoff kept to time, it meant that no-one disagreed with me...

There followed a list of scientific questions. Did, for example, the turbidity disappear when you looked at it between crossed polarizers?

Lehmann answered this and other questions curtly in a letter written on 21 June. Yes it did. But he concludes the letter by remarking that although he is grateful for the opportunity of achieving understanding in a spirit of friendly cooperation, he cannot but suspect that maybe Tammann has not had actually had the opportunity to make the observations himself.

The emotional effort of all these exchanges was clearly telling on Lehmann, as he was confiding to Schenck in a parallel set of letters. Schenck, in what we might think of as pusillanimous mode, remarks on 17 June:

> *I was very interested by the letter from you which I received today. The shady tricks by Dr T which you recount are even worse than I had expected...You mustn't make any concessions. This so-called honourable man, who denies all facts...must be offered no forgiveness. As far as negotiations with T are concerned, in order to prevent later shady tricks, one must establish a protocol, which he must sign. The presence of an expert in theoretical optics and a chemist is an absolute precondition...*

War, indeed! Meanwhile, insult or no insult, Tammann's mind was also not for changing. Lehmann was turning his attention to the relationship between liquid crystals and the life sciences, a story which we shall relate in Chapter 6. We note here only that the discouraging interaction, and the emotional frustration and effort which accompanied it, led to two further books, one of which is phrased as a Socratic dialogue.

Lehmann had been exhausted by their public disagreements, but retained an emotional commitment to the subject matter. Tammann, however, although unpersuaded, was on the point of retiring hurt. He made but one further assay into the liquid crystal literature, the following year (1906), again in *Annalen der Physik*. This time it had been Lehmann who had fired the first shot.[20]

Tammann contrived a weak, if complex, reply.[21] No longer did he refer to 'so-called' liquid crystals. Unsure of his ground, he simply entitled his paper 'On the nature of liquid crystals III'. The III was understandable; it was indeed the third in the series, but normally one would expect it to have the same title as I and II! The article contains a number of excessively complicated binary liquid phase diagrams, but continued to lack a convincing explanation of the huge birefringence.

As it turned out, even the commission of experts also turned out badly, with Lehmann and van't Hoff finishing daggers drawn, but Schenck was able to retrieve something in the form of an extensive review article in 1909 in the *Jahrbuch der Radioaktivität und Elektronik*.[22] None of these personal problems impeded the progress of liquid crystal science, as we shall see in the next chapter.

TECHNICAL BOX 2.1 Optical anisotropy, the optic axis, and the director

The problem that confounded Lehmann and others in understanding liquid crystals was the optical anisotropy they observed in the polarizing microscope. Anisotropy could be easily understood for crystals in terms of lattice theory, since if the lattice dimension along one direction was greater, or less, than that in another direction, then it was obvious that the properties of the crystal, that is, the refractive index, would differ along these two directions.

However, there can be another source of anisotropy other than lattice anisotropy. If we want to locate ourselves, or indeed anything else, in the map of the Universe, then we certainly need position, as on a lattice, but we also need to know which way up the map is, that is, the direction. On a microscopic scale, a lattice provides position, but what about direction?

On Earth, we know which way is up and which is down thanks to the force of gravity (a force in technical terms is called a field). Direction on the surface of the Earth is given us by the Earth's magnetic field, which runs from the North Pole to the South Pole in lines of force, which more or less follow the lines of longitude. But we can also gain directions from looking at things—the sun, the stars, and even a boundary marker in the distance, and we can do this because light has direction.

Thanks to Newton and Huygens a great deal was understood about the directional properties of light, and all scientists working at the beginning of the twentieth century knew everything about macroscopic optical anisotropy: refraction, the polarization of light, double refraction, and birefringence.[23] If the optical anisotropy of liquid crystals didn't come from a lattice, however, where did it come from? The crystal optical anisotropy that Lehmann and others could understand was a consequence of the positional anisotropy associated with a non-cubic lattice, but nobody was thinking about other indicators for direction on a microscopic scale.

The director

The answer is clear, but it took a long time to demonstrate it. Under suitable conditions of density and temperature, molecules having an extended shape will pack together such that they minimize their collisions with each other. Such an arrangement is often likened to a shoal of fish swimming in a river. There is a difference between fish and liquid crystal molecules, which is that the latter mostly don't care whether they are facing forwards or backwards, but otherwise they will flow like a shoal of fish. The local direction is determined, and we now give it a name, the *director*, but it need not be fixed in space. The director may wander around in a fluid, or be constrained by boundaries, or, as we shall see, it can be changed by flow or external forces. Having defined a director, the optical anisotropy is easy to understand.

The optic axis

At last the origin of the optical anisotropy in liquid crystals is explained. The speed of light is different along the director from the speed of light perpendicular to the director: in fact it is faster. The direction of maximum speed of light was something that optics specialists could easily determine for crystals and liquid crystals, and it was called the optic axis. It was a special direction, since not only did light travel fastest along it, the polarization of the light was unchanged. If you looked at a crystal or a liquid crystal film between crossed polarizers down the optic axis, then you saw nothing: it was as though you were looking at an isotropic liquid...or nothing. But slightly tilt the crystal, or liquid crystal away from its optic axis and light was transmitted by the crossed polarizers: a clear signature of a birefringent material. *The optic axis is the director.*

3
Liquid crystals, where do they come from?

———————◦◦◦◦———————

... no branch of research has permitted the explanation of the inner relationships between molecular shape and crystal forms better than the investigations into the liquid crystalline state.

D. Vorländer, *Zeitschrift für Physikalische Chemie* **105**, 211 (1923)

Traditionally natural philosophers, whom we now call physicists, observed and studied the things around them: the heavens and the earth, and what lay beneath the surface of the earth. The opponents of Lehmann were mostly physicists, and could rightly claim at the time that there were no liquid crystals to be found in nature.* Lehmann's magic materials must be an artefact of some description. Of course Lehmann himself was a physicist, and by training and instinct would have shared the scepticism of his peers. But he had been seduced by the would-be chemist, masquerading as a botanist, Reinitzer, who had introduced Lehmann to the problematic compounds that were the cause of so much trouble. Chemists were a different breed from physicists. They were descended from the magi—magicians, tricksters, followers of the occult—and they were certainly not to be trusted. True, their potions could cure, or kill, but no-one knew why. And they could manufacture fireworks for entertainment, bombs, shot, and powder for destruction, but the processes involved remained a mystery to decent men.

The only so-called liquid crystals came out of the chemist's laboratory. They did not flow in rivulets; liquid crystals could neither be grown nor mined. Quite clearly they were not a product of nature and so had to

---

* We now know that the bright multi-colours of scarabaeid beetles are due to cholesteric liquid crystals, which have special light reflecting properties. More importantly, liquid crystals form a vital part of all living systems, animal and plants, in providing the special material for cell membranes. They are able to protect the biochemistry of the cell from the sometimes hostile environment, and also provide a support for proteins and other biomolecules which manage and regulate the activity of the cell.

be treated with extreme caution. Experimental physicists investigating the properties of liquid crystals had to rely on chemists to provide them with samples, and in the early 1900s there were precious few materials to study. In 1905, as we have seen, just 21 compounds were known which exhibited properties that could be described as liquid crystalline. In his book,[1] Rudolf Schenck, a chemist turned physicist or physical chemist, was easily able to list them all.

In the decades that followed, due largely to the efforts of a single research group, led by Daniel Vorländer at the University of Halle, this number would increase 50-fold. Not only would the range of compounds available for study increase, it became possible to speculate on the particular features of molecular structure necessary for the formation of liquid crystals. This in turn enabled theories of the structure of liquid crystals to be developed.

Making new materials

Let us now return to 1905. The search for the microscopic origin of liquid crystallinity was now really on, as the number of research groups studying the phenomenon began to increase, although still not transcending the boundaries of the German-speaking world. A prerequisite for any detailed picture involves knowing what kinds of materials are able to form liquid crystals. New materials are provided by chemists, whose speciality is to synthesize new compounds from existing and available chemicals through a series of chemical reactions. This is a procedure akin to cooking, and is often still disparagingly referred to as such. The chemicals that form liquid crystals come under the umbrella description of organic compounds, i.e. they contain a substantial proportion of the element carbon and no metals.*

The chemistry of carbon compounds is known as organic chemistry, and has developed as a distinct subject of study. The chemistry of all the other elements traditionally goes under the title of inorganic chemistry, although in modern times these barriers between the subdivisions of chemistry are beginning to disappear. Why should compounds of carbon be so special as to warrant their own branch of chemistry? Partly this is a result of the shear diversity of carbon-containing compounds that can be formed. But more particularly it is a result of the close links between carbon-containing compounds and those chemicals extracted from plants and animals. The essential components of living things are compounds of carbon, oxygen, and water,

---

* Modern chemistry has now come up with metal-containing liquid crystals, but attempts to incorporate particular metallic properties into liquid crystals have so far failed.

with other elements, particularly nitrogen, sulphur, and phosphorus, also playing important roles. These days, liquid crystals can be synthesized that have structures similar to, if not the same as, almost any compound extracted from living tissue, although the chemists of Lehmann's time had no inkling of this.

The liquid crystals Lehmann received from Reinitzer, which started the story, although inspired by earlier studies of carrot juice, had in fact been prepared from an industrial source of cholesterol. The chemical reactions that Reinitzer used to convert cholesterol into compounds exhibiting liquid crystal behaviour were trivial, even by the standards of the late nineteenth century. It is doubtful that he would have claimed to have *synthesized* the materials. For chemists, as distinct from pharmacists, the difference between preparation and synthesis is important. A preparation is a simple process that may, although not necessarily, involve a simple chemical reaction. The desired product is then extracted and purified and ready to use: the equivalent in modern times of flat-pack assembly from IKEA. Synthesis needs a professional, who works with basic chemical building blocks and from them constructs a new compound. This process in general needs a series of chemical reactions, from which it may be necessary to isolate and purify the intermediate compounds.

The rising star Ludwig Gattermann was very much a professional organic chemist and, as we have seen in Chapter 2, he had accidentally synthesized a new liquid crystalline compound, PAA. Although Gattermann had no way of knowing it, this compound was to become one of the most studied materials over a period of 50 years or more. Surprisingly Gattermann lost interest in liquid crystals, but a new champion emerged in Daniel Vorländer (1867–1941) of the University of Halle (Figure 3.1). A century later, there is still a liquid crystal group in Halle enjoying a direct historical link with Professor Vorländer.

To begin with, liquid crystals (or *crystalline liquids*, as Vorländer insisted on calling them right through his long career) cannot have seemed very promising. The first paper which appeared from the Halle group on this subject was signed only by his graduate students F. Meyer and K. Dahlem. The paper was entitled 'Esters of azo-and azoxybenzoic acid' and was published in the *Annalen der Chemie* in 1903.[2] In a rather detailed commentary on the oxidation of an intermediate chemical, they mention, almost in passing, that they synthesized a material that:

> ... exhibits a double melting point... [which]... reminds one of the strange materials investigated by O. Lehmann and R. Schenck.

Indeed, just to check, they asked Schenck to repeat the observation. However, these remarks occupy less than half a page in a 15-page article, whose main thrust concerns more technical aspects of organic synthesis.

Figure 3.1 Daniel Vorländer (1867–1941).

What is noteworthy in this article to the historian of science is that it is not co-authored by Vorländer himself. It was an unwritten but nevertheless extremely rigid convention of the day that the director of an institute would publish the results of his co-workers under his own name. Only in passing would the director have mentioned that these results had been obtained (in this case) 'together with F.M and K.D.'. It stretches plausibility to suppose that Vorländer omitted his own name out of a spirit of generosity, in order to promote his subordinates' careers. Much more likely is that Vorländer missed the significance of these observations and, furthermore, aware of the controversial nature of the crystalline liquid hypothesis, was simply keeping his distance from a possibly dangerous scientific controversy.

Whatever his motivation in 1903, the Halle experiments continued, and Vorländer finally realized that he was tapping a rich vein. Three years later he returned to the subject, this time in an article authored by himself alone[3] in the premier German journal *Berichte der Deutschen chemischen Gesellschaft*. So influential was this journal in its day that references to articles in it were often abbreviated just by the letter *B*!

It was his next article,[4] published in the following year in *Berichte*, which was to have the really long-term influence on the subject. In it he discerned the first important clue to the real nature of liquid crystals. This very short article is entitled 'On the influence of molecular shape on the crystalline liquid state'. Referring back to Gattermann's liquid crystal compound PAA, Vorländer realized that he could make two other varieties of the same compound having the same chemical constitution but with different shapes. These different varieties are conventionally labelled as *ortho*, *meta*, and *para*,* the last being the variety that Gattermann had identified as liquid crystalline. Vorländer was able to show that only the elongated molecule (denoted *para*) was a liquid crystal. Clearly molecular shape was a vital factor in determining whether or not a compound might exhibit liquid crystallinity.

Shape had already been noted by Lehmann as a feature of liquid crystals, but he was referring to the variety of shapes seen in the microscope. Although small on a human size (fractions of a millimetre), these

---

TECHNICAL BOX 3.1 Molecular shape and structure

Chemical constitution—Compounds are made up of atoms of elements joined together in particular ways. The proportions of different elements in a compound are indicated by the chemical formula, which labels the relative numbers and types of atomic elements in the compound.

Chemical elements—Although over 100 different elements have been identified, those involved in liquid crystal compounds are relatively few. In chemical formulae they are represented by capital letters as follows:

carbon—C; hydrogen—H, oxygen –O; nitrogen—N.

In a chemical compound the atoms are joined by bonds, denoted on paper by a dash (–). Sometimes two bonds (=) link certain atoms, known as a double bond.

Chemical structure—This is the arrangement of atoms and their connections (bonds) which defines a molecule. The chemical structure is a representation of a molecule, but using modern methods of X-ray crystallography it is now possible to determine precisely the structure of a molecule in its crystalline state. In a fluid state (liquid or liquid crystalline) and a gaseous state, the molecular structure may not be rigid and so can adopt different shapes.

Organic chemistry is concerned with the compounds formed by carbon with other elements. Since carbon can link with up to four other atoms at any one time it provides a rich source of many compounds. It is also the building block for all living things. The arrangement of other atoms around carbon depends on the nature of the bonding. If all the bonds are single, then a tetrahedral structure is formed.

Joining a number of carbon atoms together forms a chain, and these are often vital for the formation of liquid crystal phases. The atoms that fill the vacant bonds of the carbon chain are hydrogen atoms, and such chains are referred to as *alkyl chains*.

---

* The Greek prefixes *ortho* (straight, right, normal, or correct), *meta* (with, after), and *para* (beside, beyond, parallel to) lose most of their linguistic significance in chemistry and physics, where they tend to be used as designatory letters (*o, m, p*). They appear in various contexts with different definitions.

In fluid phases, such chains are not rigid, and the various segments may rotate. However, for liquid crystals to be formed it is believed that the most extended forms of the chain are to be preferred. Rotation of chain segments will shorten the length of the chain and so tend to disfavour liquid crystal formation. If a carbon atom is joined to just three other atoms it tends to form a planar structure, and six carbon atoms joined together can form a planar ring molecule. If all the other atoms filling the vacant bonds are hydrogen, then the familiar compound known as benzene results.

This ring structure appears in almost all liquid crystal molecules, but is often represented without the hydrogen atoms for convenience. It is also conventional to omit the carbons, since they will be assumed. To keep the tally of bonding correct, some double bonds have to contribute to the ring structure, and these may appear or not depending on the preference of the chemical artist.

The liquid crystal compound prepared by Ludwig Gattermann that was the preferred material for experimental study for many decades was *para*-azoxy-anisole (PAA), the designation *para* is significant. The position of attachment of a single group to a benzene ring is immaterial: all the six positions are equivalent. However, if there are to be two attachments, then their relative position does affect the overall structure of the molecule. There are three possible ways in which two groups can be attached to a benzene ring, and these are denoted as *ortho*, *meta*, and *para*. The designation is arbitrary, but the consequence is very different chemical structures.

ortho-   *meta-*   *para-*

The chemical formula of PAA is $C_{14}H_{14}N_2O_3$, and its chemical structure is as depicted below:

In many circumstances it is possible to generate a variety of chemical structures from one chemical formula, that is the same proportions of atoms connected in different ways can result in different molecular compounds having different shapes and properties. These are called *isomers*. This is the case with azoxyanisole, and Vorländer, or at least his laboratory assistants, were able to prepare the *ortho* and *meta* isomers of PAA, and also showed that they were *not* liquid crystalline. It was this more than anything that persuaded Vorländer that molecular shape was all-important in determining the stability of liquid crystals.

ortho- azoxyanisole   *meta-* azoxyanisole

The *ortho* and *meta* isomers have shorter overall lengths, and more particularly they are less linear than the liquid crystalline *para* isomer.

images were at least a million times bigger than a molecule. It was the chemist Vorländer that made the connection between what he saw in the microscope and the shapes of the molecules that constituted the liquid crystals. From the standpoint of twenty-first century science, it is clear that the connection between molecular shape and the phases they form is very complicated, but Vorländer's intuition was largely correct, as this comment from 1923 shows,[5]

> *no branch of research has permitted the explanation of the inner relationships between molecular shape and crystal forms better than the investigations into the liquid crystalline state.*

It was sufficient to stimulate physicists to think about the molecular structure of liquid crystals, even before they were sure that molecules existed.

The exotic forms and shapes adopted by liquid crystals are illustrated in Figure 3.2. These were generated by allowing a liquid crystal phase to form in suspension in olive oil. What was fascinating to observers was how the micro-forms combined. Here was the evidence for Gattermann's copulation phenomenon, but there were other significant dynamics. For example, rod-like or worm-like 'liquid crystals' would merge when they contacted, but the resultant shape often depended on the way in which the combining liquid crystals approached one another. If two rods came together alongside, then merging would take place without a substantial change in shape—the resultant rod was just thicker. From Vorländer's perspective, it was reasonable to believe that the shapes seen in the microscope reflected the way in which the molecules were organizing themselves. The molecules were coming together in parallel rows and bundles, and the optical evidence was plain to see. The approach of Vorländer was in fact a very modern one. Even today chemists enjoy

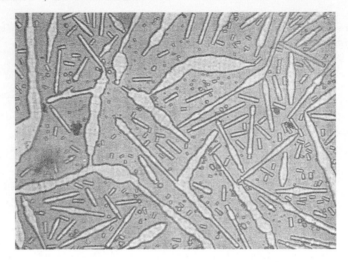

Figure 3.2 Formation of a liquid crystal phase from a solution. The rod-like forms and their coalescence encouraged Vorländer to believe that the molecules should be rod-like in order to form liquid crystal phases. From D. Vorländer, *Kristallinisch-flüssige Substanzen*, Ferdinand Enke, Stuttgart 1908 (Figure 21).

building conceptual models of how molecules might aggregate to form bigger functional structures. However, in the present day there are many ways to test the model building that were unavailable to Vorländer.

The one obvious fact for Vorländer was that to form a liquid crystal, a molecule must be extended, made rod-like, and the more rod-like the better. Introducing bends or kinks in the structure worsened the liquid crystal-forming performance, but making the molecules longer enhanced it:[3]

> *From the linear shape of the molecules follows their parallel orientation and hence the anisotropy of the liquid.*

There were many subtleties, and one suggestion,[6] subsequently withdrawn, was that if molecules were built with alternating charges along the rod axis, this might help the formation of the liquid crystal phase. As we shall see later in this chapter, this particular unsubstantiated observation was to have unfortunate consequences in the development of a theory of liquid crystals.

Over the years Vorländer and his students synthesized hundreds of liquid crystalline compounds. An interesting discovery was that amongst the slimy liquid crystals were many soaps and soap-like compounds. In 1907, he detected for the first time that a given substance may exhibit more than one liquid crystal phase. Already by 1908 he had enough material to feel able to report his accumulated results, not just in a series of papers, but in a book, which he entitled *Kristallinisch-flüssige Substanzen (Crystalline-liquid substances)*.[7]

Figure 3.3 The collection of cigar boxes in which Daniel Vorländer stored his liquid crystal samples. Courtesy of the University of Halle. (See colour plate 7.)

Vorländer's book, like that of Schenck, was aimed at the practical scientist wishing to know about and work in the field. Not for him the philosophical speculations of Otto Lehmann. In general it is very readable. It was of course necessary to cite the relatively few papers in the field (some of which, as we shall see, were by this time in French). Nowadays, no-one's career ever suffers by a little overenthusiastic self-citation. so it is interesting to observe how Vorländer refers to the Meyer–Dahlem paper from 1903, from which his name was so ostentatiously and surprisingly missing. By the time 1908 came around, Daniel had forgotten! The work had been done in his laboratory by his students. He must have been a coauthor! As a result we read of the paper by 'Vorländer, Meyer and Dahlem'.

By the time of his retirement in 1937, more than 1000 liquid crystalline compounds had been synthesized in Vorländer's institute. With the approaching hostilities, all the samples, carefully labelled, were packed in cigar boxes and stored in a cupboard in the Institute (Figure 3.3). During World War II much of Germany was destroyed, but a bizarre sequence of events (described in Chapter 8) meant that Vorländer's samples survived. More than a decade later these liquid crystals were unearthed and used to identify new varieties of different liquid crystal phases, suspected but not proved by Vorländer. A collection of these compounds in sealed glass tubes and kept in attractive cigar boxes can still be inspected in the Institute of Physical Chemistry of the Martin-Luther-University in Halle.

Professor Lehmann on the warpath again

Vorländer's book was a significant development because it drew attention to the potentially vast number of chemicals that might be expected to form liquid crystals. His chemical judgement was impeccable, but he was socially unwise enough to ask 'who was the *real* discoverer of liquid crystals?' He concluded, almost in a throw-away sentence, that, given the circumstances, it was right and proper to consider both Reinitzer and Lehmann as discoverers. One might think that the publication of such an impressive tome would have had the consequence that thousands of synthetic chemists all over the German-speaking world and even beyond would be reaching for their test-tubes and Bunsen burners in their eagerness to emulate and surpass Vorländer.

The immediate result was more prosaic. An unintended consequence of the book was an unedifying battle, fought out on the pages of the *Annalen der Physik*, concerning priority over who had *really* discovered liquid crystals. Within weeks, maybe even within days, of the appearance of Vorländer's book, Otto Lehmann, fresh from his intellectual victory over the unbeliever Tammann, reached for his pen.[8] But it was not the chemical content of the book to which he took offence. His problem was the matter of *priority*. We quote the first paragraph of Lehmann's opening gambit in his battle to defend *his* discovery:

> **The history of liquid crystals**
> **O. Lehmann**
>
> *In a book published recently, the origin of the liquid crystal concept has been described. The book contains many new observations and is a valuable addition to the chemical liquid crystal literature. However, it presents a picture of the origin of the liquid crystal concept which is seriously misleading. One might suspect from this book that it was Herr Fried. Reinitzer, who is professor in the Botanical Institute at the University of Graz, who had actually discovered the phenomenon in 1888, and I had merely renamed it. I present here, for the first time, the full story.*

Reinitzer replied some months later in *Annalen der Physik*. He set the record straight as follows:[9]

> *From these arguments it is indisputably clear that the unambiguous concept of flowing crystals was recognized by Lehmann first of all by studying my derivatives. Furthermore, it is obvious that the perception is due to Lehmann, but that I also contributed considerably in this matter... One should admit that the credit of discovering the phenomenon ought to be attributed to me.*

Reinitzer had not continued his investigation of cholesteryl derivatives. This plea for scientific recognition was his last contribution to the liquid crystal story. His career continued as a university administrator and director of the Botanical Institute in Graz.

It was not only against Reinitzer that the barbs of Lehmann's pen were applied. Here he is again, in 1914, in a direct reproach to Vorländer:[10]

Mr D. Vorländer believes he should reproach me for some errors, but in reality these errors do not exist. However, his remarks are of value. Misunderstandings between the points of view of Vorländer and myself have often arisen by confusion. Hopefully these misunderstandings can be put behind us. Vorländer has obviously only concerned himself with the investigation of which chemical constitution of a substance is necessary for the appearance of a liquid-crystalline modification. In this endeavour he has achieved great success. However, he has only rarely been involved with physical and crystal optical investigations. As a result it is not surprising that my ideas seem extremely unfamiliar to him.

The molecular science of liquid crystals

By the end of the nineteenth century, mathematics had been applied to optics and sound, to elasticity in liquids and solids, and perhaps most triumphantly to electricity and magnetism. However, the application of mathematics to Lehmann's liquid crystals was hard. The picture developed by Lehmann and Schenck was persuasive, but actually the evidence for liquid crystals was only circumstantial. The alternative mixture picture from Tammann and colleagues had been seen off qualitatively, but the underlying doubts about liquid crystallinity remained. Without a detailed mathematical picture no-one could feel confident that the advances in the frontiers of liquid crystal science would hold, and that a strategic retreat would not be necessary.

Where was one to start? At this time there was only the 1873 model of the fluid state due to van der Waals, which has already been referred to. This leads naturally to the idea of a liquid–gas phase transition, but of course says nothing about liquid crystals. Theories of the crystalline state seemed less useful, but a model for permanent magnets recently developed by Pierre-Ernest Weiss (1865–1940) in 1907 was to prove a valuable starting point.[11] This theory is now known as the Curie–Weiss theory of ferromagnetism.*

The lack of a mathematical theory was keenly felt by the experimentalists in the field. The first serious attempt at a mathematical theory of liquid crystals was made by Emil Bose (1874–1911) (see Figure 3.4) from the physical chemistry department of the University of Danzig (now Gdansk in

---

* The Curie is Pierre Curie (1859–1906), husband of the Polish-born physicist and chemist Maria Skłodowska Curie, the discoverer of radium, usually known as Marie Curie. The Curies were a talented pair: Marie and Pierre shared with Henri Becquerel the 1903 Nobel Prize in Physics for the discovery of radioactivity, while Marie received another Nobel Prize in Chemistry in 1911 for her discovery of radium. Marie was the pioneer of the use of radiation in medicine, and died of cancer in 1934.

Figure 3.4 Emil Bose, pioneer liquid crystal theoretician.

Poland, but then a German-speaking city) over the period 1907–1909. Bose's interest in the field had been aroused by discussions he had with Daniel Vorländer. As a synthetic chemist, Vorländer's expertise in mathematics was strictly limited, but he was aware that he had no overarching picture of the crystalline liquid phase in which to fit his experiments.

It was Vorländer's influence which led Bose to use the term *crystalline liquids* rather than *liquid crystals*. His contribution over this period amounts to three papers in the *Physikalische Zeitschrift*. The first of these compared the Lehmann and Tammann pictures of liquid crystals, and was entitled 'For and against an emulsion structure for crystalline fluids'.[12] The basic conclusion (not quite for the first time, but certainly not for the last time, in the history of the subject!) is that rather than talk of liquid crystals or crystalline fluids, it would be better to refer to *anisotropic* fluids. The evidence, he stated, only stretched to anisotropy, rather than crystallinity.

The second paper, which was perhaps the most influential, is entitled 'On the theory of anisotropic fluids'.[13] Bose tried, not entirely successfully, to draw eclectically from both the van der Waals and the Weiss theories. He was led to introduce the idea of molecular *swarms* (*Molekülschwärme*). With the benefit of hindsight we can see that the theory was not really successful, even as a theory, but in the land of the blind, the one-eyed man is king! For many years the literature contained earnest but awkward discussions about the difference between a *chemical* molecule (the real one!) and the *physical* molecule. The latter was supposed to be one of a swarm of molecules pointing more or less in the same direction. In any event the *swarm theory*, as it became known, passed, albeit temporarily, into the canon. As late as 1957, it was quoted approvingly in the major review article of the day in *Chemical Reviews*.[14]

Although Bose's papers were not seen as completely successful, his career was on an upward trend. The University of La Plata in Argentina was opening a physics department. At this period of history, Argentina enjoyed a high standard of living and possessed the fourth largest gross national product per capita in the world. They were able to hire the best young people, and where better to turn than Germany? Advice was sought from the distinguished physical chemist Prof. Dr Walther Nernst (a future Nobel Prize winner) at Göttingen. Nernst recommended his former Ph.D. student, the 34-year-old Bose.

So it was that Easter 1909 found Bose in Lisbon, setting sail for the New World and a new life with his wife, the chemist Margarete Heiberg. Together they worked tirelessly to build up the Institute of Physics at their new university. Then, in early May 1911, Bose fell ill with typhus. Within two weeks he was dead. He was only 36 years old.

TECHNICAL BOX 3.2 Swarms, magnets, and electric dipoles

The two elements of van der Waals theory for liquids and gases were that molecules had a size (excluded volume) which kept them apart, but also there was an attractive force between them that encouraged aggregation, similar to that between magnets. It was the balance of these effects that resulted in the properties shown by liquids and gases. At high temperatures in the gaseous state, the molecules were in rapid motion and tended to keep apart, so effects of size were of less importance than at low temperatures. As the temperature is lowered, the speed of the molecules decreases and they begin to experience the attractive forces of their neighbours. They cannot get too close, however, because of the excluded volume of the molecules: the compromise results in the liquid state.

From his experiments on liquid crystals, Vorländer concluded that the shapes of the molecules were important in determining whether or not a liquid crystal would form. So how could this extra influence be factored into the fluid theory of van der Waals? Emil Bose realized that giving a shape to a molecule meant that its dynamic behaviour would depend on its relative orientation with respect to its neighbours. Indeed he used the analogy of Vorländer's cigar box. It is possible to fit more cigars into a box if the cigars are parallel to each other—so it is with extended rod-like molecules. They will interact more effectively if they have the same orientation, that is they point in the same direction. The driving force for the molecules to come together is their mutual attractiveness, while their shape forces the molecules to line up to create a molecular swarm. This is the liquid crystalline state.

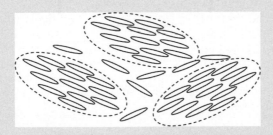

Representation of liquid crystal swarms.

There was still a bit of a problem. Liquid crystals are fluid, and so the molecules were in rapid motion: would not this destroy the swarms? This is where the doubts about the theory creep in. Perhaps, argued Bose, the molecules were in swarms locally, but there were lots of swarms moving around in the liquid crystal fluid. The boundaries between the swarms could be constantly changing, but individual molecules would each feel themselves to be a part of a single swarm. Not only did this idea solve some of the conceptual problems, it could also explain the turbidity (cloudiness or scattering of light) that was characteristic of liquid crystals. Undoubtedly this came from the various boundaries of the swarms. If all the swarms were lined up in some way, then one would surely have an anisotropic liquid.

Like many ideas in science, Bose's model for liquid crystals was adapted from another problem in physics where it actually works. Magnetism is a great phenomenon of nature. The earth is a big magnet with a north pole and a south pole. Everyone is familiar with smaller magnets, compass needles, etc., all distinguished by having north and south poles. Unlike poles attract one another, like poles repel. The theory of magnets (ferromagnetism) formulated by Pierre Weiss was that all magnetic materials, usually compounds of iron, consist of lots of small magnets arranged in a parallel fashion such that the north poles point one way and the south poles the other. However, it is a fact that not all magnets are permanent, that is the magnetism could apparently be destroyed by heating or even hitting with a hammer. It could be recovered, however, by heating and cooling, or by gently tapping the magnet in the presence of another magnet. For Weiss, the explanation was obvious. All the little magnets were indeed lined up in domains, but, all things being equal, the domains could point in different directions. What separated the domains were domain walls, and careful experimentation revealed the existence of these walls in real magnets. To produce a permanent magnet, the domain walls had to be removed, which could be achieved by controlled heating in a magnetic field or gently knocking them out.

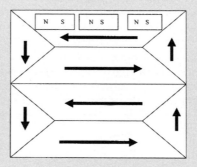

Domains and domain walls in a ferromagnet. The direction of the magnetic field in the different domains is represented by the broad arrow, which corresponds to the alignment of contributing magnets depicted schematically in the top domain with north and south poles.

The analogy with liquid crystals was so close that it had to be true, at least so Emil Bose thought. But he was wrong. There are no swarms or swarm boundaries. True, there is molecular organization in liquid crystals that results in parallel orientation of neighbouring molecules, but on a molecular scale liquid crystals are homogeneous or continuous.

Bose knew that the molecules of liquid crystals were not magnets, but that didn't affect his overall picture of the molecular organization; what was lacking was any idea of the attractive force that caused the molecules to come together. There are no north poles or south poles, but Vorländer pointed out that his molecules carried electrical charges. Within any molecule they were balanced, i.e. equal numbers of positive and negative charges, but they could still interact with charges on neighbouring molecules. In the language of the chemist, the molecules possessed electric dipoles, and such dipoles with positive and negative ends behaved in similar ways to magnets with north and south poles. Like charges would repel and unlike charges would attract. Eventually, careful experiments by Vorländer's assistants disproved the idea that electrical charges on liquid crystals were important in determining liquid crystal phase behaviour. In twenty-first century liquid crystal science we know that electric dipoles can influence the properties of liquid crystals, but not in the dramatic way initially envisaged by Vorländer and translated into theory by Born. For purposes of illustration, we show below the structures that Vorlander investigated with the likely charges represented as positive (+) and negative (-).

Vorländer's liquid crystal molecules with alternating positive and negative charges.

It is difficult to understate the sense of heartache that Bose's death evoked. Even nearly a century on, his efforts are still warmly remembered in Argentina. For his friends, the news was devastating, particularly because he was young, so far from home, and he had already tragically lost a first wife from illness. His fulsome obituary concludes painfully that he 'died in a foreign land, but lives on, both there and in the homeland, through his works and in the hearts of his friends'.[15] There can be little doubt that, had he not died, Bose would have found his way reasonably quickly to a more-or-less correct molecular theory. As it is, the wait for this turned out to be rather long.

In 1916 the liquid crystal problem came to the notice of Max Born (1882–1970), Professor of Theoretical Physics in Göttingen. While Bose had borrowed primarily from van der Waals, paying lip service to Weiss, Born concentrated more closely on adapting Weiss's theory of magnetism to the liquid crystal context.[16] His basic assumption was that liquid crystalline molecules carry an electric dipole (a separated positive and negative charge, the electric equivalent of a localized magnet) pointing in the direction of the molecule, and that this quantity drives the liquid crystallinity. In 1910, Vorländer had suggested that molecules with dipoles might have an enhanced tendency to form liquid crystals.[6] Perhaps this suggestion from the pre-eminent liquid crystal chemist further encouraged Born to develop his theory.

The theory is great. Great in this context means that the maths is right, and the assumptions and conclusions are clear. This wasn't the case for Bose's swarms, which were less easily amenable to experiment. The problem with theories is that although they provide explanations, they also make predictions. If a single prediction is proved to be false, then the theory is fatally flawed, at least until someone comes along to patch up the theory. Born's elegant theory failed in one predictive respect. By analogy with the magnetic theory on which it was based, it would be expected that liquid crystals would end up as the electrical analogues of magnets, with a large positive charge at one end and a large negative charge at the other: rather like a mini battery. This demonstrably does not happen, except in some very exotic liquid crystal phases discovered in the 1970s and now known as ferroelectric liquid crystal phases.[17]

There was worse to come. Vorländer had already gone back on his original suggestion, and subsequent work in his group had showed that electric dipoles were not essential for liquid crystal-forming molecules, but it was too late, Born's theory had already been published. The *coup de grace* for Born's theory was delivered in 1925, when Szivessy in Budapest made a nematic liquid crystal consisting of molecules *without* any electric dipole in the structure.[18] This demonstrated that the basic assumption behind the theory must be wrong. One experiment had killed the theory stone dead, and Born's paper has been relegated to the status of an interesting historical footnote.

Born himself went on to greater things. Around him in Göttingen gathered the bright young things who in the years after 1923 created quantum mechanics,[19] which changed the whole way we think about the physical world. Later still, Born, who was of Jewish extraction, fled Germany to escape the Nazis. For many years he was Professor of Theoretical Physics at the University of Edinburgh in Scotland. Uncommonly, later still, he and his wife returned to Germany on his retirement.

Lehmann's career

We have met Otto Lehmann as the prime mover in the discovery of liquid crystals. We shall meet him again in subsequent chapters as a significant player in a number of different dramas. Notwithstanding his very human frailties, he is *the* hero of this part of the story, *the* researcher who, in the face of considerable scepticism by his contemporaries, established the new field of liquid crystals. He is, unjustly in our view, little known outside the narrow academic field he made his own. Before we leave this pioneering period, it is right that we should record the high points in his career.

The post in Technische Hochschule to which Lehmann was appointed in 1889 was originally primarily a teaching post. Having authored a textbook on molecular physics, and later several other texts, Lehmann was ideally qualified. His first task was to organize laboratories for the students. As we have seen, Lehmann always emphasized to his students the importance of personal observation in the development of scientific culture. As we have commented at the outset of this book, in Anglo-Saxon culture in recent years there has been a separation (some would say even a divorce) between general and scientific culture, but Lehmann was the antithesis of an ignorant man. He enjoyed wide interests outside science, and cultivated a panoramic view of scientific culture. This enabled his teaching career over his 30-year period in Karlsruhe to be extremely successful.

A particular highlight occurred in the academic year 1900–1901, when Lehmann was Rector of the Technical High School. To the Rector fell the task of delivering the commencement lecture to graduating students. For the graduating students, this lecture would be their final memory of their years at university. Lehmann was determined that it would present a message that they would not forget.

Lehmann's Rector's Lecture,[20] entitled 'Physics and Politics', was an intellectual *tour de force*. It ranged over two millennia of cultural history, from Aristotle and Alexander the Great in ancient times, to firearms and the (then contemporary) steam engine. He invoked the example of Benjamin Franklin (1706–1790), whose early career was devoted to science, and whose later life was dominated by the struggle for American independence. For Lehmann, as for scientists in modern times, the link between physically based arguments and politics was self-evident. Nowhere was this more obvious than in the considerations governing the energy which powers the modern economy.

Given the urgency of this scientifically based political discourse in the early twenty-first century, it is worth repeating some of Lehmann's arguments in detail, for there is a prophetic ring to Lehmann's 55-page narrative. Where, Lehmann asked, does our energy come from?

> *Coal originates from primeval plants, through the reduction of atmospheric carbon dioxide. Using a procedure due to the most distinguished of present-day physicists, Lord Kelvin, it is possible to estimate the world's coal reserves from the quantity of oxygen in the atmosphere. Let us make the most favourable assumptions, namely that: (a) all this coal will be found, including inaccessibly deep coal, as well as coal under the ocean bed; (b) it is possible to extract it without inordinate cost; and (c) consumption remains at present-day levels. Then the reserves will last another four to five hundred years. But these hypotheses are not valid. It can therefore be predicted with some certainty that the reserves will be exhausted in one to two hundred years. What then?*

What is interesting from this paragraph is that this argument for future coal supplies relates to *origin*, rather than to *extrapolation*. In fact, sensible contemporary extrapolations from current usage and the rate of discovery give very similar orders of magnitude for the time-scale on which coal will be exhausted, thus providing considerable support for the physical mechanism for the origin of the coal.

There then follows a discussion of the political consequences of his deduction. Exhaustion of fossil-fuel reserves in one to two hundred years was not an immediate crisis in 1901.

> *What if no alternative is found? There would be unimaginable misery, terrible revolutions, and the larger part of humanity would experience the convulsions of a dying civilization! Probably there will be other energy sources, such as the movement of the earth, the internal heat of the earth, or solar radiation. But we have no means of exploiting them in significant quantities.*

> *The most likely alternative to coal comes from the exploitation of the sun's heat. The plants are able to use up some six thousandths of a per cent of this in the form of the heat from the burning of wood. Only a tenth of these six thousandths of a per cent can be converted into useful work by our steam engines. It does not seem inconceivable that physics might succeed in learning from plants the secret of energy storage. In such a case it would be possible to replace the technology based on iron and steel and the steam engine. This new technology would use soft and half-fluid materials,*

products of modern chemistry, to achieve much greater efficiency than the organic activity of the plants.

The unexpected very recent discovery of liquid crystals suggests that something new may well yet turn up in this respect...All these are conjectures! Only one thing is certain, and that is that the replacement for coal, if it is found, will never be found by chance! Never is a physical discovery made by chance!

Thus as early as 1901, an energy crisis had already been predicted to occur in the twenty-first century. Solar energy had already been predicted to replace coal, but for all his erudition, Lehmann was unable to predict the importance of oil to late twentieth century economies! Lehmann also foresaw that his beloved liquid crystals would be a route to energy salvation. One cannot help but despair at the continued lack of response from the scientifically untutored governing classes, given that this material was in the public domain more than a century ago.

Turning from scientific to moral discourse, Lehmann then appealed to the better nature of his student audience. His final appeal rings through the ages and serves as a reminder (if one were needed in the present troubled times!) that it is possible for idealism to defeat ignoble self-interest:

If the opportunity presents itself, cast a glance at the history of physics. It is a treasure trove of precious stimuli and spiritual pleasure. Here you will see intelligent and idealistically-inclined people devote their whole life enjoyably seeking out nature. Not for them the pursuit of money, but rather the pursuit of the common good. They are ignored, despised, laughed at, and fought against by those for whom they work. We are filled with admiration in the face of such greatness of character!

The Karlsruhe *Technische Hochschule* had no research focus and was not authorized to award Ph.D. degrees, so Lehmann had to develop the research programme from scratch. Part of this effort involved close relations with industrial firms, for example he maintained close relations with the chemical firm E. Merck of Darmstadt. In an exchange of letters in 1905, he can be found encouraging this firm, better known for its medicines than its chemicals, to manufacture liquid crystals for research work. Lehmann was keen to exploit his scientific expertise in the wider commercial world, and his microscopes were marketed by the well-known photography firm Carl Zeiss of Jena, as shown in Figure 3.5.

In the spring of 1909 Lehmann was invited to speak in France and Switzerland. This was a wonderful opportunity to proselytize, which he accepted with alacrity. The eventual result of the subsequent visit would be a major shift in the geographical centre of gravity of liquid crystal research. We shall relate the story of that visit, and its consequences for progress in liquid crystal science, in the next chapter.

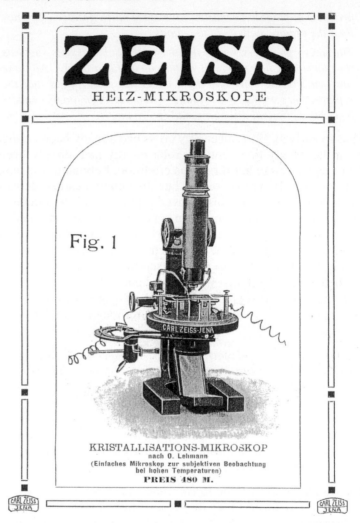

Figure 3.5 Advertisement from the well-known Zeiss company in Jena, Germany, for a hot-stage crystallization microscope designed by Otto Lehmann, dating from about 1906.

Lehmann was much honoured, receiving the title *Hofrat* in 1892, and *Geheimrat* in 1902, as well as picking up various knighthoods along the way. These titles went back to the times of the Holy Roman Empire, and are concerned with the duty of advising the Emperor (they fell into disuse after World War I). To be a *Geheimrat* was a great honour. To omit the title when addressing Lehmann would be a significant breach of protocol. It would be as though in Britain, Sir Otto Lehmann were addressed as Mr Lehmann and not as Sir Otto! A mark of his scientific

Figure 3.6 Top: Lehmann's hammer mill in the Black Forest, advertised as a summer school and a sanatorium. 'In a specially reserved room he demonstrates and explains on rainy days the most important features of liquid crystals to guests of the sanatorium who are interested in this topic.' Bottom: a photograph of Lehmann in his private laboratory with his children.

reputation is that on several occasions between 1913 and 1922 he was nominated to the Swedish Academy as a candidate for the Nobel Prize in physics, but on each occasion was just pipped at the post.

There is another somewhat surprising way in which Lehmann was a pioneer.[21] He was a great lover of the outdoors. Much of his leisure time was spent in the Black Forest, South of Karlsruhe. In 1910 he bought an old hammer mill 730 m up in the Black Forest Hills (Figure 3.6). This he fitted out as a laboratory, in which to offer summer schools on liquid crystals in a peaceful mountain environment. He made strenuous efforts to raise funds for a foundation to support the summer schools, approaching colleagues both in Germany and abroad. The result of these efforts is not recorded, but it is likely that whatever success he might have had would have been blown away by the war which started soon after in 1914.

Whether or not Lehmann was successful in prosecuting his 'Physics in the Mountains' schemes, his concept is now well-established in modern science. Nowadays summer schools, winter schools, conferences, and seminars abound in attractive venues around the globe. The physicist only has to mention Les Houches (in the French Alps), Erice (on the top of an isolated mountain in Sicily), or Aspen (Colorado, in the middle of the Rockies, just convenient for skiing), and colleagues will lock eyes knowingly as if to ask, 'How do I get an invitation?'

A good part of Lehmann's later scientific activity was spent on the applications of liquid crystals in the life sciences. We postpone an examination of this part of his work until Chapter 6, as it is really part of another parallel story. In 1915, Lehmann turned 60, and one might have expected some slow-down in his activity, but even after World War I, his scientific work continued at a furious pace.

He combined lecture courses and university administration, and also continued to publish prolifically. His health, by all accounts, remained good, and his colleagues remarked on his strong constitution. His sudden death, on 17 June 1922, came as an unexpected shock and surprise to his family and colleagues. His obituarists were united in their praise for a far-seeing and committed scientist and warm family man, who had opened up a new field.[22] Just how far the new field would travel, both intellectually and geographically, we shall see in the chapters to come.

4
La Gloire Française

⎯⎯⎯⎯◦◦◦⎯⎯⎯⎯

> It has been in France, above all, that the main steps in understanding
> the principal concepts have been made...
>
> Georges Friedel 1922

Languages and science

The last quarter of the nineteenth century was a period in which Germany
enjoyed unparalleled economic and political strength.[1] The *Länder* of
the modern German republic were originally separate entities. In 1834
they were linked together in the so-called *Zollverein*, a customs union
that removed many barriers to economic expansion. This customs union
was the first step towards the politically unified German state, achieved,
under Prussian domination, by Bismarck in 1871.

During this period Germany surged ahead of both Great Britain and
France, whose industrial revolutions had preceded that in Germany.
The chemical firms of Bayer and Merck, and the electrical firm of
Siemens are but three examples of successful industrial innovators
whose names have survived to the present day. The Darmstadt-based
(and family-owned) E. Merck, founded in 1668, initially concentrated
on pharmaceutical products, but later diversified into industrial chemicals. This company played, and continues to play, an important role in
the unfolding story of liquid crystals. A particular strength of the reju-
venated Germany was the close relationship enjoyed between the technical high schools (really, in contemporary language, not high schools
at all, but technical universities) and the industries in which their grad-
uates worked.

Hand in hand with the economic advance came a period of major
scientific progress in Germany. The physicists Weber and Ohm gave their
names to electrical units, and Hermann von Helmholtz (1821–1894) formulated the theory of conservation of energy.[2] Of chemists, the names of
Liebig (who constructed the first university laboratory in Giessen in 1827)
and Bunsen (whose burner decorated all school chemistry laboratories

when the present authors were young) are easily called to mind. With the scientific progress came an added importance of the German language for scientific discourse.

In medieval times, almost all scientific discourse was in Latin. Indeed you couldn't *do* science—or natural philosophy as it was known in those days—at all, if you didn't know Latin. Isaac Newton's major work on the dynamics of astronomical bodies, *Principia*, was published in Latin in 1686, and not translated into English until 1729.[3] However, sometime in the seventeenth century this universalism seems to have broken down, for Newton's other great work—his *Opticks*—was published in the vernacular (that is, in Newton's own language, which in this case was English), in 1704.[4]

So by the beginning of the twentieth century, individual nation states in Europe sustained a flourishing scientific literature, all in the vernacular. Various German journals appeared naturally enough in German. One particularly prominent journal was the *Berichte der Deutschen Chemischen Gesellschaft* (Reports of the German Chemical Society). Likewise, the *Proceedings of the Royal Society* of London appeared in English, and the French Academy of Sciences' flagship journal, the *Comptes Rendus de l'Académie des Sciences* (C.R.A.S.) appeared in Paris in French.

This state of affairs continued for many years, beyond indeed the beginning of the scientific careers of the present writers. Only in the mid 1970s did the global dominance of English begin to squeeze out other languages, as papers in European journals began to appear in English rather than in the relevant national language. Nowadays, of course, a scientist trying to make a career without an adequate command of the English language will cut a very sorry figure.

The news leaks out

Reinitzer (from Austria) and Lehmann (from Germany) were both native German speakers, albeit that they spoke that language with different hues. Naturally their work, and indeed that of almost all the pioneers in liquid crystal research, appeared in the German language. The introduction to Rudolf Schenck's 1905 volume *Kristallinische Flüssigkeiten und flüssige Kristalle* (*Crystalline liquids and liquid crystals*)[5] includes a list of 41 'works which are concerned with liquid crystals'. Of these 41, 39 are in German and two are in Italian.

It is perhaps fitting that the first intellectual external port of call for liquid crystals turns out to be in Italy. Since the 1970s Italy has developed a formidable degree of expertise in the mathematics, physics, and chemistry of liquid crystals, but as far as we can gather, there is no intellectual continuity between the Italian liquid crystal papers of 1902 and the current Italian interest in the subject. The two Italian authors are the mineralogist Carlo Maria Viola (1855–1925), later a professor of mineralogy at the University of Parma, and the physicist Alessandro Amerio (1876–1965). Amerio was writing at this stage of his career from Florence, where he was working at the Institute for Higher Study. Later he was to become full professor of physics at the prestigious Politecnico di Milano. Much of his career was concerned with studies of solar radiation, which he started not long after the appearance of his only paper in the liquid crystal field.

Viola tried to use crystallographic rules, with which as a mineralogist he was very familiar, to figure out the possible crystal structures in liquid crystals.* As we shall see later, this was in fact a scientifically flawed idea. It is not therefore surprising that this paper died the death that is the fate of much scientific work.[6]

The paper by Amerio is more fruitful.[7] It is entitled '*Sui cristalli liquidi del Lehmann*' ('On the liquid crystals of the Lehmann')—Italian has a quaint way of referring to someone important using a definite article. Amerio makes abundant use of the definite article, using it to refer not only to Lehmann, but also to Gattermann, Quincke, Reinitzer, and Tammann. In modern language, the key result of Amerio's work, later more carefully re-measured by Schenck himself, is that the transition between the liquid crystalline and liquid phases involves a heat change (known technically as a *latent heat*), similar to that which occurs when, say, ice turns into water. In other words, it is a *real* change; something real is going on inside the materials over and above the fact that there is a change in appearance.

Perhaps more interesting, however, are the comments Amerio makes in his final section, entitled 'General observations: are the liquids crystalline or anisotropic?' The speculations are interesting because Amerio, unlike his contemporaries, finds himself constrained to make a distinction between the two. By contrast, other workers seem straight-jacketed at this stage by Lehmann's choice of language.

Apart from Tammann, who denied the existence of the liquid crystals altogether, other workers had noticed that the liquid crystals were anisotropic, but assumed that they just did not yet understand the details of the structures involved. Tammann (correctly) was just at the same time realizing that the idea of a crystalline liquid was a contradiction in terms, but had assumed that the whole idea was bunkum and that Lehmann and everyone else had blundered in some, as yet undetermined, stupid way. Here Amerio is anticipating the later resolution of the problems:

> *(This) hypothesis... induces us to think about a special state of matter, intermediate between a liquid and a solid, a state which*

* Viola may have learnt of liquid crystals through spending summers with fellow crystallographer Paul Groth, one of Lehmann's teachers, who subsequently moved to Munich.

might possess in common with a solid a connection between the molecular orientations, and with a liquid, the capacity for molecular translation...

At that stage, the Italians had no claims to major power status (Mussolini's advent to power was more than 20 years in the future). Italians understood that they needed to read the foreign language literature. Amerio's paper includes 12 references, nine in German and three in French, but it is probably the fact that Amerio himself was writing in Italian, a minor language, which inhibited the influence of Amerio's scientific insights. Apart from the encyclopaedic reference in Schenck's book, Amerio's paper also died a death. Meanwhile, his ideas seem to have awaited rediscovery by the scientific mainstream; a similar point was to be made by Bose seven years later.

The remaining 39 German language papers referenced in Schenck's book came from 19 separate authors or sets of authors, with the pioneer Lehmann contributing no fewer than 11, and Schenck himself as many as 9. Of course, not only German native speakers read and published in German-language journals. Schenck mentions in his list of liquid crystal papers work by the young American George Augustus Hulett (1867–1955). Later a distinguished physical chemist at Princeton University, he was completing a Ph.D. under the supervision of Wilhelm Ostwald at Leipzig. In Leipzig he completed a paper for the *Zeitschrift für physikalische Chemie* (Journal of Physical Chemistry) entitled '*Der stetige Übergang fest–flüssig*' ('The sharp solid–liquid transition'),[8] published after he had already sailed back to the USA to a position at the University of Michigan.

There are also two Dutch contributions, including one on Gibbs's phase rule from the distinguished physical chemist Hendrik Willem Bakhuis Roozeboom (1854–1907),[9] who had succeeded van't Hoff as professor in Amsterdam. Finally there are two contributions from the Pole Tadeusz Rotarksi (1874–1912),[10] describing work carried out in St Petersburg. Rotarski had been a student of Tammann in Dorpat until 1900, and remained a faithful adherent to his school of thought. Indeed, the title of both papers insistently referred to 'so-called liquid crystals'!

Liquid crystals reach France

Thus we see the beginnings of an international interest in the new liquid crystal materials, as news began to spread beyond the boundaries of the German-speaking world. The famous 1905 Karlsruhe meeting even included a speaker from Paris, albeit speaking in German and on a topic unrelated to liquid crystals. The first French interest in the subject seems to have come from the mineralogist and crystallographer Frédéric Wallerant (1858–1936), who had been appointed to the chair in mineralogy at the University of Paris in 1903.* This ancient and highly respected university, usually known informally as *La Sorbonne*, is one of the most prestigious academic centres in Europe. It was founded in the eleventh century, and predates the universities of Oxford and Cambridge by a century or more.

At that time the subjects of mineralogy and crystallography were strongly interlinked, the idea being to understand the physical and chemical properties of minerals, many of which were, of course, crystalline.† There was also a national network of *écoles des mines* (mining schools), including one at the relatively minor regional centre of St Étienne in the centre of France. These mining schools prepared mining engineers for exploration for materials and their subsequent exploitation all across France's (by now considerable) overseas empire. The range of interests of the faculty at these schools was not as narrowly focused as might be supposed, however, and extended seamlessly and unapologetically to what we might now call the pure physics end of materials science. These mining schools, as well as a network of *écoles polytechniques*, were the French counterpart of the German *Technische Hochschule*.

Wallerant's interests in the years after he was appointed to his Sorbonne chair tended toward what we now call the thermodynamics of solids. One paper, for example, concerned how one might achieve solid solutions, by analogy with solutions in liquids. Another concerned transitions between crystalline forms of the same material, and a third focused on a 'modification' of a crystal (i.e. a particular crystal type) which was stable over two separate and non-overlapping temperature intervals. This series of papers was published in the *Proceedings of the French Academy of Sciences*, of which he was, of course, a member.

Wallerant's papers appeared in French, but many German scientists did not read French, and scientific journal circulation was not as international as nowadays. However, German summaries ('abstracts') of all

* A substantial part of this chapter, concerning the relationship between the original German pioneers and their French counterparts, is based on the account given by H. Kelker and P.M. Knoll (1990).

† The strong French tradition in the subject, dating back to René-Just Haüy, still continued in the early part of the twentieth century. Haüy had been Professor of Mineralogy at the National Museum of Natural History and founder of the Museum of Mineralogy in Paris. Another major French contributor was Auguste Bravais (1811–1863), who in 1845 had developed an exhaustive theory of the different types of crystalline symmetry allowed in a crystalline lattice. This theory underlies much of modern solid-state physics. Indeed 'Bravais lattice' is a key term introduced right at the beginning of any university solid-state physics course. Given that the concept of a truly 'liquid crystal' was still taken seriously at this stage, an obvious theoretical starting point would be Bravais's classification of different possible lattices.

chemistry-related papers, including those in French and other languages, were published in the journal *Chemisches Zentralblatt*, published by the German Chemical Society. Daniel Vorländer was a beneficiary. In his 1908 book on crystalline fluid substances, for example, Vorländer refers to no fewer than five of Wallerant's papers.[11] But he always includes not the original C.R.A.S. reference and title, but rather the German translation and *Chemisches Zentralblatt* reference.

Wallerant's research focus gradually crossed the threshold from solid to liquid crystals. Interestingly, despite the fact that a number of influential voices in the German literature were clearly still not persuaded by the idea of crystals that were simultaneously liquid, Wallerant seems to have been completely won round at a very early stage, as he indicated in 1905:[12]

> *This salt therefore behaves like soft crystals and crystalline liquids, whose existence can no longer be questioned following the work of Lehmann and Schenck.*

By 1906 he was publishing three papers on the subject,[13] culminating in a review in 1907 in the influential Italian review journal *Rivista di Scienza*.[14] By this time his enthusiasm knew no bounds, for we find him commenting:

> *Lehmann's discovery is definitely one of the most important of the last century. Its consequences are numerous and of the first order. They permit (us) in particular to sharpen our knowledge of crystalline matter...*

Wallerant seems to have been the first in the French crystallography community to interest himself in liquid crystals, but he did not remain alone for long. He was soon joined by Paul Marie Benoît Gaubert (1865–1948),[15] who was director of the Museum of Natural History in Paris, and Georges Friedel (1866–1933), professor (and after 1907 director) at the School of Mines in St Étienne. Both Gaubert[16] and Friedel[17] were to publish papers in the liquid crystal field in 1907.

News of the researches of the French school soon reached Lehmann in Karlsruhe. His private notes, in which reference is made to the work of Wallerant and Gaubert, have also been passed down through the Lehmann family to the present day. The scientific correspondence led to an invitation to visit France. So it was that in April of 1909, Lehmann, by now 54 and an established figure, set out for Paris. The visit also took in Geneva; at each venue he gave a long seminar accompanied by experimental demonstrations.

The excitement was great, and the reporter from the Berlin newspaper *Das Berliner Tagblatt* reported in graphic and elegiac style on 20 April 1909 on the lecture which had taken place four days previously:[18]

GERMANY AT THE SORBONNE
From our correspondent

There is a large physics lecture theatre where Marie Curie used to deliver her public lectures. Recently, in the very same lecture theatre, as we have also reported elsewhere, there occurred an event remarkable in a different respect. A German professor, Otto Lehmann from Karlsruhe, spoke about his discovery of liquid and apparently living crystals. He spoke to a completely French audience, at the invitation of a French scientific society, in the lecture halls of the Sorbonne. These lecture halls had only recently been the site of a nationalist rally against a professor whose offence had been to accompany a number of students on a study trip to Germany. The lecture was a great success. For Parisians, this lecture had been arranged at an uncharacteristically early hour in the morning. Notwithstanding, come the hour, a select audience found itself arranged in the tiered rows of the large lecture theatre. In the first row sat professors from the Sorbonne itself, mainly scientists and mathematicians, a whole gallery of erudite eggheads. Behind them could be seen youthful students, elegant ladies, numerous officers from the Artillery and the Marines, and an extraordinary number of churchmen. Not a seat was empty, although the lecture was to be repeated the day after. The scientific première beginning at 10 a.m. was exerting the same attraction as the evening theatre dress rehearsals beginning after 9 p.m.

What an enormous impression! The moment that the professor from Karlsruhe appeared, the assembly greeted him with an applause that let him know that they knew and valued his scientific merit. Then everyone followed the lecture with great concentration. At the end repeated cheering and applause seduced the unassuming professor from his place. This triumph for German science in Paris also carries a cultural and even a political significance. Those who so readily listen to this German, and who so sympathetically consider his discovery, belong to the intellectual elite of their people and form a vanguard of its youth. They all, teacher, churchman, soldier and student alike, communicate and by so doing affect each other. Probably some of those who came to the Sorbonne this morning expected something else. Near me sat some experienced students. In the best case scenario they had expected to be bored by the German Doctor, but they seemed not averse to a little scandal. But later, they were the first to applaud the cinematographical performance of the apparently living crystals. Faced by the modest strength of the German researcher, there was no evidence of displeasure, nor of resistance, nor of fatigue. After an hour, the German professor had acquired a mass of French friends and admirers. This should not be overstated, but neither should it be underestimated. In this land a German professor would have for a long time been mocked as a learned monster. It certainly is of some significance for an intellectual from this community to be greeted here so sincerely and sympathetically.

Of course, the friendly manner of Geheimrat Lehmann also played no small role in this success. At the beginning he was somewhat shy. Perhaps he was surprised by the applause that greeted him. But he soon relaxed. His head, with a grey-blond beard and a high forehead, betrayed no trace of nervousness. His eyes gazed over the auditorium through golden spectacles in a silence that verged on humour. On one occasion the projection apparatus ordered from the skilled technical assistant broke down. The professor waited patiently until the disruption was over. He attracted an even more sympathetic response after the lecture. He gave some favoured individuals an intimate view of his discoveries, using fifteen microscopes...

Lehmann had studied the French language as a student in *Straßburg*, and his 1904 book contains frequent French language footnotes. The learning seems to have come in useful, for there is no mention of a translator and no mention of any problems in comprehension.

There was also an academic record of the lecture, for versions of both the Paris lecture[19] and the later Geneva lecture[20] were published in the scientific literature. The published version of the Paris lecture, in the October issue of the *Journal de Physique*, runs to 22 pages, and includes innumerable figures of liquid crystal droplets and textures. A footnote records that the seminar was illustrated by 114 projections, including a movie (also mentioned in the newspaper report above). In addition the speaker exhibited 64 photographs (19×28) and 124 glass slides (13×13). The experiments themselves were repeated by a large number of members of the Society (*La Société Française de la Physique*) in laboratory sessions held at the Sorbonne after the lecture.

So both the professional and the journalistic reports agree that Lehmann's visit was a great success. The result reinforced the formation of the French school of liquid crystal science, which has remained influential to this day. Amongst those most influenced by Lehmann's visit was a young crystallographer named Charles Mauguin (1878–1958). At this point he was Wallerant's assistant at the Sorbonne. Having started the liquid crystal programme, Wallerant designated Mauguin as the liquid crystal specialist in his laboratory.

Mauguin was the son of a country baker. He attended primary but not secondary school, and had as a consequence originally only aspired to be a teacher. His performance at the teaching college had been sufficiently good that he was accepted into the *École Normale* at St Cloud, from where he was elevated to the prestigious *École Normale Supérieure* in Paris. There he completed a doctoral thesis in organic chemistry. He then changed to crystallography following lectures at the Sorbonne by Pierre Curie (husband of the famous Marie, who had died in a road accident in 1907). Mauguin had been hugely impressed by Lehmann's lecture in Paris in 1909, sufficiently for the young Mauguin to visit him in Germany two years later.

In the years that followed Mauguin published several papers on liquid crystals: short notes in C.R.A.S., long reviews in the *Bulletin de la Société Française de Minéralogie* (Bulletin of the French Mineralogical Society), and even a short German version in the *Physikalische Zeitschrift* (Physical Journal). Mauguin seems to have been under similar publication pressure to contemporary scientists, for it is possible to detect a certain amount of repetition in these (but nevertheless interesting and fundamental) papers.

One of his most important papers in the *Bulletin* was entitled '*Sur les cristaux liquides de Lehmann*'.[21] Like Amerio before him, but only too aware that there was controversy in the field, Mauguin was just slightly hedging his bets about the status of the liquid crystals. To refer to 'Lehmann's liquid crystals', rather than merely 'liquid crystals' made it clear to the world outside which materials were under consideration. On the other hand, he was avoiding an absolute commitment to the concept itself, while at the same time avoiding the implicit slight of referring to 'so-called liquid crystals'. As an extra bonus, the great man himself was kept happy.

It is worth quoting from the French version of this article. The emphasis is Mauguin's own, but we have concatenated Mauguin's rather parsimonious paragraphs. Notwithstanding any discussion about his use of language, it is clear where Mauguin sits in the ongoing debate about the existence of liquid crystals:

> *Azoxyanisole and azoxyphenetole, isolated by Gattermann in 1890, provided for Lehmann one of the most interesting types of* liquid crystals. *When these materials melt, the first at 116°, the second at 138°, they give rise to liquids which are as fluid as water, but* twice as birefringent as calcite. *This birefringence, which is larger than that of almost all crystals, seems enormous when compared with the liquid birefringence caused either by electric (the Kerr effect) or magnetic (the Cotton–Mouton effect) fields. At higher temperatures (134° for azoxyanisole, 168° for azoxyphenetole), these fluids become isotropic through a transformation exactly analogous to crystal melting. When the substances are subsequently cooled, the same phenomena occur in reverse order.*
>
> *That such a high birefringence could exist in materials which were so obviously fluid excited no small degree of surprise amongst*

*physicists and crystallographers. Indeed, some of their number—
and some quite eminent—did not want to believe it. Even nowadays, twenty years after the first experiments, Lehmann has still
not succeeded in convincing everyone. However, a careful study of
these effects hardly leaves any room for doubt. I propose to present
here some new results which constitute full confirmation of the
conclusions reached by the famous German scientist.*

Mauguin carried out a large number of experiments on liquid crystals in thin cells between glass plates, using a standard off-the-shelf
microscope, and of course, like Lehmann, surrounding his cell with
polarizing plates, or polarizers. These experiments were absolutely critical to the understanding of liquid crystals, and indeed Mauguin's papers
still stand as the definitive work on the optics of liquid crystals. He
approached the interpretation of the measurements as a crystallographer, but their significance becomes more apparent when we understand
a little more of the molecular structure of liquid crystals. For this reason
we will defer detailed consideration until we have learnt just a little
more from Georges Friedel later in this chapter.

After these seminal contributions, Charles Mauguin seems to play
no further role in liquid crystal history. Using his assistantship at the
Sorbonne as a springboard, he obtained a professorship in mineralogy
at the University of Bordeaux. Subsequently he moved to Nancy, then
as a *maître de conférence* (lecturer or associate professor), to Paris in
1919, and finally to his former boss's chair at the Sorbonne in 1933,
staying there until his retirement in 1948. Having abandoned liquid
crystals, Mauguin became extremely distinguished for his work in other
areas of crystallography, such as crystallographic group theory, in which
he invented an eponymous notation, and in X-ray scattering. However
for us, Mauguin's careful researches into the optical properties of liquid
crystals, even before they were properly characterized materials, remain
as the standard work. (See Technical box 4.1).

At the same time as Mauguin's researches were getting into full
swing, the competing crystallography group at St Étienne under Georges
Friedel was also beginning work in the field. It is to the profound and
lasting work of the Friedel group that we turn next.

Georges Friedel and the St Etienne group

Georges Friedel's powerful personality, his extensive and scholarly
researches, his exquisitely phrased articles, and perhaps most of all the
prodigious power of his poisonous pen were to have a transformational
effect on the development of liquid crystal research (see Figure 4.1). In
terms of influence on the subject, only Otto Lehmann himself had a

TECHNICAL BOX 4.1 Principal refractive indices and the experiments of Mauguin

Knowing that the director is the optic axis provides (see Technical box 2.1) the key to understanding all of the optical properties of liquid crystals. Mauguin had carried out a series of experiments on liquid crystal films in a polarizing microscope (i.e. between crossed polarizers) in which the optic axis had been aligned by surface treatment of the microscope slides containing the films. At the time of his experiments, the concept of a director had not been introduced, but for Mauguin that didn't matter, since everything could be explained to his satisfaction in terms of the optic axis.

For a film that had the optic axis aligned in the plane of the film, Mauguin detected the classic signature of a birefringent film. As he rotated the film between the crossed polarizers, the image changed from light to dark every quarter of a rotation. Thus, when the polarization of the light was along the optic axis or perpendicular to the optic axis, there was no light transmitted. In between the image changed progressively from dark to maximum brightness at 45°, and then dark again. This is because the polarized light of the microscope was sampling both refractive indices, and there was interference between that part of the light polarized along the optic axis and that polarized along a perpendicular direction, giving constructive interference and light transmission through the second polarizer.

In this arrangement the light was travelling at right angles to the optic axis. When the polarization direction coincided with the optic axis, this corresponded to the largest refractive index, i.e. the slowest speed of light. Rotating the sample so that the polarization of the light was perpendicular to the optic axis corresponded to a minimum in the refractive index. These two refractive indices, maximum $n_{parallel}$ (extraordinary) and minimum $n_{perpendicular}$ (ordinary) are referred to as principal refractive indices, and the difference between them is the birefringence.

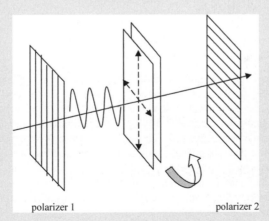

polarizer 1   polarizer 2

A liquid crystal contained by two thin glass slides is placed between the polarizers. Arrows on the front slide indicate the directions of the principal refractive indices. The vertical line corresponds to $n_{extraordinary}$ parallel to the director (the optic axis), and the perpendicular line corresponds to $n_{ordinary}$.

Mauguin was also able to prepare a liquid crystal film for which the optic axis was perpendicular to the containing microscope slides. Under this condition the light travelled along the optic axis, and the image remained black as he rotated the film between the polarizers. It was as if he was looking at an ordinary liquid.

Mauguin's twist experiment

An important experiment that Mauguin performed was to prepare a liquid crystal film in which the optic axis, or director, twisted by 90° through the sample. He did this by rubbing his microscope slides, as he had done before, but rotating one by 90° with respect to the other. He found a remarkable effect. The polarization direction of the light twisted with the director, so the emergent light polarization was at right angles to the incident polarization. So, however he oriented his liquid crystal sample, light was transmitted. This effect is known as 'guiding', and relies on having a thin sample roughly equal to the wavelength of the light being guided.

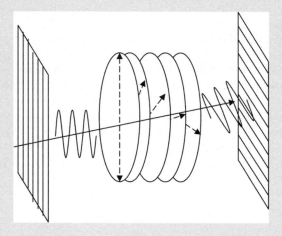

A final experiment was performed by Mauguin to apply a magnetic field along the direction of the light beam, i.e. the whole experiment was done in a magnet. The optic axis changed direction so that it was now along the magnetic field. This was now the same configuration that Mauguin had investigated with a perpendicular orientation of the optic axis at the surface of the film. No light was transmitted, as before. An optical switch could be created from a twisted state (light) to an untwisted state (dark) by application of a magnetic field. This was to have huge consequences 60 years on, when it was realized that the same experiment could be done with an electric voltage and gave rise to the liquid crystal display that we know today.

Figure 4.1 Georges Friedel (1865–1933).

comparable effect. Like Lehmann, he complemented his scientific rhetoric with a personal passion and a panoramic scientific vision. Present-day observers are fascinated and dazzled by these personal attributes, which also no doubt rendered him not an easy colleague!

The Friedels are from Alsace in the east of France, close to the German frontier, the significance of which will become clear presently. They are one of France's most distinguished scientific families.* In British terms, they can be perhaps compared in influence only to the Huxleys or the Darwins. George's great-grandfather, Georges Duvernoy (1777–1855), had been a naturalist of renown, professor at the Collège de France, where he had been collaborator of the pioneering palaeontologist and naturalist Georges Cuvier (1769–1832). His father, Charles Friedel (1832–1899), had been a well-known chemist, professor of mineralogy and later organic chemistry at the Sorbonne.†

Georges was born in Mulhouse (Alsace) at the family seat in 1865. His father's position at the School of Mines in Paris came with an apartment, and much of Georges's childhood was spent in this apartment. As a consequence he was imbued from the time he was very young with an academic spirit. Friedel entered the *École Polytechnique* in 1885, graduating in 1887 to join the School of Mines in a junior position. During the period 1891–1894 he was a mining engineer within the School of Mines at Moulins. In 1894 he became professor in the St Étienne branch

---

* For much of the personal detail on Georges Friedel, we draw on information from Jacques Friedel's (1995) memoir *Graine de Mandarin*, on the history of the Friedel family[22].

† Charles Friedel is the Friedel of the Friedel–Crafts reaction (1877). He is arguably even more famous as a chemist than his son Georges was as a physicist.

Figure 4.2 François Grandjean (1882–1975).

of the School of Mines, where he taught, among other subjects, crystallography. In 1907 he was promoted to the directorship at St Étienne.

The competition between the provincial Friedel and the metropolitan Wallerant was intense. Friedel hated and despised Wallerant with the kind of passion that scientists reserve only for their professional rivals. His copy of Wallerant's 1909 book[23] on crystallography has been conserved by the family. His grandson Jacques Friedel (b. 1921)[24] records that on it Friedel scribbled innumerable uncomplimentary comments...

> ...or Science thought of as a branch of politics...(underneath the title!)
> ..Difficult to talk more nonsense in fewer words...
> This is the crystallography they teach at the Sorbonne!

Friedel's assistant in St Étienne was François Grandjean (1882–1975) (Figure 4.2). Born in Lyons in 1882, but brought up in Paris, he was already recognized as a brilliant student in high school. He entered the *École Polytechnique* in 1902, graduating fourth in a class of 189 students in 1904. The following year he entered the School of Mines in Paris, graduating in 1908. Friedel was pleased to recruit the top Paris student as a teacher in St Étienne, and put him to work on the liquid crystal project.

In a couple of papers read before the *Académie des Sciences* in July and August 1910 (and thus published in C.R.A.S.), Friedel and Grandjean began a re-examination of the facts of the subject that would in the end lead to a proper understanding.[25] The title of the papers was 'The anisotropic liquids of Lehmann'. Note the subtle change of emphasis. Not *liquid crystals*, but rather *anisotropic liquids*, a change of description that others, as we have seen, had already made. But Friedel and Grandjean seem to be the most emphatic that this was as far as the actual evidence went; the appellation 'crystal' carried an additional inference which was not strictly there.

Having made this observation, they reaffirmed German observations that there were two types of material, which looked different under the polarizing microscope. Lehmann had distinguished two types of liquid crystal (*flüssige Kristalle*) and flowing crystals (*fließende Kristalle*). However, for Lehmann the primary distinction was a dynamic property concerned with the degree of fluidity—what we might now call the fluid viscosities.

Lehmann's liquid crystals (*flüssige Kristalle*) exhibited what we now call the *Schlieren texture*, using the description of Gatterman, as discussed in Chapter 2. These were patterns of light and dark regions. Often, as we have seen, they were brightly coloured, with so-called *noyaux* (nuclei) at which the light and dark stripes cross each other. Friedel and Grandjean called these liquids *liquides à noyaux*. The flowing crystals, on the other hand—depending on what aspect they presented to the viewer—looked either like a set of fans or a set of ellipses. Friedel and Grandjean realized that these images were really representing the same thing. What characterized these materials was the *conic sections* (i.e. cross-sections of cones) which were always visible, but which section you *actually* saw depended on how whatever was going on in the material was oriented with respect to the viewer. Friedel and Grandjean called these *liquides à coniques focales* (focal conic liquids).[26]

They then followed this up in their next paper with the essentially same theoretical reflection as Amerio:[27]

> (...both types of...) Lehmann liquids need to be thought of as representatives of a new state of matter. This new state is as different from the crystalline state as it is from the ordinary isotropic liquid. The hasty conclusion that has been drawn that includes them as classes of crystal is without justification.

The Friedel-Grandjean collaboration was extraordinarily fruitful, producing altogether seven* papers over the years 1910 and 1911. They were made of sterner stuff than Amerio, and were in a position to follow up the key insight that the Lehmann liquids constituted a new phase of matter.

However, before it was possible to push this work further a number of events intervened. In 1911 Grandjean moved to Paris and obtained a post at the *École Supérieure des Mines*. In the short run this put paid to his close collaboration with Friedel. His post in Paris was originally in palaeontology, in which he had produced a number of papers over the period 1909–1911, and perhaps it was difficult to justify further work on liquid crystals. From 1912 he was in a department of mineralogy again, but seems to have been diverted to overseas mining expeditions.

---

* Or eight, depending on how you do the counting, because Friedel and Grandjean (1910a,b) were published as two parts of the same work.

In 1914 the war started and he became an artillery officer, later with responsibilities in an armament factory.

Mainly for this reason Grandjean did not return to his liquid crystal studies until 1916. He was trying to distinguish intrinsic and extrinsic (i.e. effects due to external influences) liquid crystalline effects. His article 'Orientation of anisotropic liquids above crystals' is a careful examination of the effect of crystal cleavage planes on the liquid crystals sitting on the crystal substrates.[28] This is how (in somewhat free translation) he starts:

> *Anisotropic liquids exhibit an enormous number of different structures. In due course it will no doubt be possible to summarize these in terms of a few very simple laws. These are not yet known. However, in view of the large amount of available data, there is good reason to suppose that considerable importance will be assigned to the role of the* surface interactions *which govern the liquid orientation in the neighbourhood of the surface, in what might be called the capillary layer. These should be clearly distinguished from the* interior interactions *which fix the liquid orientation with respect to itself inside the bulk liquid. The former will govern, for example, the oriented textures at the glass surface, whereas the latter will govern the focal conics and the visible threads.*

Although most of this article is devoted to detailed observations of liquid crystalline anomalies induced by defects in the crystal surface, probably the most profound observation is almost buried in one paragraph (and, thankfully, a picture) at the top of the fourth page in an article which continues for 49 pages. There is not a little irony in Grandjean's careful study of what were then extremely abstruse materials, appearing at almost exactly the same time that the killing fields of the Somme were raging at their most furious. Scientific life was attempting, and apparently succeeding, in carrying on in apparent disregard of the human mayhem that surrounded the ivory tower. By World War II, in contrast, the lessons would have been learnt. Any scientist, on either side, worth his (and, by this time, her) salt would have his research programme hijacked in the name of the common good.

Grandjean was observing drops of his focal-conic liquid crystals. The drops were attached to crystal surfaces. He found an effect which he called *phénomène des gradins*. His drops were divided into regions of apparently more or less constant height separated by narrow steps. These have passed into current liquid crystal terminology as *Grandjean terraces*, which turned out to be the essential clue into the nature of the liquid crystals (Figure 4.3).*

---

* We note in passing that liquid crystals formed a relatively small part of Grandjean's total *oeuvre*. He published 20 papers on liquid crystals between 1910 and 1921. Like many in the liquid crystal field, he had catholic interests. After 1921, failing to persuade

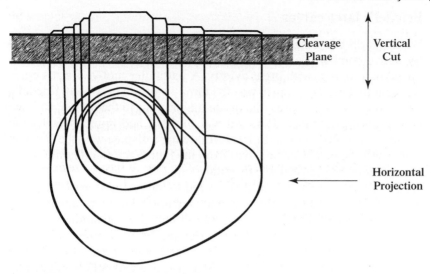

Cleavage Plane

Vertical Cut

Horizontal Projection

Figure 4.3 Grandjean terraces. Diagram redrawn from Grandjean's 1916 paper with the French legend replaced by English.

With the benefit of hindsight Grandjean's most fundamental piece of work was a modified version of Born's molecular field theory of liquid crystals. This was published in C.R.A.S. in 1917.[29] Like Born's paper this was also influenced by the magnetic theories of Curie and Weiss. Unlike Born (whom he does not cite, for this was wartime and it is unlikely that Born's paper had reached him) he recognized that liquid crystalline molecules were likely to be non-polar. Rather, he included the rod-like shape of the molecules explicitly, supposing each of the ends to be equivalent to the other. This was not *exactly* true, but sufficiently true in an idealized kind of way to give the correct answer for the molecular organization in the nematic phase. Unfortunately the paper was completely lost until the modern period, when the theory was rediscovered. We shall return to these ideas in due course.

---

his superiors in Paris that liquid crystals were a fruitful field for study, he turned to other matters. Eventually, at the relatively early age of 59, he took early retirement from the School of Mines in Paris and devoted himself to acarology—the study of mites. This small life form was another suitable topic for investigation with his microscope. Between 1928 and 1973 he amassed a grand total of 241 publications in this field. Indeed, at the time of his death in 1975, his original work in the liquid crystal field had almost been forgotten by obituary writers, such a giant was he in his adopted biological field.

Friedel's later career

When World War I started in 1914, Georges Friedel was 48 and too old for active combat. Nevertheless, with two sons on active duty, he followed its progress with great avidity. A particular motive for his commitment to the French cause was the family's origin in Alsace. Friedel's birthplace, Mulhouse, was one of the major cities in the region, and one which since the Franco–Prussian war of 1871 had rejoiced under the German name *Mülhausen*. In 1911, following a dispute with the German authorities, he had been banished from the family ancestral home, so he had a particular personal axe to grind.

The allied victory in World War I led to the restoration of the provinces of Alsace and Lorraine to French territory. The German-speaking *Universität Straßburg*, where once Otto Lehmann had studied, was no longer. In its place was a reborn Francophone *Université de Strasbourg*. A new academic staff was required, and who better than native Francophone Alsatians to people it. So it was that Friedel was invited by the Prime Minister Alexandre Millerand to transfer to the University of Strasbourg and direct its laboratory of mineralogy.

Friedel was ever a man of culture as well as an applied scientist. His inaugural lecture in Strasbourg demonstrated this, in a plea that is perhaps even more relevant today than it was just after World War I:

> *Science is not the servant of industry, but rather its mother...No, industry, applied science is not a goal in itself. It is a means. To put it at the top and make it the object of all our activity, is, as H. Poincaré rightly stated,* Propter vitam vivendi perdere causas.* *Science, just like art, has its own internal purpose. And to require it to limit itself to allegedly practical matters—i.e. those for which one can see immediately useful results—that would just be to cut it at its root. Industry as well would soon follow it to its grave.*

It was at this time that Friedel renewed his interest in liquid crystals. The following year (1922) he summarized his own researches and the work of others in a 200-page review in the *Annales de Physique* of Paris, entitled '*Les États Mésomorphes de la Matière*' ('The Mesomorphic States of Matter'), from the Greek words *mesos*, meaning intermediate or middle and *morphe*, meaning shape or form.[30]

There was another reason for producing this prodigious review, rather more prosaic than pure scientific commitment. To be director of the School of Mines, it was not necessary to possess a doctorate. The School of Mines was an arm of the state, and to hold a high position it was necessary to rise up the civil service hierarchy. To be a professor in the uni-

---

* 'To destroy the reasons for living for the sake of life' (Juvenal, *Satyricon* VIII, verses 83–84).

versity, however, was another question altogether. Legislation required certain standards, presumably to avoid the constant pressures of corruption. The French legislation in fact lasted until the 1960s. Georges Friedel did not possess a doctorate and therefore was not qualified to be a professor in the university! '*Les États Mésomorphes*' was his retort to the legislation, the *thèse d'état* to end all *thèses d'état*. In it, Friedel introduced modern liquid crystal terminology. Friedel finally talked of an optically active or chiral variant of the nematic fluid, which he referred to as *cholesteric*,* after Reinitzer's original liquid crystalline compounds. These were the substances in which Grandjean found his terraces.

Friedel's crucial insight, already formulated but repeated here in a full synthesis of the available facts, was to identify Lehmann's liquid crystals as distinct new *phases* of matter, intermediate between the solid and liquid phases. He further realized that the identifying feature of the phase was not the degree of fluidity (which is in some sense merely a derivative property), but rather the nature of the internal molecular organization. He described qualitatively the main structural properties of each phase.

The normally cloudy phase had eventually been called by Lehmann *flüssige Kristalle*, and Friedel had previously called this the *liquide à fils* (or *à noyaux*). Now, with the benefit of a grand overview, he renamed it the *nematic* phase after the Greek *nema*, for thread, recognizing the intrinsic threads which always occurred in these samples when viewed under the microscope. The threads and the cloudiness of the nematic phase are both consequences of optical properties associated with the director (see Technical box 2.1). In a sample of nematic liquid crystal in which the director or optic axis (see Technical box 3.1) varies randomly in the sample, then a light beam is constantly refracted and reflected in different directions by the changing orientation of the director. The consequence is that the sample appears cloudy. It is optically inhomogeneous, just like the emulsion mixtures of two fluids, which also appear cloudy, hence Tammann's confusion, which caused such disagreement between him and Lehmann, although neither knew the origin of the confusion.

The threads which appear in the *liquide à fils*, which occur in samples that are better aligned than Lehmann and Reinitzer's original samples, are due to defects in director orientation now known as *disclinations*. The slimy liquid crystals became *smectic* (Greek *smegma* = soap) because of the preponderance of soap-like compounds which fell into this category. This phase was characterized by a layered structure. The fans and ellipses which appeared from samples under the microscope were again the product of a non-uniform alignment of the director. Physics constrains the thickness of layers to be very closely constant. In

---

* For a description of cholesteric liquid crystals see Chapter 12.

TECHNICAL BOX 4.2 Director defects or disclinations

In samples where the director orientation is only weakly non-uniform, the light can almost follow the changes in director direction, but the director cannot change smoothly everywhere. It happens that there are lines along which the director is forced to change quickly and eventually is simply not defined. One analogy might be water running down a plug, where at the centre of the vortex the flow of water is turbulent; another might be standing at the North or South Pole, where the concept of direction loses its significance. Likewise the director in a nematic can become fixed due to a surface irregularity or a particle of dust, and cannot vary smoothly in its vicinity. Such a point is described as a defect, at which the direction orientation becomes undefined. Then there are *always* lines within the nematic along which the director is undefined, and near which it changes quickly. Near these lines the light scatters by a large angle. If you are behind such a line and you are looking through a polarizing microscope, you physically see the line as a dark thread against a coloured background.

Dark threads, appearing as coloured lines, marking fluid defects (known as disclination lines) in a nematic liquid crystal. The black areas emanating from points are known as brushes. They are areas of optical interference which connect through points on the surfaces marking the end of defect lines. © D. Dunmur. (See colour plate 8.)

There are many different types of fluid defects in liquid crystals. The fluid defects in nematic phases, known as disclinations, can be of different strengths characterized by a number s, which is the number of full rotations (360°) experienced by the director in circumnavigating the centre of the defect. The defects can be identified by viewing between crossed polarizers. If the axes of the polarizers are assumed to be horizontal and vertical, then the image will appear dark whenever the director is vertical or horizontal. Thus for s = ±½, there are two horizontal dark areas, known as brushes. For s = ± 1, there are four brushes, aligned vertically and horizontally. Both two- and four-brush defects can be seen in the accompanying image.

Figure 4.4 The fan texture of a smectic liquid crystal © I. Dierking. (See colour plate 5.)

the best-case scenario, the layers would always be oriented more or less in the same direction, but in practice the layers begin to grow with one orientation here, another there, and yet another somewhere else. Then these different regions collide with each other to form the characteristic focal conic images or textures (see Figure 4.4).

It turned out that a condition to keep the layers the right distance apart from each other was that the boundaries between the regions adopt certain specific shapes. It is known from nineteenth century geometry that these are the so-called *Dupin cyclides*, whose cross-sections are ellipses or hyperbolae. At these boundaries—grain boundaries we would call them in modern crystallography—light goes rapidly from one region to another. This causes light to be reflected, making the pattern of the surfaces immediately obvious to the microscopist. Once again the optical pattern gives the observer direct access not to the underlying phase, but rather to *characteristic irregularities* in the underlying phase.

How did Friedel know this? Again good luck: the mathematics, and in particular the geometry, that he would have learned when he studied at the *École Polytechnique* would have been very rigorous. He would have covered just this topic, although of course the pure mathematicians who taught him would have been as surprised as Friedel to discover that their beautiful mathematics had such an application.

There was also a third type of liquid crystal, which Friedel referred to as *liquides à plans*, whose optical signatures combined some features of

Figure 4.5 A microscope image of a cholesteric phase. A variety of characteristic optical textures is shown, including fans (left), Grandjean planes (centre), and schlieren (right). © D. Dunmur. (See colour plate 6.)

both of the focal conic and the thread-like liquids. By careful experiment, Friedel was able to convince himself that in these phases the director was no longer uniform, but rather had acquired a spontaneous twist. This phase he called the *cholesteric* phase because many of the materials which formed such a phase were made from cholesterol products. Because of the twist, a cholesteric phase could be left-handed or right-handed.

The helix-like rotation induces a layer-like structure, and in this way the cholesteric phase resembles a smectic phase. The separation of the layers, the pitch of the helix, is a quantity whose magnitude is determined by the material. If the direction of the layers is not uniform, then, just as in a smectic, the sets of layers may 'collide' along a focal conic surface. At the collision, the material properties change suddenly, causing light to bounce off the surfaces, so we also expect to see the focal conic (fan) texture (see Figure 4.5).

Friedel's article, then, introduces and justifies modern liquid crystal terminology. He gave much thought to his new nomenclature. As we shall see below, he hoped that his new language would replace entirely the previous catch-all 'liquid crystals' terminology. His daughter Marie was a classicist, and gave him the support he needed to seek out a suitably nuanced terminology.

Much of Friedel's review is taken up with using the structures which he had identified to provide a detailed interpretation of the dramatic optical signatures of what he was now calling the mesomorphic states. With the benefit of hindsight, it is spectacular how much of liquid crystal science Friedel already understood in 1922. He predicted, for example, that X-ray scattering on smectic phases would give rise to a pattern, and comments on the analogy between this and the Grandjean terrace phenomenon.

Occasionally, but only very occasionally, Georges Friedel made what can with hindsight be unambiguously identified as an incorrect statement. For example, he thought that all smectic phases are essentially identical (now known not be the case). He also dismissed evidence of mechanical torsion (i.e. a physical twisting effect) transmitted through a nematic, while accepting evidence of optical torsion, but these mistakes were rare in a work of genius.

No dry description of the content of Friedel's article can replace quotation from the real thing. In reading the excerpts below, it is useful to recall that the article was written in the weeks that followed the death of the original pioneer of the field, Otto Lehmann. Friedel launched straight into his subject matter, guns blazing, metaphorically speaking:

> *I use this term (i.e. mesomorphic) to designate those states of matter observed by Lehmann in the years following 1889, and for which he invented the terms* liquid crystal *and* crystalline fluid. *Lehmann had the great merit of drawing attention to these materials, but he erred greatly in naming them wrongly. The terminology is inapt, despite having been ceaselessly repeated over the last thirty years. His terminology might lead many people to suppose that these substances are just crystals, albeit differing in degree of fluidity from those hitherto known. In fact, it is entirely otherwise, and indeed something infinitely more interesting than would be the case if we were considering simple crystals exhibiting some unexpected degree of fluidity.*

Friedel is clearly not a man to mince his words, nor one who suffers fools gladly—'inapt terminology', 'erred seriously': these were intemperate remarks then and would be now. Scientists frequently disagree strongly about matters of interpretation, but nothing fires them more than arguments over notation and nomenclature. It was the words that Lehmann used to describe his observations, not the observations themselves, that Friedel scorned so vehemently. To make such adverse personal remarks about a fellow scientist is unwise, and more often than not the remarks rebound with unexpected consequences. Peer review of papers and research proposals can be influenced by personal feelings, although of course they should not be, and the public forum of a scientific conference is occasionally the scene of unseemly exchanges between overwrought scientists. Usually none of the combatants emerge with honour.

Friedel clearly did not fear a fight. Not least because Lehmann, whom he is careful to damn with faint praise, was recently in his grave. Lehmann he would not have the misfortune to meet at another conference! Perhaps Professor Friedel starts strongly for effect, but will tone down his comments in the bulk of the article? Not a bit of it:

> *It is not without hesitation that I propose new terminology. For the*
> *example of 'liquid crystals' gives us the opportunity to observe the*
> *influence of a word badly chosen. Lehmann did not really under-*
> *stand clearly how the substances he was studying would take their*
> *place in the hierarchy of material phases. He did, however, per-*
> *ceive, albeit in a rather confused fashion, that here indeed was*
> *something entirely new. We must be grateful to him for trying to*
> *convince his contemporaries of that fact. He was misled by the*
> *inadequate definitions of crystalline materials which were current*
> *at that time, and thus erred in classifying his liquids as crystals.*
> *But much more serious were the errors of the numerous scientists*
> *who contradicted him, especially in Germany. These workers*
> *refused to see...*

'Especially in Germany'—aha, could Professor Friedel be gazing at his colleagues through tricolor-embossed spectacles? So what was it that the stubborn Germans failed (rather than the literal 'refused'!) to notice? Friedel continues...

> *As to the detailed study of the mesomorphic phases... and the need*
> *to establish any order in this jungle of new facts, the German*
> *authors seem to be little concerned. An exception must be made for*
> *Lehmann, whose large number of publications contain many*
> *observations with the microscope. However, these are confused,*
> *and rendered still more unclear by the theoretical considerations*
> *which accompany them.*
>
> *It has been in France, above all, that the main steps in understand-*
> *ing the principal concepts have been made, although understand-*
> *ing is still far from complete.*

So it's official! Friedel was a political dinosaur, a nationalist of the Genghis Khan school. Of course, where else but in France could we find such an extraordinary collection of genius? The Germans had been busily (stamp) collecting useless facts, when suddenly, from out of the West, in rode the rescuing French white knights, led by their heroic champion Georges Friedel. And still the faint praise for Lehmann, who is depicted as a heroic pioneer, but fundamentally flawed for all that. Why the faint praise? Why praise at all?

Lehmann and Friedel knew each other at least slightly. Indeed as recently as 23 January 1922, Friedel had written Lehmann a polite, even effusive, letter thanking him for his hospitality in receiving a colleague and asking him for help in obtaining samples:[31]

> *...Although Mr Schnaebele has not yet returned, I'd like to thank*
> *you for the generous welcome you were able to extend to him and*
> *for the trouble that you took in writing to me on this subject. I will*
> *follow with great interest the samples you have given to him, and*

will of course send them back after we are finished with them. I've
had enormous trouble in getting hold of these interesting materials
and am extremely grateful to you…

What was going on? We have seen Lehmann pursuing Tammann, Vorländer, and Reinitzer when they had the temerity to argue with him. Had he worked similar magic on Georges Friedel? Maybe he had tried, and Georges turned out to be made of tougher stuff? Or was Georges just on his best behaviour because he desperately wanted to get his hands on Lehmann's compounds? Still, can we ask, did not Friedel feel pangs of guilt at laying into Germans with such ferocity, knowing that the Lehmann who had helped him so generously had died so recently? How would he have phrased his article had Lehmann still been alive?

But we are not finished yet. There is more…

Lehmann understood the point, albeit in a confused way, that he could save the word 'liquid crystal' only if he discovered in his materials discontinuous properties. So he searched either for polyhedral crystal shapes with planar faces or for twinning phenomena. He thought he had found them, but we shall see that it was only an illusion. In despair he was forced to recognize the lack of a periodic structure. All that remained was to find a definition of a crystal which referred to neither discontinuous surface properties nor a periodic structure:

Thus we obtain the bizarre definition: 'A crystal is an anisotropic body which has a 'molekulare Richtkraft', and which has as a result the property of growth. The 'Richtkraft' or 'Gestallungskraft' is a mysterious force which is supposed to make the vertices stick out, the edges straight and the faces planar. It is supposed to counteract surface tension, which makes the surfaces round. In fact, it is only a nebulous new Germanic Divinity, which contradicts itself in an extraordinary collection of meaningless words,* and, as we shall see, lacks any basis other than that of an inexact view of the role of surface tension.

Friedel retained the original German of *Richtkraft*, and he was right. Lehmann had indeed used this idea—of directional strength—as a central pillar in the explanation of liquid crystal phenomena in his 1904 book. He is also right in his value judgement: it is essentially gobbledygook. *Gestallungskraft* means configurational strength; from a modern point of view an explanation in terms of 'molecular configurational strength' is seriously incomplete.

Modern scientific writing is stultified by a homogeneous passive voice and other awkward phraseology, and perhaps one reason for the alienation of scientists from ordinary people has been precisely this

---

* In the original French, *une singulière logomachie*, which is extraordinarily difficult to translate.

apparent lack of soul. This lack of soul has been deliberate, for it has provided the practice of science with a veneer of objectivity. Friedel, by contrast, was very obviously human. Yet his physics also made sense—but of course we do not want it to! Even today we occasionally meet echoes of his chauvinism. What right has he to sneer at those who do not share his nationality? What right has he to mock the genuine efforts of the pioneers of the subject? All the more so, when, as he himself admits, without a good definition of what exactly a solid *is*, it is entirely unfair to castigate the stupidity of one who is unable to distinguish between solid and liquid. By 1922, as we have seen, the molecular structure of solids was clear, whereas in 1890 there was still considerable uncertainty. Lehmann and his German colleagues were not stupid. Indeed, as we have seen, in many ways they were

TECHNICAL BOX 4.3 Liquid crystal structures

Georges Friedel had provided the basis for the modern physics of liquid crystals and clearly had a good understanding of what the microscopic structure of these materials should look like.

Now we have a wealth of techniques to probe the structures of liquid crystals. Some of our most powerful images of new materials come from the computer. So confident are we of our theories, that it is commonplace to use images generated by these theories to show microscopic structures in unimaginable detail. To help the understanding of further chapters, and as homage to the insight of Friedel, we reproduce here some computer-generated pictures of liquid crystal mesophases (courtesy of Dr M Bates, University of York).

The images are generated by assuming that liquid crystal molecules can be represented by ellipsoids, and then allowing them to interact with intermolecular forces that model the interactions we now believe to be responsible for the formation of liquid crystals. Above left is the computer-calculated image of a nematic phase, while on the right is the phase that Friedel identified as a smectic phase. A hint of the layers can be seen in the computer image.

For comparison, the following images are representations of the solid crystal phase (left) and the isotropic liquid phase (right), using the same theoretical model.

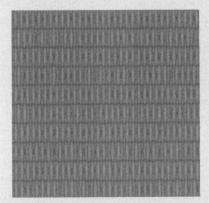 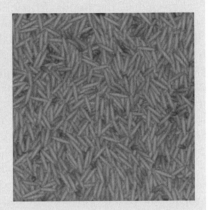

The images above are, of course, not of real molecules.

However, using a modern technique known as scanning tunnelling microscopy, it is possible to generate images of real molecules. Friedel, and indeed Lehmann and Vorländer, would have been amazed to see a picture of liquid crystal molecules organizing in a parallel fashion, just as expected in a liquid crystal phase.

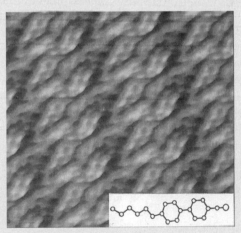

An image of a thin film of liquid crystal molecules obtained using scanning tunnelling microscopy. The alignment of the molecules is similar to that adopted in a nematic liquid crystal. Included above is a representation of the molecular structure of the molecules being viewed. Courtesy of Dr Jane Frommer, IBM Almaden Research Centre. (See Colour plate 10.)

far-seeing and insightful. They were, of course, not able to foresee the future. Born was actually rather smart, although as it turned out, unfortunately wrong. Born understood that a liquid crystal was really a liquid, as did Bose, albeit in less quantitative fashion, even a decade earlier than that.

We cannot, of course, ignore the political climate of the time, nor the background of the family Friedel. The year was 1922. Mauguin had been able, in print, to pay Lehmann the compliment in 1911 of being 'the distinguished German scientist'.* Grandjean, even while the guns of the Somme were blazing, was able magisterially to ignore the very existence of a Franco–German conflict, but the tensions seem to have exploded from Friedel's pen. The war served as *la goutte de cristal liquide qui fit déborder l'éprouvette*.† Given all of Georges's problems with the Germans, it is little wonder, perhaps, that he found himself better able to discern German scientific error than some of his contemporaries.

The article transformed the subject and remains a classic. Articles on liquid crystals refer to it even now. It is such a classic that we might suspect that in many cases the authors are referring to a paper they think they *ought* to have read, rather than one they actually *have* read. Be that as it may, an enormous number of 656 scientific authors referred to the Friedel paper over the years 1970–2008, half a century or more after it was written.

Triumph that this paper was, Friedel was destined to lose one battle, however. The idea of a mesomorphic phase, made of *mesogenic* molecules, has passed into the canon. His classification of the phases into smectic, nematic, and cholesteric has similarly been accepted. His sense of offence at the mistaken concept of *liquid crystal* was equally understood by his contemporaries. The liquid crystals were not crystals at all, but peculiar liquids with some hint of solid properties, but the epithet was vivid, and the terminology already widespread. Georges Friedel's righteous hostility to Germans and liquid crystals was insufficient to shift scientific usage of the term. The description 'liquid crystals' was here to stay.

---

* *l'illustre savant Allemand.*
† The liquid crystal drop that makes the test-tube overflow.

5

The meeting that wasn't and the meeting that was

⌒≈◦≈⌒

If the hill will not come to Mahomet, Mahomet will go to the hill.
Francis Bacon (1561–1626) Essays, 12, Boldness

The continuing liquid crystal wars

The three names which dominated the heroic early period of liquid crystal research were those of Otto Lehmann (born in 1855), Georges Friedel (born in 1865), and Daniel Vorländer (born in 1867). For each of these three, a good part of their lives was spent at war. There were literal wars, of the shooting variety, between the French and the Germans in 1871, and again, much more savagely, during World War I, but as we have seen (and will see more in this chapter), there were also personal and private scientific wars, with each other and other scientists, over scientific priority and interpretation in the liquid crystal arena.

The liquid crystal wars rarely strayed into physical violence, although here too, much later, fate would intervene. Nevertheless, they were venomous and vicious personal conflicts manifesting themselves as wars of words, which continued to the grave, and even beyond, while the science itself moved on. The bitterness engendered by Lehmann's hostility to Vorländer's partial assignment of priority to Friedrich Reinitzer could not be balanced by his kind words elsewhere. Not even Lehmann's death in 1922 could assuage the pain. Here is Vorländer in 1923:[1]

> I cannot agree with O. Lehmann's interpretation of the so-called liquid crystal characteristics of ammonium oleate hydrate and similar soaps. These materials, when melted with water, are said to yield liquid crystals. What Lehmann really observed, photographed and then described, at least in part, were probably suspensions of swollen soft birefringent phases in water. In my experience, liquid crystalline phases of fatty acid salts are only

formed at high temperatures, in the molten phase of the (incompletely decomposed) anhydrous salts. Lehmann repeatedly referred to his discovery, mainly in order to establish his priority with respect to Reinitzer...

It was only a footnote, but one could not doubt the intensity of the emotion.

During the first two decades of the century, Lehmann had done his best to promote the subject of liquid crystals. His pride often led to a confusion between this primary goal and an important secondary goal, that of emphasizing his own importance in discovering them. However, apart from proselytising, in terms of new research he did not himself make further significant progress beyond the ground-breaking observations of his early career.

Friedel, on the other hand, thought hard about teasing out the true physical basis of the new phenomena. However, in practice he tended to work only with a very small number of trusted colleagues. In particular, as we have seen in the last chapter, he was intolerant of what he felt were often less able competitors, especially those with theories not backed up by experiment. He also did not follow the literature closely, preferring (and in this respect he resembled his *bête noire*, Lehmann) evidence provided by his own eyes. Indeed, so suspicious was he of theory that he even ignored the molecular theory advanced by his respected friend François Grandjean. This suspicion had direct consequences for the development of the subject, for it took another 30 years before this correct theory was rediscovered by others.

Daniel Vorländer was an organic chemist from a classic mould. The contrast between the practical Vorländer and the ethereal intellectual Lehmann could not be more striking. Not for Vorländer the classical footnotes and sentence structures of Byzantine complexity. Vorländer's liquid crystal monograph is a primer. Lehmann's is a cross between a political tract and a work of art. Whereas Lehmann largely worked by himself, Vorländer had a whole institute at his disposal.

With a large team of scientists working in his group, Vorländer had been able to generate a huge output of work. Over the period up to 1922, 46 members of Vorländer's group submitted theses at the University of Halle. Since each of these would have taken about 5 years of work, that represented more than 200 people-years of effort. This is despite the war years, which had seen the boss, Vorländer, in action as a gun-battery commander.

In the late 1920s, with Lehmann dead, Vorländer and Friedel prepared yet again to do battle in the literature over the 'liquid crystal question', but now the existence of the phenomenon was firmly established in the scientific canon. More and more experimental details about liquid crystals were being discovered. As a result, the sophistication of the arguments likewise increased. Only the malicious rhetoric would retain its primitive unsophisticated form.

The new data were not merely the product of time, nor even of ingenuity: the progress of experimental science is often the result of technological developments in apparently unrelated fields. Key developments in seventeenth century astronomy depended on the newly discovered telescope. Likewise, biological progress resulted from the invention of the microscope. For the study of materials, the nineteenth century was the age of the optical microscope. Otto Lehmann's hot-stage microscope enabled him to study properties of liquid crystal materials inaccessible to the possessors of less sophisticated instruments.

Many variants of the microscope were developed, each with a special mode of operation. The ultra-microscope, invented by the Austrian chemist Richard Adolf Zsigmondy (1865–1929), allowed the observation of particles too small to be visually detected by normal optical microscopes. As we shall discuss in more detail in Chapter 6, Zsigmondy used the ultra-microscope to study colloids and soaps, materials which have many features in common with liquid crystals. So profound were these advances that Zsigmondy was awarded the Nobel Prize for Chemistry in 1925 for his invention.

Early X-ray studies of liquid crystals

Then, early in the twentieth century, came X-ray diffraction. This technique enabled the molecular structure of materials, particularly crystals, to be determined in detail at a previously unimaginably small scale. Now, for the first time, it would be possible, albeit with some difficulty, to learn about structures with dimensions of around a millionth of a millimetre. Since antiquity philosophers had speculated on the nature of matter, and why different forms of matter presented themselves in different guises. With X-ray diffraction it would be possible to probe directly the long-predicted—but so far unobserved—atoms.

It was Wilhelm Röntgen (1845–1923)* who had first observed the mysterious rays which penetrated matter, clouded photographic plates, and enabled the previously veiled internal details of human bodies to open themselves for inspection. In Munich, the distinguished German theoretical physicist Arnold Sommerfeld speculated that X-rays were not beams of particles, but rather electromagnetic waves—light waves with wavelengths of the order of one tenth of a millionth of a millimetre. He realized that such waves would not see ordinary matter as continuous,

---

* Wilhelm Conrad Röntgen was awarded the first Nobel Prize for Physics in 1901, for his discovery of X-rays.

but rather would respond to the atomic structure. He suggested to his junior colleague Max von Laue that a suitably designed experiment (and some sophisticated mathematical theory) might kill two birds with one stone. It would establish both the atomic structure of matter and the nature of the X-rays.

The news of the Munich X-ray experiment in 1912 by Friedrich, Knipping, and von Laue took the physics community by storm.[†2] The images of the crystal lattice seen in X-ray diffraction were not just enlarged versions of the lattices themselves seen through a massively powerful microscope. Rather, a peculiar pattern of spots was projected onto a screen. Only after some complex mathematical data processing could the structure of the crystal be deduced.

For the first time explicit evidence was produced that crystals consisted of atoms arranged in regular arrays. The crystal lattices were concrete and real, albeit on microscopic scales. The apparently unbridgeable chasm between everyday and atomic dimensions had been traversed. The importance of X-rays as a probe of atomic and molecular structure was immediately recognized by the award of the Nobel Prize in Physics in 1914 to Max von Laue.

Crystallographers the world over diverted their activity towards repeating and extending the Laue experiment. In the UK, the father and son team of William Henry Bragg and William Lawrence Bragg were able to concoct an experiment and interpret it with a simplified version of the Laue theory, for which they jointly also received the Nobel Prize in Physics in 1915. In France both major crystallography groups (led by Wallerant and Friedel) had been concentrating on liquid crystal work, but this innovation was so important that even the liquid crystal work would have to wait.

Neither group had sufficiently good X-ray equipment for their experiments on classic solid crystals to succeed. Friedel immediately made theoretical advances connecting the lattice symmetry to the symmetry of the pattern of spots of the X-rays refracted from the crystal, while Wallerant, in collaboration with the chemist Paul Villard (1860–1934), attempted to repeat the German work. However, for various reasons (not understandable at the time, but only too comprehensible later) he could not obtain a powerful enough beam of X-rays to obtain a signal. Perhaps it was not only poor equipment, but also some element of fate. Wallerant, unlike almost everyone else in the field, never believed the X-ray evidence for the existence of crystal lattices. Eventually Maurice de Broglie (1875–1960), working at the Collège de France, was able to obtain an X-ray machine and in March 1913 began to produce X-ray results on simple crystals like salt.[3]

[†] Knipping and Friedrich were even more junior colleagues who actually did the work.

Figure 5.1 Sir William H Bragg (1862–1942) in 1926.

By 1922, Friedel had returned to his liquid crystal work and X-ray crystallography was an established technique. In his overarching review of the same year, he had explained data on the phases he termed smectic in terms of layered structures. He predicted that they should exhibit a characteristic pattern of diffracted spots in an X-ray beam. This prediction was soon confirmed by none other than his own son, Edmond Friedel (1895–1972), future Director of the Paris School of Mines (1945–64). In a ground-breaking study using Maurice de Broglie's apparatus, he was able to show that the smectic phase consists of aligned molecules, neatly arranged in layers roughly 45 Ångströms thick,* but wandering freely within them.

At almost the same time, Vorländer had also realized that X-rays could make a huge contribution to the study of liquid crystals. The results of Edmond Friedel and de Broglie were published with almost indecent haste, but the insults that Georges Friedel's landmark article had loaded onto Vorländer were too savage, and too contemptuous, to ignore:[4]

> *This shows to what extent the true nature of these compounds continued to be poorly understood in Germany, even 20 years after their discovery.*

Useful collaboration between Vorländer and the Friedels was impossible. Luckily there was an alternative centre of X-ray diffraction expertise: the Braggs, in England.

On 28 May 1923, Vorländer wrote to Sir William Bragg, by now Director of the Royal Institution in London (Figure 5.1).[5] The

---

* A distance of 10 Ångströms is one millionth of a millimetre or a billionth of a metre.

hand-written letter is short and almost undecipherable, and suggests (in German) that the X-ray technique could be usefully applied to liquid crystals. There is no record that Bragg had previously heard of Vorländer, and unfortunately there is no record of Bragg's reply. Perhaps there was no reply; as we shall see later Sir William Bragg was not well-disposed towards Germany. More plausibly, he couldn't read the awful handwriting or perhaps his German was not sufficiently fluent.

However, interestingly, the records also reveal a parallel approach to Bragg from Georges Friedel.[6] The correspondence, relating to X-ray studies of liquid crystals, spans the period 1924–1925, but ultimately this exchange was no more successful than that between Vorländer and Bragg. Friedel had suggested to Bragg that X-ray measurements on compounds of fatty acids would provide a clear confirmation of his (Friedel's) model for smectic liquid crystals. With hindsight, these were probably not the best compounds to study, but the Friedels were restricted in the liquid crystals they had access to. The primary source of liquid crystal materials remained Vorländer's Halle group, but given the manner in which the review of Friedel had dismissed Vorländer's work, even Friedel realized that it would be futile to seek samples from Vorländer.

In reply to Friedel's request, correspondence from William Bragg indicated that measurements made at Friedel's instigation at the Royal Institution had failed to show any evidence of unusual behaviour:[7]

> ... all substances (studied) show sharp crystal-like reflections...

What a blow for Friedel! There must have been some mistake? Perhaps Bragg was looking at the wrong materials, replied Friedel:[8]

> It seems to me most likely that the acids (e.g. the fatty acid samples studied) crystallized, but the ether (derivatives) are smectic.

Friedel refers to his son's work with Maurice de Broglie on sodium oleate, a salt of a fatty acid, and implores Bragg just to look at the compounds under an optical microscope. However, this foray of Bragg into the field of liquid crystals came to an abrupt end in February 1925, when Bragg wrote again to Friedel,[9] insisting that he saw only crystalline reflections from X-rays. Not a good ending for Friedel, but he would get his own back in a few years when Bragg was naïve enough to tackle something else in liquid crystals.

The meeting that wasn't

In the meantime, with the new X-ray technique beginning to be widely applied, there was a lot going on in the field of crystallography. The Braggs, father and son, had plenty to keep themselves occupied without getting involved in a new field. As the father of X-ray crystallography,

William Bragg was a member of the Editorial Board of the *Zeitschrift für Kristallographie*, a publication of the German Physical Chemistry Society. Another member of the Editorial Board was Paul Ewald (1888–1985), a talented crystallographer from Stuttgart, whose method of analysis of X-ray diffraction patterns was a great advance. Ewald became a close colleague of the Braggs, and was in frequent correspondence* with Bragg the younger, who was just two years his junior.

Ewald had also become aware of liquid crystals. In a letter to W.L. (Lawrence) Bragg in February 1930,[10] he mentioned a forthcoming visit to his laboratory in Stuttgart by the liquid crystal optics pioneer Charles Mauguin. Ewald was sufficiently inspired by the reports on liquid crystals that he decided to organize a symposium, the proceedings of which would be published in the *Zeitschrift für Kristallographie*. A normal symposium is an academic meeting on a specialist subject in which the talks and subsequent discussions are often recorded for publication, but Ewald's Symposium was of a special kind, for there was to be no actual physical meeting of human beings. All that Ewald envisaged was a kind of ghostly record of what was (not) said at the (imaginary) meeting. In modern parlance we would call it a *virtual meeting*.

We can only speculate on the motives for Ewald's proposal. Probably he had realized that the application of X-rays to liquid crystals could be a significant new use for the technique that he had had such a major role in developing. In addition, it is the case that, slowly and quietly, in isolated pockets across Europe there had been an increase in the numbers of people involved in liquid crystal research. Bit by bit, without the characters in the play quite realizing it, a liquid crystal *community* was developing. Only Ewald, sitting as he did on the Editorial Board of a major journal, was in a position to take a larger overview.

During the first half of 1930, the key researchers on liquid crystals, all from mainland Europe, were invited to submit papers on their specialities. These papers were then circulated to all contributing authors and other liquid crystal scientists for comment. Finally papers, comments, and replies to comments by authors were published by the journal in 1931. It was a substantial publication, running to 347 pages, including 78 pages of discussion, and it included all the main players. Judging from the written discussion, it was probably a good thing that the contributing authors, in particular Vorländer and the Friedels, did not have a face-to-face encounter. If they had, there might have been bloodshed!

A historical survey was provided by Rudolf Schenck. By 1930 he was already 60 years old and it was many years since he had worked in

---

* Numerous letters exchanged between William Lawrence Bragg (son) and Ewald, mostly on structures of solid crystals, can be found in the Bragg Archive, Royal Institution, London.

the field. But his affection for liquid crystals remained undimmed. We left him in Karlsruhe in June 1905, having just presented a talk which comprehensively undercut the foundations of Gustav Tammann's emulsion picture of liquid crystals, and subjected Schenck to the consequent wrath of Tammann himself. We may recall that van't Hoff had intervened from the chair to call for a commission of experts to examine liquid crystal questions and to resolve the raging controversies. In his introduction to Ewald's publication, Schenck recalled the drama of the Karlsruhe meeting and further related how the commission fared: inconclusively, as we have seen in Chapter 2. His main protagonist from 1905, the by now elderly Tammann (he was to die in 1938), had not been invited to contribute to Ewald's collection of articles, although it is known that he had not changed his views. He had achieved greatness in another field and probably the emotional immediacy had left him.

The first article[11] was a joint effort by Georges and Edmond Friedel summarizing their understanding of the field. It was an amalgam of the key parts of Georges's famous 1922 paper, together with the recent X-ray results obtained by Edmond. Although most of the rest of the volume is in German, this paper (as with all contributions from French contributors) was written in French. It was simply assumed that readers would be educated enough to make sense of it! Although much shorter than '*Les états mesomorphes*', this effort still stretched to over 60 pages and was the longest contribution in the volume. The Friedel contribution was not the only one which mentioned X-rays; an article from Herrmann and Krummacher at Berlin-Charlottenburg also reported an X-ray diffraction study on a liquid crystal material.

The second article was by Daniel Vorländer. No doubt Ewald lost a little sleep over the order in which the articles appeared, for this order in itself makes a statement about the relative importance of the protagonists. That an expert such as Vorländer had been relegated to the second place behind the hated Friedels was no doubt a humiliation to the proud Vorländer.

Vorländer was annnoyed by Friedel's new 'mesomorphic' classification. He had his own classification,[12] dating all the way back to 1907. In this classification, he used the term *pl-phases* for the higher melting, low viscocity liquid crystalline phases, derived from the first and last letter of the German word for Gattermann's *p*-azoxyaniso*l*. The lower melting liquid crystalline phase he called the *Bz-phases*, derived from the letters *B* and *z* of the German word *Benz*oesäure (benzoic acid) because he first observed these phases in benzoate esters. He saw no need for new nomenclature, as is witnessed in the following sarcastic extract:[13]

> *If anybody beginning to work in this field wants to introduce new terms—please, with pleasure, go ahead.*

Vorländer's contribution to the Ewald symposium was a forceful restatement of his own position, both academically and as a pioneer in the field. By the time he had received copies of the papers and it was time to contribute to the discussion, his blood pressure was reaching seriously elevated levels, and what his contribution lacked in formal rigour, it gained in richness of language. His words were clearly carefully chosen and add much to our historical understanding of what it felt like to be a liquid crystal scientist during this period. Unfortunately their emotional content meant that they had less of a persuasive effect on his colleagues. Thus, in spite of Vorländer's high prestige in the field, the terms *pl* and *Bz* were relatively rapidly replaced—within a decade or so—by Friedel's new terminology. Nevertheless, we do find Vorländer's terminology employed by his own former collaborator Conrad Weygand in 1939,[14] and even as late as 1955 in a review article by Wilhelm Kast.[15]

Vorländer's excoriation of the Friedels, father and son, for their new crystalline liquid nomenclature reached new heights. Not only did he feel that the data were not well described, but he was also seriously offended (with good reason) by the omission of any reference to his own valuable work. We can get an idea of Vorländer's strong feelings from the following short extract:[16]

> There are experiments for recognizing correct and incorrect names for scientific phenomena. A name is correct when it is based on fact and experimental observation, but incorrect if it has at its core theories or hypotheses...
>
> Mesomorphism, mesophases, or intermediate phases for crystalline-liquid phases, are incorrect names because there is not a single fact or phenomenon to prove that the liquid crystals stand between solid crystals and amorphous melts. Give me a single property of crystalline liquids that indicates their intermediate position as crucial to the phenomenon. I know of no such property! Everywhere, even in X-ray diffraction: sharp, discontinuous change; no intermediate position. In contrast, one could cite more than a dozen properties that prove that on the one hand the liquid crystals behave just like solid crystals, and on the other, just like amorphous oils. In other words, hermaphroditism! At best, one could introduce, instead of crystalline liquids, the term crystal-like, crystalloid liquids, or liquid crystalloids, but that does not constitute essential progress.
>
> I reject the words mesomorphic states and mesophases as a designation of crystalline-liquid properties or phases. I consider the expressions mesomorphism and mesophases, as well as the words nematic and smectic, completely misguided, even if the theory which led to the words at some later time turns out to be correct.

The emphasis is ours, but the drift of his remarks is clear! Notwithstanding these problems, always faced by practitioners in a rapidly developing interdisciplinary field, it is important to reiterate the debt which the liquid crystal community continues to owe to Daniel Vorländer, not only for his work during the pioneer era, but also for his continuing work in the interwar period.

Other themes of the symposium were no less controversial. Despite the fundamental controversies still raging, a number of theorists had decided that liquid crystals or crystalline liquids, or whatever they were called, were worthy of a serious look. As related in Chapter 3, the dominant theoretical framework was the swarm theory constructed by the unfortunate Emil Bose in an incomplete state as early as 1908. A new version was presented by the distinguished Dutch theoretical physicist L.S. Ornstein (1880–1941), who was awarded the third article in the proceedings. He drew a distinction between his swarms and Bose's, claiming that, unlike in Bose's theory, there was no element of emulsion included in his swarm picture. As the memory of the argument between Tammann and Lehmann was still alive, and the emulsion picture of liquid crystals known to be Officially Incorrect, it was important to establish theoretical intellectual credentials by denying any emulsion content.

Another participant in the symposium was Hans Zocher (1893–1969), from the Kaiser-Wilhelm Institute of Physical Chemistry in Berlin-Dahlem. He was unimpressed by the swarm theory, but his objection was more theoretical, for he was a supporter of an alternative theoretical viewpoint developed by the distinguished Swedish mathematical physicist Carl Wilhelm Oseen (1879–1944) from Uppsala University. Zocher labelled this picture the 'distortion theory'.

There was a vigorous exchange in the discussion between Ornstein and Zocher on the swarm theory. Ornstein provided a long list of experimental data justifying the swarm picture, but Zocher retorted, roughly speaking, that consistency and justification were different matters. Zocher's problem was that his distortion theory was unable to provide quantitative calculations, a point used to maximal effect in Ornstein's rhetoric. Zocher was really a prophet propagating Oseen's theory, but Oseen was a relative unknown in the field and the fact that his contribution was only eighth in the list, and only 13 pages long, bears witness to his relative lack of credibility in liquid crystals at that time.

Oseen's article was much less polemical than those of some other participants. The key to his theory was that he specifically avoided details of the molecular structure of liquid crystals, preferring to describe their physical behaviour on longer length scales. We devote rather more space to a discussion of Oseen's theoretical work. This is because, despite its lack of immediate impact in the symposium itself, it turned out to provide the basis for a good deal of our modern understanding of liquid crystals.

Oseen had been working in the field since 1922, without, it must be said, attracting an enormous amount of attention. His first three papers had been entitled '*Versuch einer kinetischen theorie der anisotropen Flüssigkeiten*' ('Essay on a kinetic theory of anisotropic fluids'). Just prior to the Ewald event, in 1929, he had felt sufficiently confident of his progress to publish a monograph on the subject, which was published in 1929.[17] Eventually he was to write 26 separate papers on the topic, all more or less entitled, in one language or another, 'Contributions to the theory of anisotropic fluids'.

Much, though not all, of Oseen's previous work had been in theoretical hydrodynamics, i.e. the study of fluid flows. It is from this perspective that he approached the theory of liquid crystals, which he initially referred to as anisotropic fluids. We quote from the introduction to his third article in 1923:[18]

> *The first question which strikes a theoretician in this field is the origin of the two forms in which anisotropic liquid drops appear: the sphere-forming 'liquid crystals', and the 'flowing' crystals which also appear crystalline. Lehmann gave the answer to this question, which was that in liquid crystals the structural force is insufficient to overcome the surface tension. That this explanation is valid is hardly open to doubt. The task of the theoretician is to define this concept of structural force, and connect it to the usual molecular forces...*

There follow nearly 40 pages of almost impenetrable algebra before Oseen emerged again with something we can recognize....

> *The number of independent constants is therefore four.*

Difficult as the working is, we are already seeing the four elastic constants of what has come to be called the distortion or continuum theory of liquid crystals. Despite his technical brilliance, Oseen seems to have had some difficulty with the basic physics of anisotropic liquids. He seems, for example, to have believed that anisotropic fluids no longer obeyed Newton's laws of motion. Oseen discussed the balance between structural force (Lehmann's *Gestaltungskraft*) and surface tension, a contrast already made in qualitative terms by Lehmann himself in 1906. Georges Friedel, however, almost contemporaneously, had pointed to the *Gestaltungskraft* as an example of woolly and imprecise thinking, as an exemplar of how *not* to make scientific explanations. Friedel had also dismissed one essential of the structure of liquid

TECHNICAL BOX 5.1 Elasticity in liquid crystals

The concept of elasticity in solids is familiar, for example the stretch of a rubber band or the extension of a spring. The application of a force (stress) changes the shape of the body, and the relative amount by which the shape changes is referred to as the strain. This type of elasticity can be called *extensional elasticity*. The shape of a body can also be changed by applying a twist or torque, and this type of elasticity is called *torsional elasticity*. Normal liquids cannot support either type of elastic strain: a column of liquid cannot be stretched, nor can it be compressed, nor can it be twisted. Liquid crystals on the other hand can be twisted—they do exhibit torsional elasticity—and this above all else is what identifies liquid crystals as special fluids and defines them as new states of matter.

The unique direction in a liquid crystal that determines its optical properties and indeed most other properties is called the director. The director in a liquid crystal can be aligned by specially treated surfaces or by electric fields or magnetic fields. Even the flow of a liquid crystal will align the director sometimes. Furthermore, if we can apply a mechanical force to the liquid crystal fluid then it is possible to change the director orientation—the liquid crystal will support a *torsional strain*.

The great triumph of the theory of Oseen was to formulate this, if not in an easily understandable fashion, at least in a way that made some predictions. The great conclusion of his work was that four parameters were required to define the elasticity of a liquid crystal. Actually we now know that for most purposes three elastic constants are enough, and these determine the ease with which a liquid crystal, or more properly the director of a liquid crystal, can be distorted. These are known as elastic constants for splay, twist, and bend, and a representation of these deformations is given below.

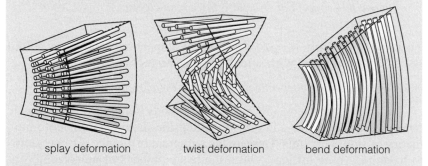

splay deformation   twist deformation   bend deformation

In an unaligned liquid crystal the director will wander around in the liquid crystal, and so there will be a combination of splay, twist and bend deformations. If the liquid crystal is aligned, then these deformations can be forced out and the director lines are everywhere parallel to each other.

Nematic liquid crystals can support splay, twist, and bend deformations, but if layers form, as in smectic phases, then only a splay deformation is possible. For tilted smectic phases (see later) the situation is more complicated since there are now two directors.

crystals, i.e. torsional elasticity, which is the structural force at the root of Oseen's theory.*

In the long run the paper Oseen submitted to the Ewald publication was of great significance, but to his contemporaries there was insufficient time to reflect on the work. It was all too confusing, the mathematics was simply too hard. Indeed, it seems that it was beyond the comprehension of most, for it occasioned no written comments from other participants at all.

The general discussion was a collection of written exchanges moderated by the editor, and these focused on the rights and wrongs of the swarm theory. With hindsight there was a rather pointless exchange of views on what to call the different manifestations of liquid crystalline behaviour, but the debate about terminology, overlain as it was with emotional significance, continued to drag on.

It seems that the symposium excited interest outside the immediate circle of invitees, for the discussion includes comments from some who had not been specifically asked to contribute articles. Two articles were included by Ewald in the *Symposium Proceedings* as the result of extended comments. One such came from the Russian physicist Vsevolod Konstantinovich Frederiks,[†] of the Ioffe Physico-Technical Institute of Leningrad.[19]

Frederiks had worked in Göttingen between 1911 and 1918, partly as an assistant to the famous mathematician David Hilbert. Both he and Ewald had worked for Hilbert in Göttingen, and so they knew each other quite well.[20] Frederiks had worked in the solid state physics laboratory in Leningrad since 1918, and set up his own liquid crystal laboratory in 1924. He had been able to use his German contacts to obtain anisotropic liquid samples from Daniel Vorländer in Halle.

Frederiks's work is of sufficient importance in the history of liquid crystals to give it some attention here. With two graduate students, Alexandra Nikolaevna Repiova and later Valentina Vasilevna Zolina, he discovered a phenomenon now known as the *Frederiks effect*. In this

---

* The present-day liquid crystal professional is intrigued to find the basic elements of current theories in use as long ago as 1923, albeit expressed in a somewhat archaic way, but the historian is more struck by Oseen's use of language. Here was Oseen, referring to exactly the same ideas as Friedel, but from an entirely opposite standpoint. Where Friedel was destructive, Oseen was constructive. For Friedel the concepts were imprecise and hence more or less useless. Friedel's physical cup was at least half-empty. For Oseen, the concepts were imprecise, but nevertheless experimentally motivated. Hence they needed to be made precise. Oseen's physical cup was half-full. What, he seemed to be saying, were theoretical physicists *for*, if not to provide a proper framework for ideas imperfectly articulated by their experimental colleagues?

† The literature contains a number of spellings of Frederiks' name. Transliteration to the German literature appeared as Fréedericksz (sometimes without the accent), and this is often the form used in English translations. In this book we have used the direct transcription from the Russian.

effect a magnetic field reorients liquid crystal molecules in a thin slab, but only above a threshold field, which is dependent on the thickness of the slab (see Technical box 5.2 for a more detailed discussion). At the time the work seemed interesting, but the real significance only became apparent later. In part this was because the magnitude of the threshold field provided a key theoretical test between competing mathematical theories of liquid crystal behaviour. It also turned out (as we shall see in some detail in later chapters) that all modern liquid crystal displays

TECHNICAL BOX 5.2 The Frederiks effect: distortion of liquid crystals by magnetic and electric fields

We devote a separate box to the explanation of the effect of magnetic and electric fields on liquid crystals, since this is at the heart of modern displays. Many scientists contributed to the discovery and interpretation of the effect, but the name of Frederiks will always be associated with it.

The concept of a director being a characteristic variable in a liquid crystal is central to the way scientists think about liquid crystals. The director itself doesn't have any real existence, we can only observe the manifestations of the director in terms of optical properties, electrical properties, and elastic properties. The torsional elastic energy of a non-chiral liquid crystal will be minimized if the director is everywhere parallel. For chiral (cholesteric) liquid crystals the minimum energy state is one with a uniform twist. A state of uniform director configuration can be achieved by aligning a liquid crystal film between two glass surfaces that have been treated in such a way that the director at the surface aligns in a particular direction through surface forces. The time-honoured method of creating an aligned liquid crystal film is to rub the glass plates in one direction, although other more sophisticated methods are also possible. In principle, if the director orientation at the boundary of a liquid crystal film is defined, then this will be propagated through the film to give a sample of uniform alignment, no matter how thick the sample. In reality in thick samples, more than a few hundredths of a millimetre, the director becomes disordered through defects and thermal fluctuations.

Forces other than surface forces will also align the director of a liquid crystal: in particular we consider magnetic and electric forces. Depending on the chemical makeup of the liquid crystal, the magnetic or electric field will align the director along the field direction or perpendicular to it. Much of this follows from the work of Mauguin and was well known to liquid crystal scientists. The innovation due to Frederiks was to consider a situation where the liquid crystal was subject to two different aligning forces: how would the director respond?

To explain what happens, we will use an analogy of a vertical cone sitting on a plane surface. Any object would do, but we use a cone because it has a unique direction, the axis of the cone, which we can liken to a director. The cone is stable on its base because of the force of gravity. Imagine that a perpendicular force is applied to the cone which causes it to tilt. If the force is not too strong, then on releasing it the cone will return to its upright position. If, however, the force exceeds a certain strength, then the cone tips over and the director has realigned with the external force. The analogy is flawed because after the threshold force has been exceeded, the director (cone in our analogy) progressively realigns as the force is increased, not instantaneously, as would happen to the cone.

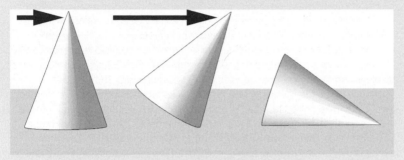

The experiment of Frederiks was to apply a magnetic or electric field to an aligned thin film of liquid crystal and then, by optical means, determine what happened to the director in the liquid crystal. He found that for fields below a certain strength nothing happened, but then above a threshold value for the field the director of the liquid crystal film realigned progressively along the direction of the field. Actually, Frederiks also discovered that the magnitude of the threshold field depended on the thickness of the liquid crystal film. This was done in a clever way by containing the liquid crystal film between the plane surface of a microscope slide and a lens placed on the slide. This gave a liquid crystal film, the thickness of which varied with position, depending on the radius of curvature of the lens.

(a)

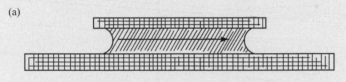

(b)

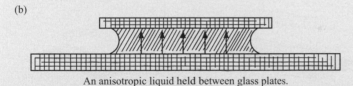

An anisotropic liquid held between glass plates.

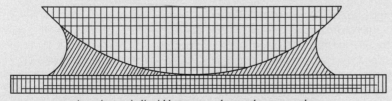

An anisotropic liquid between a plane and a convex glass.

The images on the previous page are taken from Fredericks and Zolina (1929). The samples were viewed from above in a polarizing microscope: in *a* and *b* the arrows denote the direction of the magnetic field. The sample having a curved lens enabled the effect of variable thickness to be viewed.

Having established the effect, it was Zocher who provided a sound mathematical interpretation of the experiment using essentially Oseen's distortion theory. There are many possible configurations of liquid crystal film alignment and external field directions, but variants on the theory can be used to interpret these. The experiment and its interpretation were important for the future development of displays and other electronic devices using liquid crystals. The simplest illustration of what is now known as the Frederiks effect is to imagine an aligned thin film of liquid crystal to which is applied an electric voltage (the origin of the electric field). In the absence of the applied voltage, or for voltages below the threshold value, the film remains undisturbed: left below. When the voltage exceeds the threshold value, then the director (and molecules) begins to turn into the direction of the electric field: right below. The direction of the electric field is vertical, between the two electrodes top and bottom of the cell. Realignment starts at the centre of the film, but as the voltage increases the alignment eventually fills the film, except for a very thin layer at each surface.

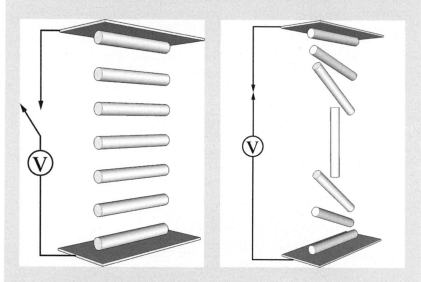

In the figures above, the Frederiks effect induced by an electric field is depicted. In the *off state* (left) the director is aligned by the surfaces of the top and bottom plates. When a voltage exceeding the threshold is applied between the plates the director realigns with the electric field between the plates to create the *on state* (right). The alignment of the director parallel to the electric field depends on the liquid crystal having the correct electric properties (positive dielectric anisotropy). These properties are determined by the chemical structure of the liquid crystal. It is possible to make liquid crystals with negative dielectric anisotropy, in which case they align perpendicular to an electric field.

depend for their mode of operation on some development of Frederiks's eponymous effect.

His group reported some initial results at the Congress of the Russian Physical Society, held in Moscow in December 1926. A longer paper was published in the physics section of the *Journal of the Russian Society of Physics and Chemistry* the following year, entitled 'On the problem of the nature of the anisotropic liquid state of matter'.[21] There, for the first time, amongst other results, could be found what was to become the Frederiks effect. It is interesting to note that Frederiks never refers to liquid crystals, nor even crystalline liquids, despite his Halle contacts. Already in his 1927 Russian paper he is regularly using the term 'nematic'. As was the custom, a substantively similar paper was published in German in the *Zeitschrift für Physik*.[22]

This work was sufficiently noteworthy to earn Frederiks an invitation to the 55th annual meeting of the American Electrochemical Society, held in Toronto in May 1929. One imagines that it was the German rather than the Russian version which had attracted the attention of the American electrochemists. By now Repiova's place in this project had been taken by Zolina, who carried out further experiments in an attic room just beneath the roof of the Ioffe Institute, increasing the maximum magnetic fields eventually up to 25 kG, some 10,000 times the magnitude of the earth's magnetic field. They also tried using electric rather than magnetic forces to realign the liquid crystal molecules. Frederiks did apply for an exit visa to deliver his paper in person, but by 1929 the political situation in Russia was becoming difficult and he was unable to obtain permission. Nevertheless he was able in March to submit his manuscript to be read at the Toronto meeting.[23] The German version published in the Ewald *Symposium Proceedings* summarized this material.

After the discussion, Ewald once again sent out all the manuscripts so that some final remarks might be made. At this stage the Friedels realized that they had never seen the intemperate contribution from Vorländer, and had not yet had an opportunity to respond. In particular they had not had the chance to address the claim by Vorländer to have identified more varieties of liquid crystal phases than the Friedels. The delay in sending Vorländer's article might have been intentional by the editor. Ewald clearly was a wise man, hoping that by the time the offending article was finally transmitted, the Friedels' passions might have died down. Not a bit:

> *Professor Vorländer's note 'Discussion on articles by G. and E. Friedel etc.' situated in the chapter 'General Discussion Notes', was not communicated to us during the course of the the Discussion. In fact, perhaps it is just as well that this violent diatribe against our ideas was thus shaded from all controversy. It self-destructs on its own.*

Professor Vorländer thinks that he sees three, four or more 'crystalline fluid' stases in some instances...We still believe that very probably this is a case of erroneous observation...It is more and more likely that a single material never offers more than one smectic and one nematic stase*.

What is striking from the Friedels' final contribution to the discussion is the articulate plea they make for more experiments, drawing attention to controversy on theoretical questions which they regard as unimportant and which can in any case as yet only be inadequately resolved on the basis of current data.

This timely survey of the field of liquid crystals was published in Volume 79 of the *Zeitschrift für Kristallographie* in May 1931. It should have become a milestone in the development of research. All the principal liquid crystal scientists had contributed to the proceedings. Talented theoreticians such as Oseen, Ornstein, and Zocher had been attracted to the field. The extensive discussion records not the usual 'off-the-cuff' remarks that might have been made in a conference hall, but the carefully considered responses made as a result of thorough study of the written presentations.

The papers do indeed stand as a significant development in our understanding of liquid crystals, but by a bizarre twist of events, history has consigned this volume to the oblivion of library stack rooms. A mixture of personalities, ambition, and the background of the political scene in Germany conspired to place the research successes recorded in Ewald's symposium into the canon of British science.

The meeting that was

Sir William Bragg's position on the Editorial Board of the *Zeitschrift für Kristallographie* gave him easy access to the papers published in Ewald's *Symposium Proceedings*. It seems likely that it was Herrmann's article on X-ray diffraction from some liquid crystal materials which excited his interest. For when Bragg had examined Friedel's materials using X-rays, he had not noticed anything of interest. Bragg clearly felt that these liquid crystals needed a second look. The papers of Ewald's *Symposium Proceedings* were not in English, but Bragg's interest was sufficiently aroused that he attempted English translations of some of the

---

* The uncertainty surrounding the nature of liquid crystals resulted in the introduction of the term 'stase', used by both the German and French Schools to denote a liquid crystal phase. Few were brave enough to nail their colours to the mast and proclaim all the different types of liquid crystal as phases. There was even serious academic discussion about the differences between stases and phases, although now we make no distinction. Liquid crystals are a new state of matter and can exist as a number of different phase types.

symposium papers.[24] He found the theory papers particularly interesting. He wrote out by hand the translation of Oseen's paper, and then turned to translating Zocher, but Zocher's (native) German turned out to be harder to construe. Eventually his efforts petered out and he gave up:

.... the next four sentences are too difficult...

If the mountain will not come to Mahomet, Mahomet will come to the mountain, although in Sir William's case it was: if I can't understand German papers on liquid crystals, then I will have them presented in English.

Bragg decided to restage the virtual symposium on liquid crystals in his institute, the Royal Institution in Albermarle Street in Central London. In keeping with the dignity invested in its name, the Royal Institution has an illustrious history. It was the academic seat of the British pioneering scientists Sir Humphrey Davy (1778–1829) and his student and successor Michael Faraday (1791–1867). It was there that much of Faraday's ground-breaking experimental work on electromagnetism was performed. It would be a fit and proper place to introduce liquid crystals to an English-speaking audience.

The problem of translation into English would be solved by the simple expedient of engaging a competent translator. Bragg duly invited all the contributors to Ewald's edited volume to a meeting entitled 'Liquid crystals and anisotropic melts'. The meeting was to be held on 24 and 25 April 1933, under the auspices of the Faraday Society* of which he was a distinguished member.

That Bragg should have wished to restage a German event in London is extraordinary. The wartime loss of Sir William's younger son, Lieutenant Bob Bragg, killed by Turkish and German forces at Gallipoli in September 1915, had inspired in him a long-lasting antipathy towards Germany and all things German. Indeed, in June 1920, he had refused point blank to attend the Nobel Prize award ceremony in Stockholm, deferred from June 1916, because:[25]

I believe that several Germans are going.

He had also persuaded his son William Lawrence, co-recipient of their Nobel Prize, to decline the invitation.

Sir William's personal interest in liquid crystal science must have been a powerful motivation. Stronger than that was his desire, as Director of the Royal Institution, to put his institute at the forefront of modern world science. Perhaps the deciding factor was the strong personal link between Paul Ewald and both Braggs, but especially with Lawrence.

---

* The Faraday Society was founded in 1903 to promote the study of electrochemistry, electrometallurgy, chemical physics, and kindred subjects. It became the *de facto* British society for physical chemistry and in 1971 merged with the Chemical Society. This and related societies became the Royal Society of Chemistry in 1980.

It would not be long before Ewald was forced from his academic position in Germany by Nazi persecution. The Braggs helped Ewald to obtain a post, first at the University of Leeds, where Sir William had been Professor of Physics, and then at Cambridge, where Ewald became a colleague of Lawrence Bragg. In due course Ewald moved on to a permanent post in the USA.

The actual organization of the meeting was entrusted to Bragg's former research student at the Royal Institution, the charismatic John Desmond Bernal (1900–1971), a rising young star of the crystallographic world. Bernal was to become an outstanding but controversial figure in science and politics, and he will figure strongly in future chapters. He had been an undergraduate at Cambridge, and his prodigious intellect had earned him the sobriquet of 'Sage'. This nickname was to stay with him throughout his life.

For the liquid crystal meeting, Bragg invited all the major figures in the liquid crystal world (although at least one died before he was able to come and others, for a variety of reasons, did not show up). The list of invited speakers for the Faraday Discussion was almost a replica of the contributors to the Ewald virtual meeting, although for some reason Ewald himself did not attend. The topics of the papers presented (this time in many cases in person) at the London meeting similarly echo the pages from the *Zeitschrift für Kristallographie* published only two years previously.

Repetition notwithstanding, the record of that Faraday Discussion is the first major document in English on liquid crystals. As we have already emphasized, the peculiar status of English, as the modern *lingua franca* of the scientific world, has conferred on the papers read at the 1933 London meeting a greater historical importance than might otherwise have been the case. The organizers of the meeting went to great trouble to ensure that papers originally written in other languages were translated so as to be accessible to the Anglophone audience. To ensure that the science was faithfully reproduced in English, most of the translations were entrusted to one of Bernal's crystallographic students, Helen Megaw (1907–2002).

One hundred and fifty members and visitors attended the meeting, which is an impressive number given the relative novelty of the field. Among eminent foreign visitors, the proceedings record, were Professors Leonard Ornstein of Utrecht and Hans Zocher (now in Prague, who, the proceedings record, brought his wife), and Professor Dr Rudolf Schenck, now of Berlin (incorrectly stated in the Proceedings, since Schenck was at the time at the University of Münster). Schenck was a particularly important visitor, not only because of his seminal contribution to the field (in which he was no longer an active participant) but because he was the current president of the *Deutsche Bunsen Gesellschaft*. Not present, except in spirit—for papers were presented on their behalf—were the theorist Oseen from Sweden, and the experimental physicist Frederiks from Russia, whose increasingly vulnerable position as a member of an aristocratic family in the Soviet Union again prevented him from leaving the country.

Particular highlights of the Symposium echoed the 1905 Karlsruhe meeting, at which Lehmann's demonstrations had been such a dramatic success. The introduction to the Royal Institution meeting records that:

> *After Professor Oseen's Introductory Paper had been taken as read and discussed, very beautiful demonstrations were given by Professor Vorländer, Professor Van Iterson and Dr A.S.C. Lawrence. Further experimental demonstrations were given on Tuesday by Professor Van Iterson and Mr Bernal.*

The fact that the first paper in the Discussion Proceedings was by Oseen is of significance.* Oseen (number 8 in the Meeting that Wasn't) had been promoted to first place in London. What had changed between 1930 and 1933? Partly the absence of the Friedels required a new keynote contributor. Georges Friedel had been invited,[26] but by this time was so ill that he could not leave his bed and was certainly too sick to prepare a new paper. His son Edmond, whom he had hoped to send in his place, seems to have been detained by other commitments. Probably the major factor was Sir William himself. He had written out in longhand (and presumably at some trouble) a translation of Oseen's Ewald symposum paper and been impressed. Because of the format of the meeting, Oseen's physical absence would not have been a major inconvenience. In 1933 Oseen had been working on anisotropic fluids for 12 years, and had already published 20 papers. In 1934 he switched from German to French as a protest against the rise of Nazism in Germany. Oseen's contribution to the London conference is his only paper in English, and judging from the published proceedings (in which the translators were acknowledged on the title page), he nevertheless wrote it himself with presumably only minor editorial help. By May 1933

---

* The format of the Faraday Discussions (they continue to this day) is different from other scientific meetings. The usual format of a scientific conference involves the presentation of talks, which are then followed by some (often disappointingly desultory) discussion. The Faraday Discussions, by contrast, in keeping with the spirit of the distinguished thinker after whom they are named, are designed to make lasting and profound scientific contributions. The real contribution, it is hoped, will come not so much from the talk itself (for the basic meat will have been published elsewhere as well), but from the expert discussion that it elicits. Often in other meetings this discussion does occur, but in the theatrical wings, so to speak, and hence inaccessible to the historical record. So in the Faraday Discussions, the presentations are written up beforehand, then circulated to participants, but then taken as read at the meeting, with perhaps only five minutes of introduction from the author. The discussion is minuted live, but with the opportunity for participants to change their views or add extra points *ex post facto*, so as to give a snapshot of the usually hidden ongoing intellectual process.

his ideas were somewhat mature, although not yet quite in their modern form. The article was a wide-ranging review of Oseen's current views on liquid crystal science. To a modern eye the article looks quaint. For example, the concept of the *director* (see Chapter 2, Technical box 2.1), so central to the modern paradigm, is absent. However, Oseen, even as a theorist, tried to deal with all the major aspects, such as defect structure, optics, dynamics in nematics, and smectics.

The proceedings then include fully five separate articles by the 66-year-old Vorländer, including his demonstrations. Once again, Vorländer had been relegated to number two in the proceedings, but perhaps permission from the organizers to present five separate papers mollified what could otherwise have been significant indignation. By this time he seems not to have had much new to say, but at least the controversy over naming the liquid crystal phases had died down.

The paper from the absent Russian invited participant Frederiks and his collaborator Zolina was an update on work presented at the Ewald symposium. Frederiks had previously described experiments on the competing effects of the alignment of liquid crystals by surfaces and magnetic fields. In the paper submitted to the Faraday Discussion, he gives results obtained using electric fields rather than magnetic fields. He correctly makes the connection between the experiments using magnetic and electric fields. This significant discovery was that liquid crystal films could act as optical switches, on or off, when activated by magnetic or electric fields. This would prove to be the principle behind modern liquid crystal displays.

Not only that, but a later paper in the meeting from Professor Hans Zocher, entitled 'The effect of a magnetic field on the nematic state', provided a theory for the effect. Again, related work from Zocher had appeared in the Ewald *Symposium Proceedings*, in German, but had defeated the translating efforts of Sir William Bragg. This time at the Faraday Discussion the paper was in English, translated as noted above by H. D. Megaw.

Zocher's presentation is essentially the same paper which appeared in 1929 in the *Zeitschrift für physikalische Chemie*.[27] This is not the paper the translation of which had defeated Sir William Bragg, but rather one which directly addressed the experiments of Frederiks. Zocher's title in itself was making a political statement. Notwithstanding his German background, he was using the French-inspired term *nematic*, rather than the more esoteric notation that Vorländer had been vociferously promoting. In less than a decade, Friedel's terminology (and hence his physical picture) had achieved clear a victory.

The two major schools of thought emphasized different aspects of liquid crystalline behaviour. For the swarm theorists the most important property to be explained was the turbidity of the liquid crystal fluid, whereas for the distortion theorists the essential feature was the optical

Inhaltsverzeichnis des 79. Bandes.

Heft 1/4. (Ausgegeben im Mai 1931.)

Seite

G. et E. Friedel (Strasbourg), Les propriétés physiques des stases mésomorphes en général et leur importance comme principe de classification. (Avec une figure et deux planches) . . . 1
D. Vorländer (Halle a. S.), Chemie der kristallinen Flüssigkeiten. (Mit 6 Textfiguren) . . . 61
L. S. Ornstein (Utrecht), Experimentelle und theoretische Begründung der Schwarmbildung in flüssigen Kristallen. (Mit 4 Textfiguren) . . . 90
H. Zocher (Berlin-Dahlem), Die Optik der Mesophasen . . . 122
K. Herrmann und *A. H. Krummacher* (Charlottenburg), Röntgenuntersuchungen an kristallin-flüssigen Substanzen. II. Phenetolazoxybenzoesäureallylester. (Mit 10 Textfiguren) . . . 134
W. Kast (Freiburg i. Br.), Dielektrizitätskonstante. (Mit 10 Textfiguren) . . . 146
W. Kast (Freiburg i. Br.), Magnetische Suszeptibilität. (Mit 7 Textfiguren) . 161
C. W. Oseen (Upsala), Probleme für die Theorie der anisotropen Flüssigkeiten 173
C. Hermann (Stuttgart), Die Symmetriegruppen der amorphen und mesomorphen Phasen . . . 186
Wo. Ostwald (Leipzig), Über mesomorphe und kolloide Systeme. (Mit 3 Textfiguren und 4 Figuren auf 1 Tafel) . . . 222
V. Fréedericksz und *V. Zolina* (Leningrad), Über die Doppelbrechung dünner anisotrop-flüssiger Schichten im Magnetfelde und die diese Schicht orientierenden Kräfte. (Mit 4 Figuren auf 2 Tafeln und 2 Textfiguren) . . 255

Diskussion über die vorstehenden Aufsätze.

Vorwort . . . 269
Allgemeine Diskussionsbemerkungen . . . 269
R. Schenck (S. 269), J. J. Trillat (S. 274), D. Vorländer (S. 274).
Diskussion zur Arbeit Friedel . . . 289
G. Foex (S. 289), C. W. Oseen (S. 291), G. et E. Friedel (S. 293), H. Zocher (S. 296), G. et E. Friedel (S. 297), P. P. Ewald (S. 299), G. Friedel (S. 307).
Diskussion zur Arbeit Ornstein . . . 308
L. S. Ornstein (S. 308), K. Herrmann (S. 308), L. S. Ornstein (S. 311), H. Zocher (S. 312), L. S. Ornstein (S. 313), H. Zocher (S. 314), M. Jeżewski (S. 314), W. Kast (S. 316), M. Jeżewski (S. 317), W. Kast (S. 318).

Figure 5.2a Contents page of the Ewald symposium. The article of Fréedericksz and Zolina (in at no. 10) was added as an afterthought. Ornstein is at no. 3, as might be expected for a well-known and well-respected theoretician.

Contents.

PAGE

Liquid Crystals and Anisotropic Melts. *A General Discussion* . . 881
The Theory of Liquid Crystals. By C. W. Oseen . . . 883
I. Remarks on Systems with Mixed Dimensions. By D. Vorländer . 899
II. Supracrystallinity of *p*-Azoxybenzoic Acid. By D. Vorländer . . 902
III. Remarks on Liquocrystalline Resins and Lacquers. By D. Vorländer 907
IV. Cyclo-Pentanone and Cyclo-Hexanone. By D. Vorländer . . . 910
V. The Polymorphism of Liquid Crystals. A Demonstration. By D. Vorländer . . . 913
A Simple Arrangement to Demonstrate Liquid Crystals. By G. van Iterson, Jr. . . . 915
Forces causing the Orientation of an Anisotropic Liquid. By V. Fréedericksz and V. Zolina . . . 919
New Arguments for the Swarm Theory of Liquid Crystals. By L. S. Ornstein and W. Kast . . . 931
The Effect of a Magnetic Field on the Nematic State. By H. Zocher 945
Magnetic Properties of Mesomorphic Substances. Analogies with Ferro-Magnetics. By G. Föex . . . 958
Inclinations of Molecules in some Crystalline-Fluid Substances. By Karl Herrmann . . . 972
Rotating Molecules and the Liquid Crystalline State. An X-Ray Examination of Long Chain Esters. By T. Malkin . . . 977
Alterations in the Nature of a Fluid from a Gaseous to Liquid Crystalline Conditions as shown by X-Rays. By G. W. Stewart . . . 982

Figure 5.2b Contents page of the Faraday discussion. Oseen has overtaken Vorländer in the hierarchy, while Fréedericksz has jumped from no. 10 to no. 4. Ornstein is down from no. 3 to no. 5. His downgrading is an early indication of his loss of prestige in the field.

anisotropy. The swarm theorists therefore postulated the existence of swarms of aligned molecules, and the literature of the day is filled with diagrams of these swarms allegedly in constant thermally induced Brownian motion. A precise mathematical model of the phenomenon was proving elusive, but Ornstein had used this kind of a picture elsewhere in his successful description of critical point phenomena in classical fluids.

Zocher was seriously grumpy that anyone was taking these swarm theorists seriously. His paper reads like a modern theoretical paper on the continuum theory of nematics. He calculated, in business-like fashion, the dependence of the couple (i.e. the turning force) transmitted by a nematic between two walls of a cell, as a function of magnetic field. He (or rather his assistant Eisenschimmel) then tested his theory experimentally.

Of course, to begin with Zocher could not but admit his irritation, He starts:

> *Two fundamentally different hypotheses have hitherto been used to account for the changes in nematic systems in a magnetic field; these are the swarm theory and the distortion theory* (Verbiegungstheorie). *It is extremely important to decide which is the correct one, not only for the particular problem dealt with in this paper, but for the whole general question of the structure of these phases.*

There followed a mainly sober paper, but with occasional digs at Ornstein or Kast, and the inability of the swarm theory to explain this or that piece of data. In the discussion, however, Zocher was more forthright:

> *As to the swarm theory, it is* impossible *(authors' emphasis) to find out the diameter of the swarm…*

he averred, pointing out that one kind of experiment indicated a swarm diameter of about 0.1 microns (10 millionths of a centimetre), whereas another indicated a hundred times that. His scepticism concerning the swarm theory was backed up in the discussion by a long and detailed intervention by Bernal.

The importance of Zocher's article lies in the progress that had been made since the Ewald symposium. In 1930–1931, Zocher had been fending off demands by Ornstein not merely to criticize the swarm theory but actually to provide some viable alternative. The rhetoric is unconvincing, even if Zocher himself was sure of his ground, but the Frederiks experiment provided a specific procedure to calculate the elastic constants. Now the distortion theory was up and running, and the writing was on the wall for the swarm theory. As we have seen, soon the Frederiks group also was to begin to conceptualize not in terms of Ornstein's swarms but in terms of the Oseen–Zocher elastic constants.

We also note a contribution entitled 'Lyotropic mesomorphism', by A.S.C. (Stuart) Lawrence (1902–1971), a former research assistant to Sir William Bragg and also to the previous Director of the Royal Institution, Sir James Dewar. At the time of the London meeting, Lawrence was based in Cambridge, but later would become a professor in Sheffield and a colleague of one of the present authors. This is our first encounter with so-called *lyotropic* liquid crystals, yet they were investigated by Lehmann and others in the very earliest period of liquid crystal research. This new form of liquid crystal behaviour provides a connection with biological systems, which will be explored in more detail in the next chapter.

However, for the participants at the 1933 Discussion on Liquid Crystals, Lawrence summarized the analogies between the liquid crystals formed by heat (*thermotropics*) and liquid crystals formed by the action of water (*lyotropics*) as follows:

> It is interesting to recall that Lehmann's liquid crystals of ammonium oleate, which formed the foundation of the subject under discussion, were deposited from solution; and also that the amount of water present profoundly modified their form. Since that time, however, most of the substances studied have been in the form of anisotropic melts of single compounds. The action of heat and of a solvent are not dissimilar insofar as they both loosen the directive forces holding the molecules in their normal crystal lattice. For a mesoform to appear, it is necessary for these forces to persist in either one or two dimensions after loosening of the third. But even when this occurs and a mesoform exists it is unlikely that heat and all solvents will bring this about equally well.

Lawrence's work on the connections between soaps, colloids, and liquid crystal phases started a process which was to yield insights into the relationship between liquid crystal phases and life.

The final two papers presented at the 1933 discussion were part-authored by J. D. Bernal. The second of these was concerned with the structure of water. This was an obsession of Bernal at the time, as he was convinced that therein lay the secrets of living systems, which of course all need water. For us, however, it is the paper with his research student that is more interesting.[28]

The young Dorothy Crowfoot (1910–1994), to become Hodgkin, would later be celebrated for determining the atomic structure of vitamin B_{12}, an achievement for which she was awarded the 1964 Nobel Prize in Chemistry. In 1932 she had joined Bernal at Cambridge to learn crystallography from 'Sage' and was soon captivated by his knowledge and charm. For his part Bernal, despite being her supervisor, couldn't but notice her fair hair which 'when caught by the sun, stood around her head like the halo of a stained glass window mediaeval saint'.[29] The pair inevitably became lovers, but we do not know if the contribution to the conference of 1933 was a product of this union. Whether it was or not, it bore the marks of both: Dorothy's careful experimentation and Bernal's confident conclusions:

> ...the mesophases, far from being an anomalous manifestation, take their place in a regular procession from the disorder of the ideal liquid to the regularity of the ideal crystal...

The general discussion on all the papers presented ranged widely. The transcript of the discussion is not a true and accurate representation, of course. There are inserted contributions and participants are able to edit their words so as to make it appear that what they had said is more intelligent than had appeared on the day. By contrast with the Ewald symposium, however, underlying the published proceedings is the record of a human occasion.

Epilogue

That was not quite the end of the proceedings. The relationship between the optical forms (focal conics) seen under the microscope and the layering was crucial to Friedel's interpretation of the smectic phase as a layered structure. The mathematics of this explanation had not appeared in the scientific literature. Sir William saw an opening and was moved to append to the published record a short paper giving his detailed mathematical derivation. As far as we are aware it is indeed the first time this appeared in the literature, but it drew a decidedly mixed response from Friedel himself.

Letters from Friedel written to Bragg provide some sort of epilogue to the saga. Writing (in French) on 22 May 1933,[30] from what was to become his death bed, Friedel apologized for having failed to attend the meeting in London, but points out that:

> For several years, my health has prevented me from leaving my bedroom. I had hoped that my son Edmond, who is well-acquainted with all of my work and ideas, would have been able to attend instead, but pressure of his work prevented him. I imagine that at least we would have been able to convince our English colleagues to abandon the unfortunate description 'liquid crystals'.

Furthermore, in the letter Friedel cannot restrain himself from criticism of Vorländer, who, in his view:

> ...in a Teutonic frenzy (an important enough phrase for Friedel to use Latin—furor teutonicus) threatens to use these compounds (liquid crystals) to break heads.

In this battle against the liquid crystal terminology, Vorländer was at least partly an ally of Friedel, although as we have seen Vorländer had his own arcane nomenclature, and insisted on referring to 'crystalline liquids'. Finally Friedel thanked Bragg for having explained to the meeting the structure of focal conics. This refers to Bragg's contribution to the discussion, which was written up for the proceedings, but there is a sting in the final paragraph of the letter:

> *This (*the explanation of the focal conics*) has been known for more than 20 years, and taught all over France for more than 10 years, but most Germans still don't want to take any notice.*

So in this one sentence Friedel had dismissed Bragg's work as trivial and insulted the German nation. At least Sir William Bragg and George Friedel could agree on one thing: the Germans were awful!

One final letter exists from Friedel to Sir William Bragg dated 7 June,[31] in which he revisits the issue of nomenclature. More in fond hope than expectation, Friedel refers to Bragg's reply to an earlier letter:

> *However, it is good news for me to learn that you intend to observe the effects for yourself. I am convinced that much better than my telling you, is that the observations of an experimentalist such as yourself will determine a rational nomenclature.*

There was no trickery over the staging of the Faraday Society general discussion at the Royal Institution in April 1933. However, it was a bit surprising that in the introduction and acknowledgements there was no mention of the pioneering work of Paul Ewald in bringing together, albeit on paper, all the principal liquid crystal scientists of the age in the publication in the 1931 issue of *Zeitschrift für Kristallographie*. That the 1933 Faraday Society meeting is a much quoted source in the liquid crystal literature is a fact, but one due in larger measure to the language of the publication than to the merit of the publication.

As a final postscript to the meeting, Sir William Bragg wrote a review of liquid crystals for *The Naft*, a magazine for the Association of Oil Producing Countries.[32] In this he speculates:

> *It then appeared that no one in this country (*Great Britain*) had paid any attention to the matter for many years......It is difficult to explain this strange and rather disconcerting fact.*

He goes on to encourage his fellow scientists, or at least those coming from oil-producing states (not, at that stage, the OPEC countries of the Middle East, but Europe and the Americas), to take up the study of liquid crystals.

> *Perhaps the bewildering complexity of the observed effects (in liquid crystals), and the difficulties found in correlating them have prevented some from taking up the subject...Fortunately our Continental friends have not been quite so cautious...We should not be afraid to follow them.*

This entreaty to follow up research into liquid crystals fell on unreceptive ears, as the scientific community was to be turned upside down by political events in the real world. To a large extent the liquid crystal story stalled, and science and scientists turned their attention to preparations for war. The fates of the participants in liquid crystal science were varied, ranging from tragic death, through ignominy, to great career triumph. However, we postpone relating these stories in order to take a trip along the byway predicted by Otto Lehmann to be of the greatest scientific importance of all. The next chapter explores the connection between liquid crystals and life.

Friedel died in December 1933. He would be appalled to know that we still refer to his mesophases as liquid crystals. He would be more pleased to know that we do also use the terms nematic and smectic, if disappointed to learn that the term 'cholesteric' has recently gone out of fashion. It is an irony that we also refer to molecules that are capable of forming mesophases, i.e. liquid crystals, as mesogens. Thank goodness that Friedel did not get to hear about that bastardization of nomenclature.

6
The threads of life

What's this about your liquid crystals. Can they even eat?

In September 1906 Otto Lehmann was given an opportunity to address the 78th Meeting of German Scientists and Physicians in Stuttgart. He had recently published a two part paper[1] which drew analogies between liquid crystals and living organisms, and a multi-disciplinary audience of chemists, physicists, and doctors was just the forum for Lehmann to explain the far-reaching implications of liquid crystals. The controversy surrounding the discovery had already reached the ears of the wider scientific community, and there were plenty of sceptics ready to challenge Geheimrat Lehmann. The lecture was scheduled towards the end of the meeting, and delegates anticipated the occasion with curiosity.

Two old scientific friends, the chemist Müller and crystallographer Schulze, were conference delegates, but were less than enthusiastic about Lehmann's forthcoming presentation:

Müller

I see in the programme that on Friday there will be a plenary lecture, entitled 'Liquid and apparently living crystals'. Isn't this ridiculous? Everyone learns already in school that crystals are rigid bodies, built according to strict mathematical rules, the direct opposite, so to say, of liquids, and as dead matter, also the direct opposite of material that is alive. Just recently I happened to be looking at the latest big physics text book by Chwolson. Inside I found the comment that Tammann's classic studies have established that 'hard' and 'crystalline' are in principle equivalent. Of course, it does say later that there is some difficulty in reconciling this statement with Lehmann's hypothesis about the existence of liquid crystals. However, this can only mean that Lehmann's work isn't much good! And now he wants to tell us more about 'apparently living' crystals! What do you think about all this?

Schulze

As far as I know, if the proposal for the lecture came from the Conference Scientific Committee, then the material should be OK. In my opinion, Lehmann must be making a mistake somewhere. The Scientific Committee clearly didn't realize. The first I heard about Lehmann's apparently living crystals was in an article in the Karlsruhe Zeitung last year, in which they are described as a 'New Wonder in Physics'. More like an April Fool's joke, but the date wasn't right... Since then Lehmann himself has published a pile of articles on the subject. Of course I haven't read them, since they certainly have nothing serious to say. I can only feel sorry for him because of this scientific blunder.

Müller

Normally, I wouldn't give the time of day to something so ridiculous. But it is true, isn't it, that even in 500BC the philosopher Heraclitus pointed out that everything flows? Presumably that includes crystals too! I recall reading something in my youth General Morphology *by Haeckel (1866), in which a relationship is claimed between crystals and lower organisms. Let us at least have a look at Lehmann's demonstration in the Exhibition Hall......*

......later....

Schulze

How remarkable! His crystals really do look like worms! Let's ask the demonstrator how he prepares his samples!

(The pair now encounter Professor Lehman himself, who is very pleased to explain the experiments to them, and gives them a further demonstration.)

Müller

This is even more interesting than the things we saw earlier! I can see an unending diversity of forms, rows, worms, simple and double spheres, all in great confusion moving too and fro and then suddenly colliding again like insusorien [?ed.microbes] in a water droplet.

This extract of dialogue is from a text invented by Lehmann. He adopted the Socratic polemical method to publicize ideas on liquid crystals and life because Lehmann knew the potential for ridicule that his proposals carried. By hiding behind the figures of Müller and Schultze, Lehmann could present his arguments at arm's length. He was being cautious, but he was convinced that liquid crystals were something very different from anything that had previously been identified. Lehmann's attempts to justify his points of view led him to write his 1907 polemical dialogue *Die scheinbar lebenden Kristalle* (Apparently Living Crystals).[2] He would also publish in 1908 a monograph entitled *Flüssige Kristalle und Theorien des Lebens*[3] (Liquid crystals and theories of life), which we shall discuss in more detail later in this chapter.

Up until the nineteenth century it was generally believed by ordinary people, by religious intellectuals, and even by scientists, that dead stuff (inorganic matter) and living stuff (organic matter) were fundamentally different. To transform one to the other required some kind of *vital force*. The Maharal of Prague (1520–1609) is said to have created a clay man to defend the Jews of Prague against anti-Semitic attacks, and to have breathed life into it by means of Hebrew prayers and incantations. By the nineteenth century more modern transformational techniques had come into vogue, and in Mary Shelley's 1818 novel,[4] Victor Frankenstein's monster needed a hefty bolt of electricity to get him going.

The special characteristics of living organisms are nevertheless such that we might label living material as a distinct state of matter, to go alongside the 'dead' states of gases, liquids, and solids. This is where liquid crystals come in. Although the speculations of Lehmann about living liquid crystals were contrary to contemporary opinion, it is clear that he had thought deeply about his observations under the microscope. He saw many phenomena that recalled the processes of life. For Lehmann the obvious explanation was that something was occurring that closely resembled creation and growth.

There is a yet deeper question about liquid crystals and life, since the complexity of molecular organization in liquid crystals closely mimics that found in biology. The explanation of the complex structures found in living matter presents a considerable challenge, and even today some suggest that the so-called 'physical laws of nature' are not enough in themselves to explain life. There is a need for something extra: 'a law of organized complexity'. As Paul Davies explains in his book *The Fifth Miracle*:[5]

> .... it (biogenesis)* may involve some non-local type of physical law, as yet unrecognized by science, that explicitly entangles the dynamics of information with the dynamics of matter.

This is very much a contemporary issue, especially as the creationism/evolution 'debate' moves once again into the public arena. The distinguished theoretical biologist Stuart Kauffman, former Director of the Santa Fe Institute, has spent a lifetime searching for the laws of

---

* The descriptive term for living organisms giving rise to other living organisms, i.e. life from life.

self-organization and complexity.[6] Without going beyond the accepted laws of thermodynamics, Kauffman now believes[7] that in addition to natural selection, there may be a mechanism for evolution through a combination of self-organization of processes and selection.

Liquid crystals combine molecular self-assembly, molecular organization, and organizational complexity; they are also intrinsic to living systems. This is the aspect of liquid crystals that we will explore in this chapter.

Films, micelles, and fibres

In the last chapter, we introduced the idea that liquid crystals could be formed by mixing soap and water. Indeed Lehmann had made much of this, only to be castigated by Daniel Vorländer for stretching the bounds of (his) credibility, but it turned out that Lehmann was right. By the 1930s, the existence of lyotropic liquid crystals—formed by dissolving solids in liquids (usually but not always water)—would be firmly established. Early studies of lyotropic liquid crystals were directed to the common mixtures of soap and water. Traditional soaps are salts of fatty acids, although there are many varieties, all with a common origin in the tissue of once-living matter. Mixtures of soaps with water produce a number of different preparations having different properties, depending on the proportion of soap and water. These preparations were well-known to the soap-makers of old, who had even given names to them such as 'curd soap', 'neat soap', and 'middle soap'. It transpires that the liquid crystal structures found in soaps are fundamental to our understanding of plant and animal cells.

The method of making soap would have been well-known to the next key figure of our story, Benjamin Franklin, whose father Josiah Franklin was a tallow chandler and soap boiler trading in the city of Boston, Massachusetts. Born in January 1706, Benjamin Franklin (died 1790) was in many respects a man before his time. He is remembered both as a statesman and, for his early experiments with electricity, as a scientist. It was his scientific experiments on Clapham Common that earn him a place in the liquid crystal story. As a regular transatlantic traveller, Franklin had become interested in the phenomenon of 'pouring oil on troubled waters'. He found the effect of calming seas by emptying barrels of oil fascinating, and requiring further investigation.

So it was that Franklin found himself on a particularly windy day in 1772 by the pond on Clapham Common, at that time just to the south of the conurbation of London:[8]

> 'I fetched out a cruet of oil, and dropt a little of it on the water.
> I saw it spread itself with surprising swiftness upon the surface;

*but the effect of smoothing the waves was not produced; for I had
applied it first on the leeward side of the pond......I then went to
the windward side....and the oil, though no more than a teaspoon-
ful, produced an instant calm over a space of several yards square,
which spread amazingly, and extended gradually till it reached the
lee side, making all that quarter of the pond, perhaps half an acre,
as smooth as a looking glass.'*

The oil had certainly calmed the waves, but Franklin was astonished
by the huge area that a small amount of oil had covered on the surface
of the pond.

*'In these experiments, one circumstance struck me with particular
surprise. This was the sudden, wide and forcible spreading of a
drop of oil on the face of the water, which I do not know that any-
body has hitherto considered.'*

This experiment was repeated all over England for the benefit of
Franklin's scientific chums. One of these was a doctor, William
Brownrigg, who lived at Ormathwaite, overlooking Derwentwater in
the Lake District of north-west England. Brownrigg was a Fellow of the
Royal Society and persuaded Franklin to publish an account of his
observations in the *Philosophical Transactions* of the Royal Society,
which duly appeared in volume 64 in 1774.[9]

In fact, Franklin had already established his reputation as a scientist
through his experiments with electricity, and for this research had been
awarded in 1754 the Copley Medal of the Royal Society. This was a rare
honour for a person from the colonies, and was followed soon after by
his election as a Fellow of the Royal Society.

The chemical community remembers Franklin because of his
Clapham Common experiment, which marked the dramatic beginning
of the science of surfaces and colloids, but it was some decades after
Franklin's death before the significance of the thin film of oil on the
pond would be appreciated. What he had produced was a membrane
just a single molecule in thickness. A simple calculation showed that
the film thickness and hence molecular length was just a few billionths
of a metre. Why and under what circumstances such monomolecular
films form is another story. It is enough for us to know that membranes
consisting of two such films, known as *bilayers*, are the protective walls
of living cells. They are also liquid crystalline.

The apothecaries of old had discovered that adding water to materials such as gelatine (extracted from animal skin and bones) or agar-
agar (from seaweed) produced jelly-like substances now known as
gels. The characteristic property of a gel, as every child knows, is that
it is a peculiarly wobbly solid that liquefies extremely easily on heating
or stirring. What possible structure might cause this behaviour? The

Swiss natural scientist Karl Wilhelm von Nägeli (1817–1891)* proposed, on the basis of his microscopic observations, that gels consisted
of assemblies or clusters of molecules separated by water. He named
these clusters micelles (from the Latin word for 'diminutive cells or
grains') and their birefringence led him to assume that they were crystalline. Nägeli also proposed that micelles give the gel its particular
properties, but the trapped water allows the material to be mobile. The
existence of micelles is now very well established in colloid science,
but micelles are much too small to have been visible using the microscopes available to Nägeli. His intuition was good, but the real structure of micelles would take much longer to be confirmed. They are now
recognized as being components of many lyotropic liquid crystals.
Nägeli has a more important claim to fame: his microscopic studies led
him to discover sperm cells in ferns, and he is credited with the first
identification of chromosomes in cells, although he had no idea at that
time of their nature or function.

The prominence of Nägeli was such that Gregor Mendel (1822–1884),
the inventor of modern genetics, sent the results of his experiments on
cross-breeding of peas to von Nägeli for comment. However, the great
professor replied very snootily to the unknown Mendel, suggesting that
he try to replicate his results on hawkweed (*Hieracium*), a plant that
Nägeli had studied for decades. Mendel tried, but was unsuccessful, and
for good reason: peas reproduce sexually like us, but hawkweed is different and mostly reproduces by cloning. Mendel's rules do not apply.
Despite this setback Mendel published his original data on peas, but in
an obscure journal where they remained undiscovered for 40 years.
Nägeli's unhelpful response to Mendel perhaps contributed to a hold-up
in our understanding of the laws of heredity, and certainly deprived
Nägeli of an even greater legacy if he had made the connection between
chromosomes and inheritance.[10]

It was the microscopic observations of a well-known German pathol-
ogist, Rudolf Virchow (1821–1902)† that gave Otto Lehmann his idea
that liquid crystals might have something to do with the propagation of
life. Virchow was a very successful Prussian doctor-turned-scientist with

* Nägeli was born into a medical family and began his professional studies as a
medical student in Zurich, but soon changed to botany, obtaining a Ph.D. for a disserta-
tion on a Swiss thistle. Apparently not satisfied with botany, Nägeli then went to Berlin
in 1841 to study philosophy under Hegel, and this clearly broadened his interests, which
became much more concerned with life processes. Nägeli did, however, return to botany, which he taught at universities in Zurich and Freiburg-im-Breisgau, finally becoming a distinguished professor at the University of Munich until his death. An unsigned
obituary appeared in Nature **44**, 580–583 (1891).

† See http://www.clendening.kumc.edu. A well-reviewed biography of Virchow
exists, but is not readily accessible: E.H. Ackerknecht, Rudolf Virchow: Doctor,
Statesman, Anthropologist, Madison NY 1953.

Figure 6.1 Rudolf Virchow.

wide-ranging interests and a rebellious streak (Figure 6.1). However, he was vehemently against theories of evolution. Perhaps prejudiced by these beliefs, he famously misidentified the remains of the Neanderthal man found near Dusseldorf in 1856 as those of a diseased Cossack soldier. Despite having traditionalist views on many things, Virchow was a political liberal who argued that improved social structures amongst the poor could help prevent disease. From time to time he fell foul of the authorities, but in due course he founded his own political party (the Peoples' Progressive Party) in opposition to Bismarck.

Virchow's contribution to the liquid crystal story, well before liquid crystals were actually identified, was his study in 1854 of human nerve fibres; these he labelled as *myelins*, from the Greek word for marrow. Nowadays the term is used to describe the insulating layer that forms around nerves and enables transmission of electrical impulses along them. For our story, the important point is that these myelin fibres were found to share some of the optical characteristics of Lehmann's liquid crystals.

A few years after Virchow's studies of nerve fibres, Christof Freiherr von Metterheimer (1824–1898) found these myelins to be visible in a polarizing microscope, and so they shared the optical characteristics (double refraction) of crystals. Why the apparently fluid-like fibres should behave optically like crystals was a mystery. George Quincke, Professor of Physics in Berlin, whom we have already met as one of Lehmann's adversaries, carried out extensive microscopic studies of myelins and related structures. He was able to produce artificial myelin-like fibres from mixtures of certain soaps with water, and concluded that the crystal-like optical properties of myelins were due to a solid crystal core.

Figure 6.2 Reproduction of line drawing of Virchow's nerve fibres.[12]

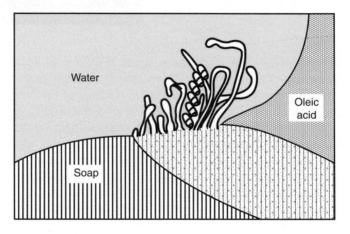

Figure 6.3 Redrawn from Lehmann's Woodcut[13] showing artificial fibres grown from soap solutions.

Lehmann, no doubt keen to undermine Quincke's work, repeated the latter's studies, but he came to a very different conclusion:[11]

> *I repeated Quincke's experiments, and it turned out that the myelin forms are nothing but flowing crystals [i.e. liquid crystals]. Because of differences in the surface tension at different areas, they had been distorted into strange forms similar in appearance to organic shapes.*

The similarities between Quincke's artificial fibres, as reproduced by Lehmann, and the nerve fibres originally observed by Virchow are remarkable (see Figures 6.2 and 6.3), and reinforced Lehmann's idea that perhaps liquid crystals had a life of their own.

Thus it had been demonstrated that under appropriate conditions it was possible to form molecular films, micelles, and molecular fibres, and the liquid crystallinity of the latter had been demonstrated, at least to Lehmann's satisfaction, by the fact that they exhibited birefringence in the polarizing microscope.

Liquid crystals and the theory of life

This was the title of Lehmann's lecture to the Joint Meeting of German Scientists and Physicians held in Stuttgart in September 1906, with which we opened this chapter. For Lehmann, unsurprisingly, the usual 1-hour lecture was insufficient. His exposition spread into two sessions, one delivered on 17 September and a second delivered in a plenary session on 21 September. He used the opportunity to give what he saw as a mature exposition of his current views on the liquid and flowing crystal problems. As he was talking to doctors, and as in any case his interests had been moving in a biological direction, his title was suitably ambitious. The opportunity was also ripe for him to discuss the resemblances between properties of liquid crystals and some aspects of the behaviour of lower organisms.[1]

Perhaps the most important intellectual influence for his interest in biological matters was the biologist, natural philosopher, and free-thinker Ernst Haeckel (1834–1919). Haeckel was a committed follower of Darwin who believed that dead matter (inorganic) and living matter (organic) were both consequences of the same set of physical laws; there was no need for a 'vital force' breathed by the Almighty or anyone else to bring things to life. On the other hand, he did believe that atoms in living systems were endowed with 'atomic souls', the interactions between which constituted 'life'. The 'death' of any living system would then be a consequence of the breakdown of the interaction between these atomic souls.

Lehmann, even as a boy, had been attracted to Haeckel's ideas. When Haeckel, as an old man of 70, heard about Lehmann's liquid crystals, he believed them to be the missing link between inorganic and living systems. There began an intensive correspondence between the two men, which continued until the end of Haeckel's life (although, surprisingly, there is no record of them ever meeting).

In his lecture to the Physicians of Stuttgart, Lehmann embellished his current views on liquid crystals with further philosophical speculations on life, the universe, and everything. He began, suitably for a talk with such an ambitious title, with some comments on what it means to be alive, and what are the rules which govern living processes.

> *Physics and chemistry are of interest to the medical man, because they offer him valuable aids, in the form of instruments and medicines. But there is a deeper reason, which underlies the fact that this is the 78th*

Annual Meeting of practitioners of the exact sciences with medical men and biologists. This is that the ideas, the materials and the forces which act in organic nature, are the very same as those which occupy physicists and chemists. In addition, their behaviour and properties are subject to the same rules as in the inorganic world.

This beginning already put him in the modern, materialist, reductionist scientific camp, in which the processes of life are in principle explicable by the laws of physics and chemistry. Lehmann had abandoned the medieval view of life as a vital force, so he was expressing above all a philosophical point of view, which he hoped later to link to his observations in physical chemistry. Even if he failed, his commitment to the philosophical doctrine was unshakeable.

But if Lehmann was sure of his philosophical ground, he was living in a world in which religious views about life were still extremely influential. He felt he had to address this viewpoint head on:

Despite much scientific study, life is admittedly just as much of a puzzle as it has always been. Phenomena of the mind cannot be derived from physical or chemical principles. According to the usual dualist point of view, a person consists of body and soul. Suppose, however, we want to endow each living being with such a soul. Religious dogma requires us to believe that this soul is immortal, can be separated from its body, is indivisible but at the same time distinct, and can exercise reason and free will. But now we encounter some unexpected difficulties.

We shovel up an earthworm in the garden. By accident, we cut it into two parts. Which half contains the soul? Or is a worm soul divisible? Both halves creep forward, and are cured, becoming again normal worms. Or we cut a twig from a willow tree and plant it in the earth. In time it too grows up into a tree. Has the act of cutting involved cutting a part of the tree's soul? Has the soul grown with the young tree, and from where?

Or, a nearly ripe apple falls from the tree. This is apparently dead matter, but in the cellar it ripens further. Must it necessarily contain life? Admittedly it is a very incomplete living being. Eventually it rots, and disintegrates into atoms. Are the atoms dead, or do they also in some way possess life, like the apple that fell from the tree?

Are there atomic souls? A necessary question! Who knows whether atoms actually exist, since no-one has ever seen one! That is right, but we believe them necessary. In order to make sense of natural phenomena we cannot do without them.

What is peculiar here is not that Lehmann is trying to do philosophy. All scientists will dare to express philosophical views in the privacy of their own homes. What is peculiar is that he is doing philosophy *in public* when he said he was going to do natural science. But in some sense it is progress that he is articulating his prejudices in public, *before* progressing to discuss scientific data.

Normally the people who address such issues are philosophers or theologians. But when the nature of the scientific discourse is changing, even scientists must address philosophical issues. It is noteworthy, of course, that Lehmann is making in 1906 comments about the presumed and necessary existence of atoms, which were still doubted in some philosophical circles. But within a decade these atoms would in some sense 'be seen' clearly in X-ray experiments.

The question which Lehmann is addressing is as old as human consciousness. How do we define what is alive? He is unenthusiastic, almost to the point of contempt, about the old-fashioned religious dualist point of view because in his view it leads to quite absurd conclusions. Indeed, given that the latest discoveries pointed to the existence of building blocks much smaller than the elemental atoms, he suggests that the dualist view would lead to the existence of living organisms small even on the scale of atoms. Lehmann's view was to reject the supernatural.

Later on he points to a grey area, between life and not-life, which suggests that the concept of life continuously shades into not-life, without there being quite a precise cut-off point. What exactly characterized life, he asked?

> ......*assimilation, and dissimilation and self-regulation of all functions. But let us think of a leaf, which has fallen from a tree, which lives on a little time and then withers. Or of Galvani's frog's leg: dead, but still somehow alive under the influence of an electric current. Or of a heart that has been cut out, but which beats strongly when salt water passes through it just at the point of rigor mortis. Here the self-regulation is absolutely imperfect. Then we might think of Koch's seed kernel, which were kept for a month in a highly evacuated glass tube, and which remained viable. Or of Leeuwenhoek's bear embryos, died out for a whole year, which seemed dead, but didn't actually die......*

His basic point is clear. It is extremely difficult to distinguish that which is alive from that which is dead. When it comes to humans, there are a whole range of properties that something alive possesses and something not-alive does not, but one can conceive of circumstances which include certain properties and exclude others. Although the degree of organization required for higher life is significant, for lower forms of life this is a different matter, and Lehmann agreed with Haeckel that there exist close analogies between crystals and lower life.

Lehmann noted that both crystals and living organisms grow, whereas amorphous bodies don't grow. Furthermore, there are crystals which remind us of life forms, such as those which under the microscope resemble pine cones. Crystal growth occurs from nuclei, and by the same token there are sometimes centres of a living organism from which the whole organism can regrow (and others from which it can't). Finally crystals, like living organisms, can absorb foreign material, but can be 'poisoned' if they attempt to absorb too much.

Notwithstanding these strong analogies, Lehmann pointed out that there were some significant differences. The most notable of these, he added, was that most organic material is 'soft and resembles egg-white', whereas, of course, crystals are hard—and (everyone knows!) that liquid crystals cannot exist, but (aha!) recently (would you believe it?) liquid crystals had actually been observed.

This gave Lehmann the excuse to launch into his by-now-familiar rhetorical defence of the liquid crystal concept. It might be that the existence of liquid crystals was contrary to current theory, but this was of little import, he averred, sarcastically quoting the philosopher Georg Wilhelm Hegel (1770–1831), who had said in a such a situation:

> ... *so much the worse for the facts* ......

It was only towards the end of his exposition that Lehmann again finally returned to the relationship between liquid crystals and life. We recall that he had made detailed microscopic observations of liquid crystals, which he recorded by pages and pages of painstakingly reproduced coloured figures. What he saw were mobile fluid droplets with complex and changing structures. Within these droplets were nuclei, which sometimes coalesced.

The coalescence phenomenon had been labelled by Ludwig Gattermann as copulation, a sexual analogy eagerly adopted by Lehmann. It was not clear how literally to take the analogy, and indeed it is not clear how seriously Gattermann himself took it. But certainly Lehmann took the opportunity of this lecture to reiterate the analogy between droplet coalescence and copulation in lower organisms.

Furthermore, he points out, if droplets of *different* materials coalesce, then instead of a new uniform droplet forming (as in usual copulation), one obtains 'bizarre structural disturbances', sometimes obtaining lamellae, or tori, or other peculiar shapes. This permits an analogy with cross-breeding. In the case of living organisms, such cross-breeding can lead to 'bastard formation'!

Now analogies between living organisms and liquid crystals flowed thick and fast from Lehmann's pen. Virchow's myelin forms were clearly liquid crystalline and they actually came from living organisms. Liquid crystal droplets sometimes formed buds that were analogous to the *pseudopods* formed when lower organisms reproduced, and liquid crystal drops could form long snake-like structures resembling bacteria. A clincher for Lehmann was that liquid crystals seemed to be able to move of their own accord, just like primitive animals. The *Gestaltungskraft*, the directional force later so contemptuously dismissed by Georges Friedel, found a second *raison d'etre* over and above changing the shape of a droplet. Now, apparently, it was the driving force for the transformation of chemical into mechanical energy, an absolute precondition for living beings.

Was there a real connection between his liquid crystals and real 'living', i.e. organic, systems? Lehmann was cautiously optimistic:

> *[This work provides] a superb proof of the correctness of our doctrine. [There are] some phenomena which would previously have been seen as actions of the soul because there appeared to be no physical analogy. These can now be grounded in physical and chemical effects. Thus it will be possible to remove the difficulties raised by the assumption of a soul in each bioblast or cell.*

Clearly the lectures had some impact. Not only was each lecture (separately) published in the *Physikalische Zeitschrift*,[14] but the whole thing was republished as a book under the full title of *Liquid Crystals and Theories of Life* in 1906, with a second edition in 1908. The republishing enabled Lehmann to expound somewhat on his central theme in the notes.

How seriously did Lehmann himself take these analogies between liquid crystals and life? Lehmann does seem to have been committed to the idea that life is just an elaborate combination of chemical and physical phenomena, but he was careful when pressed to avoid making the link between liquid crystals and life. Despite Lehmann's caution that his[15]

> *Apparently living crystals are obviously not to be understood as real living beings,*

he did succumb to the temptation of hype (sexing-up or spin as we would say these days) in promoting his work. Certainly he was not displeased when others, such as Haeckel, made the claim on his behalf that liquid crystals and life were one and the same.

Others clearly felt that Lehmann was verging on madness. As quoted at the beginning of this chapter, one member of the audience for his address to the Society of Natural Scientists and Physicians challenged him:

> *What's this about your liquid crystals. Are they even able to eat?*

Whereas Lehmann was cautious, the philosopher Haeckel was, by contrast, completely convinced that liquid crystals were the link between the living and the dead:[16]

> *In the year 1904 there appeared a number of highly significant works…The surprising discovery of 'liquid, apparently living, crystals' by Otto Lehmann (Karlsruhe), who had worked with dedicated concern for twenty years on this topic, came to fruition as a great work.*

So Haeckel could end his life happy in the knowledge that:[17]

> *What Goethe* suspected a hundred years ago, with his wonderfully deep understanding of nature, and what he prophetically expressed in incomparable words of poetry in Weimar and Jena, has today emerged to the radiant sunlight of scientific recognition.*

In fact Haeckel was wrong, although Lehmann's faith in the relevance of liquid crystals to life forms seems to have been justified. Nothing that we have learnt in the ensuing hundred years has suggested that any supernatural forces are required to explain biology. As we shall see, liquid crystalline phenomena do play a prominent role in the properties of the living cell.

It had indeed been tempting to think that the myelin fibres observed by Lehmann and Quincke were actually consequences of the life generation events that the former had seen in his microscope. These artificial myelin fibres formed in simple mixtures of soap and water, but what was actually special about these apparently living structures? To find out we have to delve into another area of science: that concerned with colloids, sols, and gels.

Lyotropic liquid crystals—the magic of water

The invention of the the ultra-microscope by Richard Zsigmondy has already been heralded in Chapter 5. This important advance in microscopy occurred around the beginning of the twentieth century as a result of a collaboration between Zsigmondy and an optician, H.F. Siedentopf, working for Zeiss-Jena, the famous optics company. Together they developed the ultra-microscope, which extended downwards the size limit of optical microscopes. The new principle of the ultra-microscope was to observe objects indirectly by way of the light scattered, rather than directly using refracted light, as in a conventional microscope.

The ultra-microscope enabled Zsigmondy, now Director of the Institute of Chemistry at the University of Göttingen, to carry out

---

* Goethe followed the philosophy of Spinoza. He was a determinist, and everything derived from Nature. Understanding would reveal itself through study of Nature, and so, at least according to Haeckel's interpretation, liquid crystals had showed themselves to be the origin of life. [*Reflections on the natural philosophy of Goethe*, Wolfgang Yourgrau, Philosophy **26**, 69–84 (1951)].

definitive studies of some selected gels (the technical term for jellies) and soaps. In 1912, Zsigmondy together with his student Bachmann published the results of their microscopic studies of mixtures of different concentrations of various soaps in water (e.g. potassium and sodium oleate, palmitate and stearate).[18] They observed distinct changes in properties at various temperatures, which they stated were consistent with the self-assembly of micelles into more complex structures. Zsigmondy did not know what these structures were, but he had provided some clues as to the nature of the myelin fibres observed some 15 years earlier by Lehmann and Quincke in mixtures of soap and water.

That soap and water mix in various proportions to give concoctions of different properties is obvious to anyone leaving a bar of soap on the side of the bath, where it becomes surrounded by a slimy goo. The history of soap-making goes back perhaps even as far as the early Babylonian period (2800 BC). The basic ingredients are fat or oil from animals or plants and an alkali, originally the ash from burnt trees or plants. Boiled together with water, 'neat soap' separates on top, leaving a residue of so-called 'middle soap' below. The products have very different viscosities and we now know they are just two of a number of liquid crystal phases that can be formed from mixtures of soap and water. The nature of the components of soap were unknown until chemistry came on the scene in the eighteenth century, when the fats were identified as a source of fatty acids, with the ash providing alkaline salts of sodium or potassium.

Much later, Quincke and Lehmann studied a soap-like system consisting of oleic acid* (a fatty acid) and an alkaline ammonium salt, which works just as well as salts of sodium or potassium. They were both surprised to find that it was possible to make films, membranes, and fibres from such a simple mixture that had structural integrity and strength, and which, at least according to Lehmann, had some movement and life of their own. As we have seen, Lehmann and Quincke differed vehemently in the interpretation of their observations.

As always, it was a new technological breakthrough that helped to unravel the mysteries of soap and water mixtures. The new technique of X-ray diffraction had at last revealed the regularity of solid crystalline structures, and it was not long before scientists were trying to use X-rays to probe more complex materials. The French physicist Jean-Baptiste Perrin in 1918,[†] obtained an X-ray image from a soap film formed from a mixture of potassium oleate, glycerine, and water, which showed that

---

* Oleic acid is a fatty acid found in animal and vegetable oils. It is a brownish-yellow liquid and almost completely insoluble in water. Readers will be familiar with it without realizing it, for it forms the largest component of olive oil (which is why olive oil is the colour it is). An *oleate* is a salt formed when a metal such as sodium or potassium or a group like ammonia combines with the oleic acid and water.

† Perrin (1918). Jean-Baptiste Perrin (1870–1942) was awarded the Nobel Prize for physics in 1926 in recognition of his work on discontinuity in the structure of matter. His investigations provided clear evidence for the atomic and molecular nature of matter.

it consisted of stacked layers each having a thickness of about 54 Å. This work was followed up by his American colleague Peter Wells in 1921,[19] who obtained the slightly lower figure of 42 Å. Georges Friedel without hesitation designated these soap films as examples of his (Friedel's) smectic liquid crystals. Confirmation that soap structures were indeed liquid crystals meant that if the artificial myelins of Lehmann and Quincke were definitely liquid crystals, then so also would be the living myelin nerve fibres of Virchow.

Liquid crystals formed from mixtures of soap-like compounds in water became known as lyotropic liquid crystals, while those formed by pure compounds on heating are now labelled as thermotropic liquid crystals. Despite Lehmann's identification of oleic acid, ammonium oleate, and water mixtures as liquid crystals, the background to the study of lyotropic liquid crystals lies in colloid and surface science. The origins of surface science lie in the attributes of soap. Traditional soaps clean without damage by wrapping dirt and soil in a membrane which itself is soluble with water. This process has no particular connection with liquid crystal activity, since soap for cleaning is used, or should be used, in very low concentrations. At low concentrations soap molecules can aggregate to form clusters in water solution known as micelles (see Figure 6.4). However, if the concentration of pure soap in water is increased, then indeed the molecules of soap organize themselves in a similar fashion to liquid crystals (see Figure 6.5).

The connection between the behaviour of soaps and the properties of liquid crystals has taken a lot of unravelling. The process was initiated by Stuart Lawrence, whose contribution to the 1933 meeting in London we briefly mentioned in Chapter 5. In 1927, while working at the Royal Institution as Sir William Bragg's research assistant, Lawrence

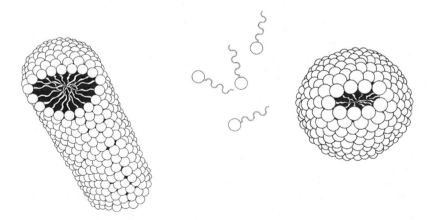

Figure 6.4 Spherical and cylindrical micelles formed by clustering of soap molecules in water solution.

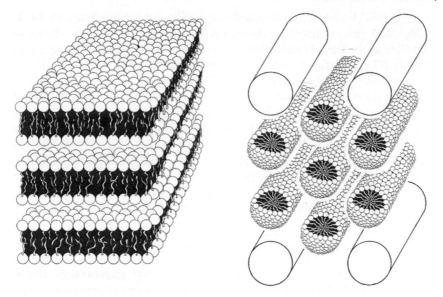

Figure 6.5 Liquid crystal phases of soap formed from micellar clusters as the water content decreases. On the left is the layered phase of neat soap, with water between the layers. On the right is the phase known as middle soap, in which water occupies the spaces between the cylindrical micelles. Both pictures show schematically the imagined internal structures of the layers and micelles.

suffered a long illness during which he wrote a book entitled *Soap films: a study of molecular individuality*;[20] thanks to him the observations of the ancient soap boilers found their place in scientific literature.

In his presentation to the Faraday Society at the meeting of 1933, Lawrence clearly identified neat soap as a smectic liquid crystal and middle soap, one of the products of soaps-making, as another liquid crystal phase of unknown structure. Although not usually recognized as a pioneer in liquid crystals, Lawrence provided the important connection with colloid science. He was to have a life-long career studying lyotropic liquid crystals of both soaps and biomolecules. His research, first at the Royal Institution, then at Cambridge and at Imperial College, was interrupted by World War II. Lawrence, however, continued to practice as a scientist in the Royal Naval Volunteer Reserve. His knowledge of soaps, oils, and thermodynamics was put to good use devising methods for de-icing warships. His ideas were put to the test during the sea trials of HMS Wrestler in a wintry North Atlantic. Presumably they were successful, since Lawrence was promoted to Commander and became a scientific advisor to the Admiralty.

In due course, Lawrence resumed his academic career. moving to the chemistry department of Sheffield University, where his success and fame grew. Always a bachelor, Lawrence managed to fill his laboratory with pretty girls and even extended his interests to liquid crystals in cosmetics. His addiction to smoking finally caused his death in 1971. Later that year, a meeting sponsored by the Faraday Society on liquid crystals at the Royal Institution, London, was dedicated to his memory, nearly 40 years after Lawrence's entry to the liquid crystal scene at the first Faraday Society meeting at the Royal Institution in 1933.

The connections between lyotropic and thermotropic liquid crystals had been established at the landmark meeting at the Royal Institution in 1933. However, papers on liquid crystals as a component of life were notably absent from that meeting, and the wilder claims of Lehmann about 'apparently living crystals' had been quietly forgotten. Despite this, there is a prophetic contribution from J. D. Bernal in the final pages of the general discussion.[21] In his discursive contribution, Bernal surveys the characteristic properties of liquid crystals and identifies them as

> ...an ideal medium for catalytic action, particularly of the complex type needed for growth and reproduction.

A few years later, Bernal would return to the question of liquid crystallinity and biochemical function in the context of proteins.

Bernal was a larger-than-life figure.[22] His dominating role in twentieth century crystallography (he supervised several Nobel-prize winners) was matched only by his wide scientific knowledge, his long-standing commitment to Socialism in general and Soviet-style Communism in particular, and his legendary prowess with women. In matters of science, Bernal tended to be cautious and he made no claims about liquid crystals being the origin of life, but in his personal life Bernal* was never far from controversy. Despite coming from a wealthy Irish land-owning family, Bernal was certainly no establishment figure. As a student at Cambridge he abandoned Catholicism and almost immediately adopted Communism in its stead. The Cambridge of the 1930s was an intellectual hot-house for Marxism, and it was through the Cambridge Communist Party, rather than as a result of a strictly academic collaboration, that Bernal met the biochemist Bill Pirie (1907–1997). The consequent academic collaboration resulted in the identification of liquid crystal behaviour in a protein solution derived from a virus.

The source of the material was sap from plants infected with tobacco mosaic virus (TMV), strains of which cause damage to species of tobacco

---

* Bernal's affair with the Nobel-Prize winner Dorothy Crowfoot (later Hodgkin) while she was still his research student was just one of his many extra-marital adventures. In 1937 Bernal became Professor of Physics at Birkbeck College, University of London. His secretary there was Anita Rimel, who was devoted to 'Sage', and well-acquainted with his wide circle of colleagues. She became firm friends with Dina Fankuchen, the wife of Isidore Fankuchen, one of Bernal's research students. Well aware of Bernal's philandering, Anita and Dina formed a society of women who had never been to bed with 'Sage', with Anita as treasurer and Dina as president. Despite his known communist sympathies, Bernal was allowed to travel to the USA in February 1949 for a World Peace Conference; he was accompanied by his secretary Anita. Not long after, Dina received a telegram from Anita saying 'You are now the president *and* treasurer of the society'.

and tomato. Extracts of infected plants, separated, concentrated, and purified, gave water solutions having the properties of liquid crystals. Perhaps for the first time, living creatures were used in the testing of a liquid crystal. Goldfish released to swim in a bowl of a dilute solution of TMV left a turbulent wake which could be observed between crossed polarizers. As before this was the optical signature of crystalline material, but it was seen in the relatively clear fluid of the goldfish bowl. These observations stimulated further experiments to determine the structure of the TMV solution.

There were four members of the scientific team responsible for the first X-ray measurements on a liquid crystal solution of the protein from TMV:[†] Bernal, Pirie, Bernal's American research assistant Isidore Fankuchen (1904–1964), and the young agricultural scientist Fred Bawden (1908–1972). Although the most junior in age, Bawden is the lead author on the letter to *Nature* that records the results.[23] As the only plant physiologist on the team, and thus the source of the precious viruses, Bawden was probably the guiding spirit for the work. It was, however, an unlikely collaboration: two communist intellectuals, a New York Jew, and the clean-cut young man Bawden, destined later for official recognition, a Government position, and a knighthood.

Still, the collaboration worked. The outcome of the research, however, was not a cure for a plant infection. Mother Nature did deliver a profound result, but not of the expected type. The conclusion of the experiment was that TMV molecules had rod-like shapes. These rods then organized themselves in a water solution to form a nematic liquid crystal. Bernal and coworkers did not themselves claim to have discovered a nematic lyotropic liquid crystal, but they did note the parallel organization of the virus rods and the absence of molecular layers (Figure 6.6).

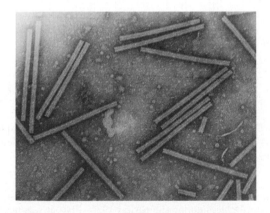

Figure 6.6 A modern electron microscope image of tobacco mosaic rods in water, showing their parallel organization. Courtesy of Seth Fraden.

[†] Tobacco mosaic virus is a complex between a protein and a nucleic acid (RNA).

Viruses, like the myelin fibres of Lehmann and Quincke, are not quite living organisms. If, however, a virus is released into a cell, it hijacks the cell's reproductive system, creating havoc for the living host. The work of Bawden and Bernal had suggested that viruses could self-organize, which would eventually be confirmed by electron microscopy, but the structure of myelin fibres still had to be explained. X-ray measurements once again were to provide the answer. The key to understanding myelin fibres was to determine the structure of soap-like liquid crystals.

Preliminary work by Edmond Friedel had confirmed the layered arrangement in smectic liquid crystals of soaps, but how did the molecules arrange themselves? It was a chemist, Joachim Stauff* (1911–), who provided the clue, at least for soaps and micelles. He took some X-ray pictures of soap solutions and found reflections characteristic of layered structures. Combining these measurements with a knowledge of the molecular dimensions of soap molecules, Stauff proposed a structure for soap films consisting of bilayers of fatty acid molecules separated by layers of water, which acts as a sort of chemical glue keeping the layers together, yet remains fluid (Figure 6.7).

This was the answer. Soap films, myelin fibres, and the walls of living cells all share the common feature of a bilayer structure. The inside of the layer is fluid and can accommodate a variety of other molecules

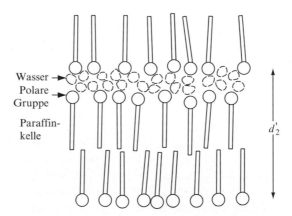

Figure 6.7 Stauff's picture of a smectic layer in a soap–water liquid crystal. The layer spacing is approximately twice the length of a soap molecule, hence the description 'bilayer'.[24]

---

* Joachim Stauff was a senior scientist at the Kaiser-Wilhelm-Institute of Physical Chemistry in Berlin-Dahlem during the period 1936–1940. It was here that he carried out the pioneering work to determine the structure of micelles and soap films. In 1976 he retired from the Chair of Colloid Chemistry and Directorship of the Institute of Physical Biochemistry at the University of Frankfurt.

such as proteins, cholesterol, and other biomolecules, but the structural integrity of the layer is preserved by water molecules bound to the surfaces of the bilayers. There is a flexible bag, or in the case of myelins a cylinder, inside of which life-supporting biochemistry can take place, protected from the sometimes hostile environment.

Cells and membranes

At its most basic level, we recognize life in the division and growth of cells which we can see under the microscope.

All organic beings originate from and consist of vesicles or cells

wrote the German natural philosopher Lorenz Oken (1779–1851) in 1805.[25] It is Oken's compatriot, the physiologist Theodor Schwann (1810–1882), who is usually credited with introducing the cell model for living organisms. He proposed that all cells, whether from plants or animals, shared the characteristic features of a nucleus surrounded by protoplasm, and the whole was contained within a cell membrane.[26]

> *The great barrier between the animal and vegetable kingdoms, viz. the diversity of ultimate structure, thus vanishes. Cells, cell membrane, cell contents, nuclei in the former are analogous to the parts with similar names in plants.*

The cell membrane maintains the integrity of the cell, yet is permeable to allow biomolecules in and out to perform their various functions, but what was its structure?

In 1925 a Dutch paediatrician turned researcher, Evert Gortner (1881–1954) and his research student F. Grendel published a paper entitled 'On bimolecular layers of lipoids on the chromocytes of the blood'.[27] In this work, they had separated lipids* from the blood of various animals and were able to show that there was exactly enough lipid extracted to coat each cell with a layer of lipid two molecules thick. This was exactly the same structure as that found for membranes of soap molecules by Stauff 14 years later using X-ray scattering. The paper by Gortner and Grendel remained undiscovered in the literature for nearly 50 years. Now the structure of cell walls consisting of lipid bilayers and containing various proteins and other biomolecules such as cholesterol is well established (Figure 6.8).

The importance of liquid crystals to living things is now well established. They contribute to the structural integrity of cell walls and

---

* Lipids are biological molecules similar to fatty acids that have partial solubility in water and oils.

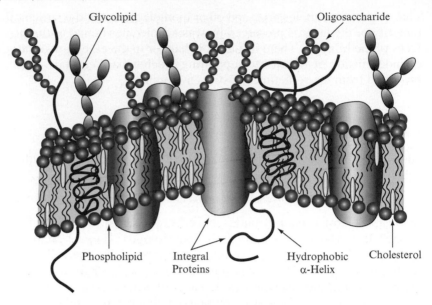

Figure 6.8 Schematic of a cell membrane with embedded proteins and other biomolecules. Courtesy of John Goodby.[28]

membranes. Lipid bilayers forming the cell membrane enclose some of the active biomolecules of the cell such as DNA and provide a support for other biomolecules within the cell walls. The most common lipids found in cell membranes are known as phospholipids. They have the basic structure of a fatty acid, but have phosphate groups attached, which form part of the water-soluble end of the lipid molecule. For cells to function there must be communication between the outside and inside, and this is provided by protein channels in the cell walls. These channels are carefully regulated, and though the cell walls are not living themselves, they are vital for all the chemical processes that contribute to life.

The special structural properties of cell walls and membranes are characteristic of the liquid crystal state. The liquid crystals provide fluidity to allow transport through the walls, but the membranes are still strong enough to contain the contents of the cells and prevent them being contaminated by their environment.

Modern molecular biology has almost exclusively focused on the molecular aspects of biological systems. Recently, however, the attention of some scientists has switched to the structural aspects of the cell,[29] specifically its mechanical or elastic properties and how these might influence or control the biochemistry of cells. It is, above all, the elastic properties of liquid crystals that impart their special properties.

Lehmann and Friedel knew something of these, but nothing of molecular biology. Their liquid crystals did in some circumstances appear to have lives of their own, and perhaps Lehmann can be forgiven for thinking, or at least hoping, that the secret of life was within their *Flüssige Kristalle*.

7
The winds of war

‹———————›

...if (our) act of terror against Stalin is successful, this will be a
signal for counter-revolutionary elements...

> Extract of record from NKVD interrogation
> of V.K. Frederiks, Leningrad, 1937

History, and the profound memories of a generation, record that the
1930s and 1940s were not normal times in Europe. The scientific actors
on the liquid crystal stage could not fail to be affected by the epoch-
making political and military events taking place around them. For
some, the preparation for war (and, of course, the war itself) opened up
new scientific opportunities, great excitement, but little personal inconvenience. For others, there followed disruption and instability which
devastated their personal and scientific lives, and in a number of cases
resulted in death. In this chapter we record the human stories of some of
the major liquid crystal scientists during this turbulent period. We do so
well aware that in many cases the human stories can be regarded as a
distraction from the course of scientific history.

For nations at war, local perspectives became constrained by national
imperatives. Scientists became the servants, or sometimes the victims,
of their political masters. The fate of our liquid crystal characters was
largely determined by their nationality, and science became a parochial
endeavour. Networks of scientific interaction ceased, and exchange of
ideas and information was regulated by the demands of 'national
intelligence'.

The last international meeting of the liquid crystal community before
the social and political maelstrom began in Europe had been in London
in 1933. The scientific importance of this meeting is discussed in
Chapter 5. Almost all the significant liquid crystal groups from the con-
tinent had contributed to the meeting, either in person or through submitted papers. After the war there were very few of the liquid crystal
pioneers still active, and several had perished. In what follows we record
the destinies of some of the key pre-war liquid crystal scientists through
the turbulent years before and during World War II.

Daniel Vorländer (Germany)

We start, as is perhaps fitting, in Germany, original heartland of liquid crystal activity. Daniel Vorländer was a survivor from the pioneering days of liquid crystal science. At the age of 66, he had been a major contributor to the 1933 liquid crystal meeting in London, having presented no fewer than four papers and mounted a lecture demonstration. Although he had retired in 1935, Vorländer had continued in the prewar years to be active scientifically. When, after only six years of retirement, he died in Halle on 8 June 1941, Germany's war prospects looked optimistic. The war in Europe was all but won, even if Britain was proving a longer-lasting irritation than expected off the far western European coast. The eventually disastrous German attack on Russia was just a fortnight in the future, while America would not enter the war until December of the same year. Vorländer's obituary,[1] which appeared after a delay of two years in Germany's premier chemistry journal—*Berichte der Deutschen Chemischen Gesellschaft*—not only provided a glowing report on his scientific career, but augmented it with some personal and philosophical reflections.

Vorländer's loyalties to family and to country were singled out by the obituarist for particular praise. He commented favourably on Vorländer's war service in the Great War, and reported that even in the present war, despite being already in retirement, Vorländer had taken over the running of the family business owned by his son, who was absent through war service. Only his eventually fatal illness, which came on in early 1941, had induced him to retire from his self-imposed national duty. At the end of the obituary, the writer quoted a poem by Vorländer on chaos and order. The great German writer Johann Wilhelm von Goethe (1749–1832) was also a scientist, and indeed had founded the science of *morphology*— the science of shape and form in plants.[2] The obituary links the work of Vorländer with that of Goethe[3] (and subtly with national socialism), finishing lyrically as follows:

> *Daniel Vorländer also belonged to Goethe's tradition, and he remained true to it right through his life.*[§]

Less controversially, Vorländer's scientific achievements are also noted:

> *...that he knew that his crystalline fluids would have a future, and his work had not been pursued in vain.*

The obituarist was Conrad Weygand, Extraordinary (i.e. Associate) Professor of Organic Chemistry at the University of Leipzig. Born in

---

[§] Roughly translated from: *Auch er, Daniel Vorländer, gehörte zu Goethes Verwandschaft und seinem Auftrag ist er zeitlebens treu geblieben.*

Leipzig in 1890, Weygand's career had been slow to take off, partly because of interruptions due to World War I.[4] His career as a textbook writer had, if anything, been more successful than his research career,[5] with books published in 1931, 1938, and 1941. His research career had blossomed after Vorländer's retirement in 1935. Vorländer needed a competent colleague to act as a vehicle for his further research, and Weygand, located in Leipzig, a mere 42 km to the south west of Halle, fitted the bill nicely.

It was no accident that Weygand had tried to link his mentor Daniel Vorländer with Goethe and morphology, for this was his favourite philosophical hobby-horse. Indeed, in 1942 he produced a whole book,[6] entitled *Deutsche Chemie als Lehre vom Stoff* (German chemistry as philosophy of matter[†]), linking chemical properties, shapes of molecules, and German-ness. Leipzig, like most other German cities, became a focus for anti-semitism,[‡] but this did not bother Conrad Weygand. His philosophy of German chemistry does not seem to have contained any explicitly anti-Jewish sentiments,[7] but he was a member of the Nazi party. According to legend in the liquid crystal community, this membership was rather enthusiastic.

In 1941, when Daniel Vorländer died, much activity in German universities was carrying on as normal. To Conrad Weygand in Leipzig, the war still seemed very far off, but after 1943, when German fortunes in the war had begun to wane, life was completely disrupted. Leipzig, in particular, was the victim of a number of extremely destructive British air raids.

As the German *Reich* shrank, young teenage boys and men of a mature age were called up in a desperate attempt to shore up the collapsing armies. By mid-April 1945, the allies were in the heart of Germany. For Conrad Weygand, a keen supporter of the regime, his duty was clear. Recruited as a storm-trooper, he was killed on 18 April 1945,[8] two days before the city was taken by American troops of the 69th Infantry Division.

[†] Weygand's *Deutsche Chemie* was a chemical counterpart to the *Deutsche Physik* of Philipp Lenard and Johannes Stark, a physical doctrine which dismissed both Einstein's relativity theory and the new quantum mechanics as 'Jewish physics'. See Gratzer (2000), p. 244ff.

[‡] In 1930 there had been more than 15,000 Jews in Leipzig, with a synagogue in the Moorish style that had been specially commissioned in 1855. The mayor between 1930 and 1937, Carl Friedrich Goerdeler (1894–1945, executed by the Nazis), was a noted conservative opponent of Hitler, but he was unable to prevent the destruction of the synagogue in officially sponsored anti-semitic riots over the night of 9–10 November 1938, the so-called *Kristallnacht*. Between 1927 and 1942, Werner Heisenberg (1901–1976), discoverer of quantum mechanics and Nobel Prize winner, was professor of physics in Leipzig, and the first German experimental nuclear reactor was built there by Heisenberg's group. By the war period these Jewish scientists, and others that could, had escaped from Germany and any Jews who remained in the university had been dismissed.

Figure 7.1 Hans Zocher (1893–1969). Image from citation in *Proceedings of the 3rd International Liquid Crystal Conference*, Berlin 1970.

Hans Zocher (Czechoslovakia)

Another prominent player in the 1933 London Faraday Society meeting was the German physical chemist Hans Ernst Werner Zocher (Figure 7.1). His major contribution to the meeting had been as a militant proponent of the distortion theory of liquid crystals. In that role he had been the main opponent of Ornstein's swarm theory. Politically, it would turn out that Zocher and Ornstein were closer together than they were in the realm of science.

Zocher was born in Bad Liebenstein, a spa town in the central German province of Thuringia in 1893. From 1912 to 1914 he was an under-graduate student in the universities of Leipzig and Jena.[9] This was fol-lowed by war service as an infantryman. He was injured in the cheek in 1916, an injury which impeded his speech and from which he never entirely recovered. After that he returned to Berlin for further university study, obtaining a Ph.D. in 1921. Between 1922 and 1931 he worked at a number of institutions in Berlin on colloid and liquid crystal problems, rising from the rank of research assistant to that of Extraordinary (Associate) Professor. His most distinguished collaborator in Berlin was the celebrated physical chemist Herbert Freundlich (1880–1941).[10]

In 1931 Zocher had been appointed as extraordinary professor at the German Technical High School in Prague. When the liquid crystal pio-neer Friedrich Reinitzer had worked in Prague in the 1880s, Prague had been a regional capital within the Austro–Hungarian Empire, with German the dominant language, but the end of the Great War had brought the dismemberment of Austria–Hungary and the formation of the new state of Czechoslovakia. The students of Zocher's Technical

High School were taught in German, but it was definitely a minority language within the new state. Czechoslovakia contained some three and a half million native German speakers (about a quarter of the whole population), mainly concentrated in the so-called Sudetenland, a semi-circular region ringing Prague, but close to the German and Austrian frontiers. In addition, there were maybe 100,000 native German speakers—half of whom were Jews—in the city of Prague itself, constituting maybe 15% of the population. The Technical High School drew its students mainly from German-speaking Czechs.

In January 1933 Adolf Hitler was appointed as Chancellor of Germany. The situation of its Jewish citizens, and even those with some Jewish descent,* immediately became more difficult. Distinguished scientists in this situation found jobs abroad. Those less distinguished soon found themselves unable to make a living.

Hitler's aggressive ultra-nationalism found an echo both in Austria (culminating in *Anschluss*—union with Germany—in 1938) and in the German-speaking regions of Czechoslovakia.† The inter-community atmosphere rapidly worsened, particularly because the mainly Czech-speaking bureaucracy in Sudetenland was seen as the agent of an occupying power. Although many German-speakers were vigorously pro-Nazi, many German-speaking academics (and, of course, the German-speaking Jews of Prague) felt rather differently.

Zocher was by origin a Protestant. In 1928 he had married Katarina Adler, who worked as a secretary in his Berlin laboratory. But Katarina was a Jewish girl, brought up in Breslau (now Wrocław in Poland). Zocher himself had long-standing left-wing sympathies, and he viewed the advent of Hitler in neighbouring Germany with alarm. He also had little sympathy with those in Czechoslovakia advocating separation for the Sudetenland. Around 1934, given the ongoing local community tensions, he took the precaution of obtaining Czech citizenship. Whether the *Technische Hochschule* management were aware of this we do not know, but in any case his growing international reputation led to his promotion to a full professorship in 1937.

But dark storms were beginning to gather on the political horizon. In September 1938, Hitler presented claims to the international community for the annexation of the Sudetenland by the German *Reich*. After much diplomacy involving Britain and France, who had treaty obligations to defend Czechoslovakia in just such a situation, the

---

* Freundlich, for example, was only one quarter 'racially' Jewish—and 50% Scottish!—but nevertheless felt forced into exile, leaving for London in 1933, and thence to Minneapolis in 1937.
† The *Sudetendeutsche Heimatsfront* (SdP), led by Konrad Henlein, agitated for a separation of the German-speaking regions from the rest of Czechoslovakia. In the 1935 Czech elections the SdP won 60% of the vote in the German-speaking regions.

Munich agreement essentially forced the Czech government to capitulate to the German demands. On 1 October, German troops occupied Sudetenland.

International academics were increasingly concerned for the future of their colleagues in Nazi Europe. The distinguished British physical chemist F.G. Donnan (1870–1956), retired professor of chemistry at University College in London, was a former colleague of Zocher's Berlin senior colleague Freundlich. Donnan had already written to Zocher, and on 19 October 1938 Zocher replied:[11]

> *Thank you very much for your kind letter of Sept 22. Since then the situation has significantly changed. Our High Schools have lost their students to Germany. For the local region there is hardly the possibility of even one of the three High Schools surviving. Most colleagues are likely to move to Germany. For me this is impossible... Finding employment in local industry is also unlikely. Firstly the economic situation means that there are not many posts, and then also it is becoming more and more prevalent, at least in large industrial concerns, to hire only Czechs, and to get rid of (*saübern*) Germans.*

> *So I expect that very soon I will no longer be able to make a living and be forced to seek an opening in another country... They say here that in some of the dominions, e.g. Canada, they need scientists and technicians. Could you perhaps advise me who I might best approach in this connection?*

> *I apologise for bothering you again with such a request, but my situation here is really very unpleasant.*

A similar letter, sent the same day, was addressed to his other British contact, Sir William Bragg,[12] reminding him of their meeting in 1933 and likewise seeking help in emigration.

In fact there was already an organization specifically set up to help refugees from Nazism. The Academic Assistance Council was set up in Cambridge in 1933 by a group of high-profile academics, including Lord Rutherford and John Maynard Keynes.[13] Behind the scenes, J.D. Bernal also played a key role. Over the period 1933–1939 this organization brought some 1500 scientists out of Nazi Europe, including a number of future Nobel Prize winners. Of the characters in our story, P.P. Ewald from Stuttgart[14] had emigrated to the UK in 1937 under the auspices of this society, by now renamed the the Society for the Protection of Science and Learning (SPSL). Both Donnan and Bragg immediately referred Zocher's letters to the SPSL, who rapidly dispatched to him their standard application form.

There followed a busy exchange of letters in the UK between Bragg and Donnan and the officers of SPSL. In January 1939 a letter of support was solicited, and subsequently received, from Freundlich, now in Minneapolis. The British academic community was not large, however, and liquid crystals was not the high-profile subject it is today. An academic job was difficult to find, particularly given the large number of scholars seeking help. In the end political events overtook them. On 15 March 1939 German forces invaded what remained of Czechoslovakia.

Over the war period itself, information on Zocher's fate is sparse. There is no doubt that it was a very difficult period. In his SPSL application form[15] he recorded that he had Czech nationality, had 'not yet' lost his post, but that he was in danger both because of his 'democratic attitude' and the 'mixed' origin of his wife. The Nazis put heavy pressure on 'Aryans' married to Jews to divorce their spouses. Refusal provoked retaliation (although it did preserve the life of the spouse). Indeed, the arrival of the German troops in March caused Zocher to lose his job.[16,17] Zocher was able to survive by doing industrial consultancy work, but his wife and children, as a result of their Jewish origin, were sent to a concentration camp.[18]

However, there was more distress to follow. The war in Europe ended on 8 May 1945, but Czechoslovakia was not on an important military route from anywhere to anywhere, and was largely forgotten in the struggle to reach Berlin. The American General Patton liberated most of Western Bohemia from Nazi Germany's control by 5 May. Soviet troops only arrived in Prague on 9 May.

The post-war situation in Prague was difficult. The economic shortages were compounded by a natural desire of the Czech population to get even with the Sudeten Germans, whom they saw as having acted as the agents of German power throughout the wartime period. The Potsdam Conference of major national leaders agreed in August that Sudeten Germans would be transferred to Germany and Austria. In practice, the expulsion of the Sudeten Germans was brief and brutal. Often people would have to leave with only a few hours' notice. In November, the German universities were summarily closed.

Thus the end of the German occupation, welcome as it was to Zocher, merely exchanged one set of problems for another. Having retrieved his family, he wrote (in English) on 7 August 1945 to SPSL in Cambridge, again seeking help.*[19]

---

* Zocher's family had survived their internment, but the experimental assistant of Zocher, Walter Eisenschimmel (b. 1902), did not survive. The work of Eisenschimmel had been acknowledged by Zocher in his paper presented at the Faraday Discussion held in London in 1933. Eisenschimmel was deported first to Theresienstadt, where he worked in scientific laboratories. In October 1944 he was transferred to Auschwitz, where he is presumed to have perished. We are grateful to Dr Michael Simunek of the Centre for the History of Sciences and Humanities in Prague for information on the fate of Eisenschimmel.

About seven years ago I wrote to you that I had no possibility to remain as a professor at the German technical high school [sic] here on account of the racial descendence [sic] of my wife according to the laws of Nürnberg. Indeed I was pensioned by the German authorities in 1939 and lived here since that time as a technical adviser of different firms.

After the liberation of the world from the oppression of national socialism I am banned of all work because I am a German. Therefore I address to you again the question if there is—according to your informations—a possibility for me to find a place of activity in England...

A hand-written letter of support from Donnan to SPSL on 24 October stresses that[20]

Zocher is in a <u>very bad</u> position, since the Czechs are closing the German university in Prague.. As a result, Zocher has lost his post, and I fear also his house and possibly his belongings. He had a bad time under the Germans... and now he is persecuted by the Czechs owing to his German nationality! Zocher is <u>certainly no Nazi</u>, and is a <u>first-rate man of science</u>....I do hope you can help!

In December a British visitor, Dr Nora Wooster, visited Prague and reported that:[*][21]

[Zocher] looks very much older than his age. His wife looks very ill and his children are undernourished. His pension has been stopped. He is now working with a Chemical Firm in its Coal Research Institute... While (I) was away in Brno after visiting the Zochers, the police called at their house, affixed a label stating that they were Germans, took an inventory of every detail of their property, which they forbad them from selling as it was now State Property. [I] saw the Minister of Information and gave him the details. He promised to intervene.

Dr Wooster thought Zocher an excellent candidate for a scientific trade union-sponsored 'respite holiday' in the UK.

However, despite a flurry of letters from Donnan and the SPSL, nothing could be arranged. The Home Office was very loath to offer

---

* Dr Nora Wooster was the wife of Dr W.A. Wooster, president of the (British) Association of Scientific Workers trade union. Both husband and wife were crystallographers and left-wing activists. Nora (née Martin) had been a research student of J.D. Bernal. She had been on a British-Council-sponsored tour of Czechoslovakia lecturing on the use of X-rays in metal industries. She also probably used the opportunity to do some union-related work.

immigration visas. The UK was just absorbing a very large number of European nationals who had been in the UK on temporary visas before the war, and was mindful of possible xenophobia.

The SPSL also tried hard to arrange a post for Zocher in Germany, where the de-Nazification process was in full swing.[22] There were explicit British army processes for finding posts for displaced persons. Zocher declared himself also willing to accept an industrial post in Germany or Austria. In fact as a former anti-Nazi with a Jewish wife, it seems unlikely that Zocher would have wanted such a post, and of course he would not have been terribly welcome in Germany either.

The secretary of SPSL, Mr J.B. Skemp, was very sympathetic about the Zocher family dilemma, referring to 'these unfortunate people', but in the end there was little he could do to bring Zocher to the UK. Luckily Zocher himself had other irons in the fire.

In 1940 his Austrian colleague Fritz Feigl (1891–1971) had managed, after a long and tortuous odyssey, to reach Brazil and hence restart his scientific life.[23] Feigl's work on coffee production brought him into contact with Dr Mario da Silva Pinto, Director of the National Laboratory for Mineral Production in the Ministry of Agriculture. It was Dr Pinto who in March 1946 sent Zocher a telegram offering him a post in his laboratory, starting on 1 August 1946. The Zocher family, with their daughter Dorothea (aged 14) and son Michael (aged 12), set sail from Gothenburg on 3 July, arriving in Rio de Janeiro on 28 July.

Zocher's career flourished in Brazil, where he was much honoured and where he died in 1969. He did not, however, resume his work on liquid crystals. When liquid crystal research took off in the late 1960s, he was also honoured by the liquid crystal community as a whole for the pioneering role he took in its early period.[24]

Leonard Ornstein (The Netherlands)

Leonard Salomon Ornstein* (1880–1941), Director of the Physical Institute at the University of Utrecht in The Netherlands, had been an important contributor to both the Ewald and the Faraday symposia. Ornstein's major contribution to these meetings, we may remember, was his advocacy of the swarm theory of liquid crystal behaviour, which was later superseded by the Oseen–Zocher distortion theory.

---

* Ornstein was born in Nijmegen in the Eastern part of The Netherlands in 1880. He studied mathematics and physics at the University of Leiden between 1898 and 1908. In 1908 he graduated with a doctoral degree, having written a thesis on statistical mechanics under the supervision of the Nobel Prize winner Hendrick Antoon Lorentz. During the period 1909–1915 he worked as a lecturer in theoretical physics at the University of Gronigen, before being appointed as professor of theoretical physics in the University of Utrecht in 1915. He transferred to the Physical Institute in 1925.

In the 1920s and 1930s, Ornstein had a major effort in liquid crystals, working with a whole set of junior collaborators. One of them, Wilhelm Kast, who would later end up in Halle, spent the year 1932–1933 working with Ornstein in Utrecht. The outcome of this collaboration was the joint paper presented to the London Faraday discussion in 1933, which both attended. Later in 1933, Kast returned to Germany and, for reasons which will soon become apparent, continuing collaboration between Kast and Ornstein on liquid crystals became difficult. Later in the 1930s Ornstein turned to other work, including nuclear physics and biophysics, the latter work being funded by the Rockefeller Foundation. He had a very good reputation as a laboratory boss, being known as 'the Professor', and taking a strong personal interest in the work of junior colleagues.

The fame of his laboratory attracted numerous visitors. One such visitor, over the period 1936–1937, was the Polish physicist Marian Mięsowicz (1907–1992). In 1935 in Cracow, Mięsowicz had made measurements on the viscosity of liquid crystals. He had found, to his surprise, that the fluid property of viscosity was not well defined in a liquid crystal. Rather it depended on the geometry of the cell and the direction in which the liquid crystals were pointing with respect to the cell walls. This first measurement of an 'anisotropic' viscosity later turned out to have profound implications on our understanding of liquid crystals. At the time, although it was possible to submit a first publication in 1935, a fuller exposition had to await the end of the war. Mięsowicz's submission to the London scientific journal *Nature* was lost in the confusion at the beginning of World War II. Interestingly, although it was Ornstein's fame in the liquid crystal field that had attracted Mięsowicz to Utrecht in the first place, Ornstein was also instrumental in changing Mięsowicz's research focus, for in the late 1930s the new field of nuclear physics was taking off. We know now that Ornstein's perspective on liquid crystal theory was mistaken and unfruitful, but this was surely not the only reason that a dispassionate and conscientious Ornstein redirected Mięsowicz's youthful passion toward the coming field of nuclear physics. It seems likely that Ornstein would have been exceedingly surprised that 70 years later the field of liquid crystals remains as alive as that of nuclear physics, and attracts less opprobrium from outside the scientific world!

Ornstein's Jewish family possessed a long Dutch heritage stretching back several centuries. Despite this, his family resisted complete assimilation. From 1918 until 1922 Ornstein was Chairman of the Dutch Zionist Society. He played a prominent role in the organization of a physics department at the new Hebrew University in Jerusalem, which opened in 1925,[25] serving on its council.[26] Despite the onset of war in 1939, he made no attempt to leave The Netherlands, thinking perhaps, that, as in the Great War, The Netherlands would remain neutral and that he would not be affected by the Nazi anti-Jewish Nuremberg Laws. On 10 May 1940 The Netherlands were struck by a surprise German attack without any declaration of war. It took only a few days for a poorly armed Netherlands to fall to the Germans.

Ornstein had an offer to go the USA at this time, but declined the opportunity, feeling that he would be letting down his colleagues in Utrecht. It took little time before The Netherlands felt the full force of Nazi anti-Semitism. The Utrecht laboratory had been funded by grants from the Rockefeller Foundation in the USA, but after the German invasion this funding was cut off. More seriously, in September, Ornstein was excluded from his own laboratory. In November he was dismissed from the university.

Following this, Ornstein unsurprisingly became depressed. On 29 November 1940, Ornstein resigned his membership of the Dutch Physical Society. He withdrew from scientific life, became ill, and by the end of May 1941 he was dead. He was still six months short of his 61st birthday.

Ornstein was a distinguished Dutchman. Important as his Jewish activity was at a particular stage in his life, it seems certain that his scientific life was much more important to him. The scientific heritage that he left is profound* and identified with The Netherlands. He had the opportunity to relocate to Jerusalem—to put the Zionist theory into practice, as it were—but chose not to do so and to remain in his native land. It is almost sure that had the Germans not invaded, and had there been no anti-Semitic laws, he would have continued his scientific life for many more fruitful years.

Ornstein's death was ostensibly from physical disease, but almost certainly, on a deeper level, it followed also from a broken heart, from the impossibility of continuing to follow the scientific life that he loved. Had he not died of natural causes in 1941, however, the prognosis would have been poor. Those Dutch Jews who did not go into hiding were transported to Nazi concentration camps and almost all died there. Even Ornstein's great scientific eminence would not have saved him.

John Desmond Bernal (UK)

All the scientists appearing in this chapter, apart from Mięsowicz, had been present in body or spirit at the London Faraday Discussion of

---

* Ornstein wrote over 200 papers in physics. One of his best-known works was the so-called *Ornstein-Zernike theory*, produced in Groningen in collaboration with the future Nobel Prize winner Frits Zernike (1888–1966) in 1914. This theory explains why near the so-called *critical point* between a liquid and a gas (at the temperature and pressure where the gas and the liquid become indistinguishable), the fluid becomes very cloudy and scatters light strongly, exhibiting so-called *critical opalescence*. This work left a permanent imprint on the theory of fluids and is known by all students who study the subject.

Figure 7.2 J.D. Bernal (1907–1971). Image from citation in *Proceedings of the 3rd International Liquid Crystal Conference*, Berlin 1970.

1933. This had been the brain-child of Sir William Bragg, but had been organized by his former research student, J.D. Bernal (1901–1971) (Figure 7.2).[27] The proceedings of this meeting, discussed in Chapter 5, have etched their way into the historical scientific record. In so doing, this publication has unjustifiably veiled previous work from the gaze of all but the most historically and linguistically committed. Bernal was also a key figure in a later Faraday meeting, held in Leeds in 1958, which marked the renewal of interest in liquid crystal work in the post-war period. We shall hear more about the Leeds meeting in the next chapter.

Over and above his organizational role, Bernal's experimental work drew attention to the relevance of liquid crystalline properties to living systems. With Dorothy Crowfoot he had contributed a short paper to the 1933 meeting.[28] His most cited work in liquid crystals was discussed in more detail in Chapter 6.[29] This concerned the liquid crystalline properties of crystalline tobacco mosaic virus (TMV), and was published in *Nature* in 1936. The interesting, and still unresolved, questions about the relationship between physical studies of liquid crystals and molecular biology in general also owe much to Bernal.

In Britain, Bernal did not fit the mould of the ruling elite,* what is colloquially known as 'the Establishment': his profound commitment to Marxism and the Communist party has already been mentioned in Chapter 6. This commitment, as well as his indefatigable energy and the

---

* This discussion of Bernal is largely distilled from Andrew Brown's biography (Brown, 2005). His immediate roots were Irish Catholic. Further back his roots seem to have been more cosmopolitan, for his mother was born American and his father had Sephardic Jewish antecedents.

almost unbounded breadth of his intellectual interests, served as threads that linked together the different facets of his career.

In the summer of 1931, Bernal met Russian delegates to an international conference on the history of science,[30] led by Nikolai Bukharin, who in the 1920s had been Lenin's major economic commissar.[†] Later the same summer, Bernal was part of a delegation of left-wing scientists and intellectuals who visited the Soviet Union, passing through Leningrad, Moscow, and Kiev.[31] Among the Soviet scientists that he met was the distinguished plant geneticist N.I. Vavilov. Vavilov was later to be a victim of Stalin's purges, following a dispute about the inheritance of acquired characteristics with the notorious scientific impostor Trofim Lysenko, who, supported by Stalin, had challenged the science of genetics.

Bernal was back in Moscow again in August 1932 as part of a delegation led by Lawrence Bragg, visiting the School of Physical Chemistry at the university.[32] It was probably during this visit that Bragg and Bernal were able to forward an invitation to the Leningrad-based V.K. Frederiks to present a paper on his liquid crystal work at the London Faraday meeting planned for April 1933.[33] A further visit of Bernal to Leningrad in December 1934 coincided with the assassination of the local party secretary, S.M. Kirov.[34] Kirov was shot and killed in what appeared at the time to be an anti-Soviet act, but much later turned out to be a pretext for further repression by Stalin's leadership. Many observers, including some of Bernal's own travelling companions, noticed the rather heavy political atmosphere, but Bernal was oblivious or unwilling to take notice.

The increasingly difficult political situation in Europe in the 1930s worried not only left-wing groups. It is probably the case that the firmly anti-Nazi position taken by the Communist Party in Britain and the Soviet Union may well have attracted some towards communism who might otherwise not have been attracted. Winston Churchill, for example, was just one rather well-known right-wing figure who advocated British rearmament in the face of a newly powerful Germany. The Munich agreement of September 1938 promised 'peace in our time' at the expense of an independent Czechoslovakia, creating, as we have seen, deep concern to Hans Zocher in Prague.

By 1938, Bernal, whose previous commitment to peace had been unstinting, was involved in a campaign to improve air raid shelters for

---

[†] A particular feature of this conference was the Russian Marxists' insistence on the relationship between the history of science and other more general historical issues, much to the discomfiture of the British delegates (see Sheehan [1993]). This strongly influenced Bernal, who later published his own encyclopaedic work *Science in History* (Bernal, 1954), and hence indirectly influenced the present authors as well.

civilian populations. Experiences in the Great War, and more recently in the Spanish Civil War, had led to fears that civilian casualties in air raids could be devastating. In January 1939, Sir Arthur Salter, Member of Parliament (MP) for Oxford University,[‡] aware of his involvement in this campaign, invited Bernal to a lunch party in Oxford at All Souls College at which these matters were being discussed with Sir John Anderson, the government minister with responsibility for civil defence.[35]

Bernal's trenchant private criticism of the current state of war preparedness at this lunch party had some immediate effects. Anderson contacted the Government's Chief Scientific Officer with a suggestion that Bernal's talents be employed forthwith. From then until 1945, Bernal's main activity shifted from crystallography to what might be termed 'war physics'. Such basic science that did continue was either exported to the USA (where the war did not begin until late 1941) or carried on by women.[36]

Bernal's initial war physics work was concerned with the civil defence problems about which he had crossed swords with Anderson. His first papers concerned the physics of explosions; not only theory, but the very real problem of how it might be possible to make measurements of key properties in the vicinity of an explosion. Later, after the war itself started in September 1939, he was involved in the clearance of unexploded bombs, dealing with delayed-action bombs, touring bomb sites during the blitz, and the calculation of the progress of bomb-induced shock waves in narrow streets.*

After attachments to Bomber Command and later to Combined Operations, by the summer of 1943 Bernal was in Canada, testing equipment to be used in the floating harbours which played such an important role in the allied invasion of Europe in June 1944. A key element in this campaign required deep knowledge of the physical geography of the French coastline. Bernal, the intellectual polymath, was in his element. For months he pored over new photographs taken from the

---

‡ Until 1948, the universities of Oxford and Cambridge returned separate MPs to the British parliament, allowing some people two votes, one as an individual living in the locality and another as member of the university.

* An important calculation of Bernal in spring 1940, which turned out unfortunately to be of prophetic utility, concerned the likely casualties from an air raid on a city. The city chosen as a model was Coventry. The calculation required sensible input parameters, such as the landmark chosen as a target (he supposed the cathedral), the statistical accuracy of the bomb-droppers, the likely whereabouts of the civilians (inside shelters?), the likelihood of a direct hit on a shelter, and so on. The most dramatic conclusion concerned the shelters. If it was desired to reduce casualties, it was most important to provide well-protected shelters. When in fact the German bombers did target Coventry, on 14 November 1940, the calculation turned out to be all too accurate, and the cathedral, as he had inferred it would, ceased to be.

air at all sorts of angles and phases of the tide, old charts (sometimes dating back to Roman times), obscure history books, local scientific journals, and even the *Michelin Guide Bleu*, in a quest to obtain the most precise details of the coastline, the slope of the beaches, and their defences. These details were crucial to the invading armies on D-Day and the days following, when the beaches had not yet been secured. And indeed, in order to check the calculations, Bernal was himself on the Normandy beaches the day after D-Day.

The general skills used by Bernal in all his war work were those he learned as a practising physicist, even though the problems themselves differed greatly from those on which he had learned his trade. There is a contrast to be drawn between the kind of work Bernal was involved with and that carried out by the contemporary cohort of nuclear physicists, all of whom had mysteriously disappeared from circulation. These physicists, by now relocated to Los Alamos in New Mexico, USA, were using precisely targeted special skills to crack the atom bomb problem.

By the end of the war in Europe on 8 May 1945, the atom bomb was near to completion. It was, of course, a closely guarded official secret. Bernal was not included on the circulation list, but any senior person who had been active in physics before the war would be able to make an intelligent guess that something was in the air. With the end of the war and the vanquishing of the German enemy, old rivalries were re-establishing themselves. Russian spies had alerted Moscow to the development of the atom bomb. On 14 June 1945 Bernal was due to fly to Moscow to celebrate the 220th anniversary of the Russian Academy of Science.[37] Late on 13 June news came through that Bernal (and some other well-known left-wing scientists) had been forbidden to travel.

The war was for Bernal a tremendously exciting and stimulating period, during which he was able to turn his peculiar skills to the service of the nation. He had been able to spend the war years in rewarding activity close to the seat of power. As we have seen, others in similar positions across Europe were not so lucky. We now turn to perhaps the most tragic victim of the war years, Vsevolod Konstantinovich Frederiks. He was not actually a casualty of World War II, but perversely a victim of the political regime that Bernal so enthusiastically and mindlessly supported.

Vsevolod Konstantinovich Frederiks (Soviet Union)

The liquid crystal researches of Vsevolod Konstantinovich Frederiks (see Figure 7.3) and his group at the Physico-Technical Institute (PTI) in Leningrad had gained the attention of the international scientific community as early as 1927. Frederiks was publishing not only in Russian language journals, but also in German (where he would be sure

Figure 7.3 Vsevolod Konstantinovich Frederiks in the 1930s at the height of his academic powers.

to be read by the major workers in the field), and even in what was at the time the less fashionable medium of English. His reputation was such that he was attracting international invitations. As we have seen, the American Electrochemical Society invited him to attend their Toronto meeting in 1929 and in 1932 Bragg and Bernal had also invited him to take part in the Faraday Society symposium in London in April 1933, but he had not been allowed to attend either meeting by the Soviet authorities.

Concerning his invitations abroad, Frederiks's biographers record laconically that 'Vsevolod Konstantinovich could not go'.[38] Indeed one abiding feature of the Soviet regime always was the harsh response to attempts to travel abroad. The feeling was that those who wished to travel abroad must always have a hidden agenda; the journey must be a cover for meetings with foreign agents whose only wish was to overthrow Soviet power. Furthermore, Frederiks was labelled as an aristocrat and was already under scrutiny by the secret police.*

In 1927 Frederiks married Maria Dmitrievna Shostakovich (1903–1976). Her father, Dmitri Boleslavovich Shostakovich, had been a chemist. More than that, he had been a co-worker of the great D.I. Mendeleev, creator of the periodic table of the elements and

---

* Vsevolod Konstantinovich's grandfather, Platon Alexandrovich Frederiks (1828–1888), had been a military man and in due course had become Governor of Eastern Siberia. V.K. Frederiks' father, Konstantin Platonovich Frederiks (1857–1918), continued the distinguished family tradition and was Governor of Nizhny Novgorod (600 km east of Moscow) under the Tsar, but died in mysterious circumstances in 1918 during the period following the Soviet revolution. Most of the discussion of Frederiks in this chapter is drawn from Sonin and Frenkel's biography (Sonin and Frenkel, 1995).

pioneering Russian winner of the Nobel Prize in chemistry. The atmosphere in the Shostakovich household was one of high culture and liberal politics. It was Maria Dmitrievna's mother, Sophia Vasilevna Shostakovich, herself trained musically at the St Petersburg conservatoire, who set the cultural agenda of the household. Of her three children, the best known was to be her son, Dmitri Dmitrievich (1907–1975), known as Mitya to the family. Dmitri's musical fame, as a composer and performer *extraordinaire*, was to stretch the world over.

By 1934, Frederiks's scientific reputation was growing. His eponymous effect had, with the quiet help of the Swede Oseen and the rather more vociferous help of the German Zocher, seen off the swarm theory. Indeed, the period between 1934 and 1936 was extremely fruitful for the Frederiks group, leading to 11 published papers in all. A new and extremely able student, Victor Nikolaevich Tsvetkov (1910–1999), had joined the group. He and Frederiks published a number of papers on the experimental determination of the liquid crystalline elastic constants. These constants were a central feature of Oseen and Zocher's distortion theory. It was Frederiks's adoption of the distortion theory in order to explain his experiments which led to the demise of the swarm theory. With Repiova, he was even beginning studies of the smectic phase.

Frederiks' star was rising. His increasing reputation did not lead to his election to the Academy of Sciences in 1934, but it was rumoured that his election in 1936 would be little more than a formality. However, this period coincided with a time of intense political ferment. On 1 December 1934, as we have seen above, the Leningrad party secretary Kirov was assassinated. On 16 December, the disgraced pro-Trotsky former politburo members L.B. Kamenev and G.Y. Zinoviev were arrested. There followed a repression of quite massive proportions. Some estimates suggest that in the year following Kirov's assassination in the order of 100,000 people were deported from Leningrad.

One instinct of the scientist in difficult political times is to keep a low profile. Much intellectual work necessarily involves expressing opinions on important issues of the age. By contrast, scientists have often been able to claim that their work is devoid of political implication. Perhaps the intense period of activity in the Frederiks laboratory was not despite, but correlated with, the external political drama. In fact neither the university nor the Frederik's Institute in Leningrad were able to escape the purges. Indeed a number of Frederiks's friends and colleagues were arrested, culminating in the arrest, on 15 October 1936, of his geologist colleague S.V. Burslan. According to Valentina Zolina,[39] the atmosphere in the laboratory in autumn 1936 seemed fraught with danger. Frederiks' group was evicted from its laboratory to a small room in which it was impossible to work. Finally, on 20 October 1936, Frederiks himself was arrested.

The spirit of the times was such that guilt was presumed and there was little that one could say in one's own defence. Interrogations would last for days at a time and suspects would be allowed little sleep, still less access to a lawyer. Some transcripts of Frederiks' interrogation were recovered from the NKVD headquarters.* Frederiks is said to have been interrogated over the period 21–28 October 1936. He was accused of counter-revolutionary and anti-state activities, and given the circumstances of his questioning, it is not surprising that he eventually admitted guilt.

Amongst the transcripts were the following surreal exchanges:

> *Q: Against whom personally was it decided to commit acts of terror?*
> *A: It was decided first to commit an act of terror against Stalin. Our opinion is that if this act of terror against Stalin is successful, this will be a signal for counter-revolutionary elements to start a whole set of similar acts. This will make it easier for us to reach our planned goal of overthrowing Soviet power.*
> *Q: What specific actions did your organization actually carry out in order to prepare acts of terror against Comrade Stalin?*
> *A: I personally don't know any plan in which acts of terror are being prepared against Stalin by participants in our counter-revolutionary terror organization.*

Frederiks was accused improbably of having a connection with a German spy ring. The closest that he could have been to connection with the alleged German spy ring was that he did attend scientific meetings at which some Germans might have been present. The accusation was absurd. More than that, it contains deep irony and pathos. Western left-wing scientists sympathetic to the Soviet Union, of whom John Desmond Bernal was a particularly well-known (but by no means unique) example, went to great lengths—often to the detriment of their professional careers—to arrange visits to the Soviet Union. At such meetings they could meet with, issue invitations to conferences to, and even try to arrange collaborations with their Soviet counterparts. For their interlocutors it was a poisoned chalice. They might be rewarded with a free trip to Siberia or worse.

Frederiks also allegedly made counter-revolutionary statements in a colleague's flat in 1935.[40] In this case, there might have been some substance to the accusations. It is possible that he made such remarks. Frederiks participated in a number of geological expeditions to remote parts of the Soviet Union. Indeed he was an expert on electric devices

---

* NKVD stands for *Naradnyy Komissariat Vnutrennikh Del*, which under Stalin was the organization of public and secret police. Its agents, as well as directing traffic, conducted extrajudicial executions and ran the Gulag system of forced labour.

used to find minerals and as such was a valued member of the team. One of the accusations was that he deliberately gave bad advice. What is true is that members of these expeditions would have travelled widely through the Soviet Union. They would therefore have known that the official propaganda about the state of life in the Soviet Union was at best exaggerated and at worst completely false.

The precise remarks attributed to Frederiks were that he compared Stalin with the character Ugrium-Burcheev. This is a stupid government official from the book *The history of a town* by the Russian author Mikhail Saltykov-Shchedrin (1826–1889).[41] There is, of course, a long tradition of Russian satirical literature. Given Frederiks' cultural background, such remarks are quite conceivable, but there is a big gap between mildly satirical remarks and an interpretation in terms of a physical plot against Stalin.

Finally, on 23 May 1937, the case came to court. There were no lawyers, no witnesses, and no due process. The judges were military judges. Frederiks was sentenced to 10 years hard labour from the date of his arrest. In addition, he was to lose five extra years of civic rights, and all his personal possessions were to be confiscated. The verdict was not subject to appeal. The court in fact established conclusively that Frederiks had been involved in counter-revolutionary activities as long ago as 1932.

Frederiks was first sent to a labour camp in Siberia, where he was made to do hard physical work. By this time he was already over 50 and his health suffered. After some time he was moved to another prison camp near to Moscow. Meanwhile, his wife, Maria Dmitrievna, had herself been exiled to Frunze (now Bishkek), in Kirghistan, in Central Asia. Their son Dmitri (Mitya) was in the care of his grandmother Sophia Vasilevna.

In these troubled times, the best that could be hoped for was that friends in high places would have enough influence to arrange one's release. When the famous physicist L.D. Landau (a much less tactful and much less likeable man than Frederiks, but one of quite inestimable scientific eminence) was arrested in 1938, another distinguished physicist and academician Peter Kapitsa was eventually able to arrange his release* by writing personally to Stalin.[42]

A similar appeal in February 1939 on behalf of Frederiks, signed not only by Kapitsa, but also by the applied mathematicians A.N. Krylov and N.I. Mukheshvili, and the physicists S.I. Vavilov, brother of the plant geneticist N.I. Vavilov, and a future President of the Soviet Academy of Sciences, A.I. Yoffe and V.A. Fock, academicians all, failed to have any effect. Similar letters from I.V. Kurchatov (a

---

* Note that Landau's close colleagues Rumer and Koretz, who were arrested at the same time, were not so lucky and served long periods in jail.

famous nuclear physicist and later director of the Soviet atomic bomb programme), Ya.I. Frenkel (the founder of the physical theory of liquids), and Frederiks's famous composer brother-in-law Dmitri Shostakovich likewise fell on deaf ears. It was risky to support victims of purges, and only those sufficiently eminent were in a position to help. Shostakovich was helping a relative, and thus doubly putting himself at risk, but by now he was so famous as almost to be immune to Stalin's purges.

In the autumn of 1940, Frederiks was moved again, to a camp at Ukht-Izhem in the Komi ASSR. This was west of the Ural mountains, closer to Moscow, but in an extremely isolated area. The camp was attached to an oil well and with its related industrial complex; later Ukht became a town in its own right. In Ukht Frederiks worked in the laboratory of an oil refinery. Among his co-workers was the physicist L.S. Polak, who survived the war and was later released. Polak and Frederiks worked in their spare time elaborating theories of elasticity in nematic liquid crystals,* using the backs of packing paper to write on.

Some of Frederiks's letters from this period have survived. It was, in principle, possible for prisoners in the camps to receive family visits. A letter dated 22 April 1941 to his wife records:[43]

> *I would like very much to see you this summer and hope that it will be possible. As soon as it is clear, I will write to you with all the necessary details. Thanks to your letters, I can now see our son better. Last night I received his photo and yours, together with your parcels from April 8. The photos of you and him together are very good. I also received his small photo in a letter which he sent himself. Thank you very much for all this. Generally I want to thank you for everything, for the warm clothes, for the underwear, for the vitamins, the coffee and the tea, really for everything. Thanks to the parcels from you and Sophia Vasilevna, all of which I have now received, I eat very well. With the help of the medicines, I will soon be healthy. I now have enough parcels. Recently I also received a letter from Natasha. Thank her for that. She writes that she has now completely recovered from her illness. Is that true?*

> *As regards me, I continue to work in my speciality, and am very satisfied with the work. My health is OK, but so that you don't*

---

* This piece of research was an almost encyclopaedic discussion of different cases of elasticity in nematic liquid crystals, anticipating work by Nehring and Saupe in the early 1970s (J. Nehring and A. Saupe, 1971, 1972). It was conserved by Polak, and finally appeared in print in Sonin and Frenkel (pp 142–170).

worry about all of this unnecessarily, I am consulting with doctors,
and then I will write about it in detail.

Recently I heard Mitya's† quintet on the radio, and of course, it
was a delight for me to listen to, as it is for everyone.

For you spring is in full flow, but here it is only just beginning.
Despite the cold, the climate here is very good. It is peculiar, but
there is no wind. For the six months that I have been here, I can
recall only a few windy days.

However, it is at least possible that Frederiks' comments about his
health are over-optimistic in order not to worry the family and that actually the climate was awful. There were, of course, many qualified doctors in the camp (the purges hit the educated classes disproportionately),
but no medicines or bandages. In the event, there is no evidence that
Frederiks's wife was in fact ever able to visit him.

Hitler's forces attacked Russia on 22 June 1941. In the early phases
of the war the Germans made dramatic advances. In autumn 1942 it was
proposed that scientists in Soviet prisons should work on war problems.
An important critical factor for the Russians was the shortage of oil, and
in Ukht this was naturally a major priority. To begin with Frederiks was
set the task of finding chemicals which would soften the earth and thereby
facilitate drilling for oil. He explained in vain to the camp authorities that
this task would be impossible without the necessary materials.

What Frederiks and his scientific prisoner colleagues were able to
suggest was an ultrasound method for separating oil into heavy and
light oils. This is useful because only light oils can be employed for
motor vehicle propulsion. As a reward for this invention it was decided
to release Frederiks and his scientific colleague Polak early in August
1946. Unfortunately, Frederiks was not to live this long.

Precise details about Frederiks' death seem to be lacking.[44] Usually
reliable informants give different stories. According to his liquid crystal
colleague Polak, Frederiks died of pneumonia in winter 1943–1944,
while being moved to work in a war industrial institute in Gorky (the
name for Nizhny-Novgorod during the Soviet period). According to
D.G. Alkhazov (later director of the cyclotron laboratory at the Frederiks
Institute in Leningrad), however, Frederiks was invited by I.V. Kurchatov
to participate in the Soviet nuclear programme and died on his way to
Moscow to join the nuclear team.

What is certain, however, is that his letters home stopped some time
in 1943. The family made enquiries but received no answer. Dmitri
Shostakovich in particular sought information. Eventually, in March
1944, his wife Maria Dmitrievna was invited to the NKVD headquarters

---

† i.e. Shostakovich.

in Moscow and told that he had died of pneumonia in Gorky, without giving the date of death. Some time later, in winter 1944–1945, photographs which Frederiks had had with him were sent back to the family.

The war ended, but the political harshness in the Soviet Union continued, with regular new waves of purges and deportations. Only with the death of Stalin in 1953 did the atmosphere relax. In 1956 the First Secretary of the Communist Party, Nikita Khrushchev, made his famous speech to the 20th Congress of the Communist party of the Soviet Union, denouncing the purges and terror that had occurred during the period of Stalin's leadership. It was now possible for families of the victims of the purges to receive news of their loved ones.

Frederiks' family made another formal approach to the authorities. Now there was some news. An official paper, dated 3 July 1957, stated that Frederiks had died of pneumonia on 6 January 1944. There was also news of his rehabilitation. On 29 October 1956, the military prosecutor of the Supreme Court of the USSR had received a statement. Apparently, all witness statements against Vsevolod Konstantinovich Frederiks had been false. Indeed, it turned out that the interrogators had used unauthorized interrogation techniques. Furthermore, the interrogators involved (six in all) had themselves turned out to be enemies of the people and counter-revolutionary agents. By way of recompense, as it were, they had themselves been arrested and executed.

There are some (more hopeful) scientific footnotes to this sad story. One concerns Frederiks' student V.N. Tsvetkov. At the early age of 26, Tsvetkov inherited Frederiks' intellectual mantle and, in the absence of any other qualified leader, his scientific group as well. Tsvetkov was able to live through the purges and build a large and influential scientific school. He was the first to introduce the key concept of the *order parameter* into the liquid crystal field, in a paper published in 1942 in German[45] and completed in Leningrad only two months before the German attack on the Soviet Union. This quantity, which plays an important role in all mathematical theories of liquid crystals, is a figure of merit describing the degree to which molecules on average point in the special director direction.

Much of Tsvetkov's later work was on the physics of high polymers. He received numerous honours and was elected to the Russian Academy of Sciences in 1968. Amongst his honours was a Stalin Prize for physics in 1951. He was also, at the age of 87, the first recipient of the Frederiks Medal, instituted by the Russian Liquid Crystal Society in 1997. This medal is the now the premier international prize offered by the Russian Liquid Crystal Society and is the object of keen competition. Tsvetkov, unlike Frederiks, was lucky enough to live into old age, dying only in 1999.[46]

Now is the time to return to the scientific story, and it will in some senses be a memorial to Frederiks. The effect that is now known as the *Frederiks effect* was first reported in 1927. An explanation of the effect is given in Chapter 5 and, like most scientific discoveries, it owes much to earlier researchers. However, it is the Frederiks effect which is at the heart of the modern liquid crystal displays viewed in computers and televisions around the world. There is still a way to go scientifically from the dismal end of Frederiks in the gulag to glittering flat screen colour televisions, but, in some deep sense, notwithstanding other intellectual narratives and intellectual property issues explored exhaustively in courthouses across the world and which we discuss in Chapters 9–11, Frederiks was the inventor of the modern liquid crystal display device.

8

Renaissance

———⟫⟨⟨⟩———

Far more marvellous is the truth than any artists of the past imagined

Richard Feynman, *Lectures on Physics*, 1963

The sometimes catastrophic disruption of the lives of liquid crystal scientists caused by the political turmoil of the late 1930s and early 1940s has been recorded in the last chapter. Research on liquid crystals had no perceived military value and so had not been a priority during the war years, although now in the twenty-first century no military force could function without liquid crystal displays. A few isolated researchers had managed to continue, and from their work slowly emerged a renewed interest in liquid crystals as a fourth state of matter.

This chapter traces the developments in liquid crystal science from devastated Europe at the end of World War II to the flowering of understanding that has resulted in a major world-wide technological revolution. We begin in Germany.

Halle, German Democratic Republic: the source of new smectic phases

In the period immediately before, and to some small extent during, World War II, it was largely in Germany that liquid crystal research was carried forward. Wilhelm Kast, whose Ph.D. supervisor Gustav Mie had been an assistant of Lehmann, was working on liquid crystals at the physics department of the Albert-Ludwigs University in Freiburg-im-Breisgau, just 100 km south of Friedel's native city of Strasbourg. In 1937 Kast returned to his home city of Halle as a Professor of Physics at the university to continue the tradition of liquid crystal research started by Vorländer. The pioneer of making liquid crystals, Daniel Vorländer, had retired from his position as Professor of Chemistry and Head of the Institute at the Martin Luther University of Halle in 1935, but continued to exert an influence on activities in the institute until his death in 1941. He had been succeeded by Karl Lothar Wolf, a physical chemist, who had come from Lehmann's university in Karlsruhe.[1]

In 1939, a new student, Horst Sackmann, was due to join the university at Freiburg to study chemistry. However, in anticipation of the invasion of France, the university was closed. Instead Sackmann matriculated at Halle, and encountered liquid crystals for the first time through his classes in physics with Kast. Following the outbreak of hostilities, the Nazi regime rapidly took control of most of Western Europe. By mid-1940 the power of the Third Reich extended from the Russian/Polish border in the east, to the Pyrenees and the Atlantic coast of France in the west. From a German perspective the situation in Europe was calm, and the University at Freiburg re-opened, allowing young Sackmann to resume his studies there. The false peace was short lived and in 1941 Sackmann was enlisted in the German army. He soon saw action, but suffered serious injury, which caused his return to hospital in Halle. Following a partial recovery, but no longer fit for action, Sackmann returned to study and research at Halle University.

As it turned out, Halle was not such a bad place to spend the war. The tenth largest city in Germany, Halle is best-known as the birthplace of the composer Handel. It was spared the devastating aerial bombardment of its near neighbours Dresden and Leipzig, and so as the war came to a conclusion, Halle was mostly intact. The arrival of the 'liberation' forces of the Americans in April 1945 could have changed that. Indeed divisions of the rival armies were ready for a pitched battle, as indicated in leaflets distributed by the Allies to the local population.[2]

MEN AND WOMEN OF HALLE
Complete destruction threatens your city. Either Halle will be surrendered unconditionally or it will be destroyed.

Surrender is the only choice. We Americans do not wage war against innocent civilians. If, however, the military commander and the party leaders do not want to prevent bloodshed, we have no other alternative but to completely destroy Halle.

Surrender was not so easy for the German Commander, General De Witt, as he was followed everywhere by SS Officers with drawn pistols to ensure that capitulation was not an option. Furthermore, 'he was afraid of what they (the Nazis) might do to his family'.[2] However, a dramatic figure by the name of Count Felix van Luckner,[3] who lived in Halle, had a secret meeting with the Allied Command and managed to limit the fighting and destruction to just a part of the city. The old buildings of the Martin Luther University mostly survived. Crucially for liquid crystal research, the late Professor Vorländer's Chemical Institute was largely undamaged (see Figure 8.1) and the vast collection of liquid crystal materials (over 1000 at his death) remained intact in their cigar boxes.

Figure 8.1 The Institute of Chemistry, University of Halle. Vorländer, Sackmann, and others researched liquid crystals at the Institute, which is pictured in the 1950s undamaged by battle.

Following the war, the city of Halle was part of the German Democratic Republic, under the protection of Soviet Russia. However, working as a scientist in a communist state was not easy. Resources were restricted, often solely at the disposal of party apparatchiks, and exchange of information with scientists from non-communist countries was beset with difficulties. The greatest disadvantage was that the work of scientists from behind the Iron Curtain was not widely circulated in the West, and much research of value went unrecognized. Exchange visits did happen from time to time, but often it was party members or sympathizers that got to travel to the West, rather than scientists with talent and ideas.

Horst Sackmann (see Figure 8.2) was no communist, yet he survived. By 1958 he had risen in the hierarchy and was appointed Professor of Chemistry and Director of the Institute of Physical Chemistry at

Figure 8.2 Horst Sackmann (1921–1993).

Halle. Nevertheless there were tensions in the Institute: some members of the university were members of the Communist Party, including the Rector. Sackmann's talent and reputation as a scientist, however, and his position as a member and eventually Vice-President of the respected Deutsche Akademie der Naturforscher Leopoldina, meant that he was able to carry on his research without too much interference. The Leopold Academy is the oldest scientific society on the planet and was not under the influence of the Communist Party. Knowing who was and who was not a party member regulated relationships at the workplace, but there were also undisclosed agents and informers of the Stasi, the hated East German secret police. Later, after reunification of East and West Germany, the discovery that some former friends and colleagues at the university had been informers for the Stasi was to cause anger and a great sense of betrayal. Those exposed suffered humiliation and condemnation, and they lost their positions in the university.

As a physical chemist, Horst Sackmann did not synthesize new compounds, but became an expert in what is called 'the thermodynamics of mixtures', that is, what substances mix with what, and in what proportions, and how the properties of the mixtures compare with the behaviour of the pure components. These experimental skills were exactly those required to sort out the controversy between Vorländer and Friedel over the existence or not of more than one smectic phase.

Vorländer had been convinced at an early stage that he had observed different variants of a liquid crystal phase, which he could not bring himself to refer to as a smectic phase. On the other hand, Georges Friedel had clearly described the smectic phase in his elegant 1922 article, and he was convinced that there was only one variety. The difference of opinion had been vociferous, sometimes heated, and continued for more than a decade. However, evidence was slowly accumulating of further smectic modifications. First Vorländer,[4] in one of his last papers, and then his collaborator Conrad Weygand[5] observed more than one smectic phase. Weygand, stubbornly resisting the Friedel nomenclature and holding to Vorländer's (see Chapter 5), discovered a substance with what he called one *pl* and three *Bz* phases. The three *Bz* phases were labelled *Bz-I*, *Bz-II*, and *Bz-III*.

Sackmann and his student, Heinrich Arnold, using Vorländer's samples, made a classic microscopic study of many compounds that appeared to be liquid crystalline. The method was one of categorization: they wanted to distinguish between different types of liquid crystal simply on the basis of what they looked like under the microscope, the so-called optical textures. Recognizing the essence of the smectic textures from the Friedel classification, Sackmann and Arnold observed that there were in fact different smectic textures. Classifying them, they found not two, but three distinct types. One, the usual texture, which they labelled smectic A, is the familiar smectic liquid crystal identified and characterized by Friedel (see Figure 4.4). There were two others,

Figure 8.3 A characteristic optical texture of a smectic B phase forming from a nematic phase. The low-temperature smectic B phase appears as butterfly shapes in the nematic fluid. There is no way of knowing if Sackmann observed this particular texture. (See colour plate 9.)

which they labelled smectic B and C. Modifications B and C seemed to contain some extra features which reminded them of the schlieren texture (Figure 2.4) so characteristic of nematics. In the phase they have labelled the smectic C, the fan-shaped textures appeared to be 'broken'. The smectic B phase, as well as showing textures similar to Friedel's smectic phase, sometimes revealed patterns in the microscope never seen before (Figure 8.3).

As well as new phases, Sackmann and Arnold had introduced a new technique: the study of mixtures of the liquid crystal compounds. There is a fundamental rule in the thermodynamics of mixtures: if two fluid compounds coalesce to produce a uniform homogeneous mixture, then the phases of the original pure compounds and the resultant mixture are all identical. Unfortunately, the converse is not true, so that if two fluid phases fail to mix, it does not necessarily mean that their phase types are different. For example, attempts to mix water and oil fail,* yet they are both simple liquids. Sackmann and Arnold made mixtures of Vorländer's liquid crystals, and so were able to determine which

---

* If oil and water are strongly stirred, then they will form an emulsion, i.e. one liquid dispersed in another, but if left the two liquids will separate. Addition of a mixing agent can stabilize an emulsion against separation, but it is still an intimate mixture of two distinct liquids.

varieties of phases were the same and which were different. Furthermore, by changing the concentration of some of the mixtures, they could show that there was never a smooth passage between what are now identified as essentially different varieties of smectic phases.

At the time of these systematic studies in Halle, other reports were appearing in the literature of different varieties of smectic phase.[6,7] Eventually, in 1959, Sackmann and Arnold published a series of four papers on the subject of 'Isomorphic relationships between crystalline-liquid phases',[8] and these established beyond doubt the existence of three different varieties of smectic phase. They were designated as A, B, and C, a notation that survives today (2010).

Smectic A was clearly the original Friedel smectic, but what of smectics B and C? For various reasons, Sackmann and Arnold understood that smectic B phases possessed a local crystal lattice-like order within the layers, but as far as smectic C is concerned, they were mystified. In their own words:[8]

> *Phases of type C should possess a common third type of structure, of which at present there is no indication of its nature.*

The resolution of this problem had to wait for another 10 years, and marks the beginning of what might be described as the modern era of liquid crystal research. Since that time, more than 20 different smectic phases have been identified, and many of these new liquid crystals have remarkable properties, as we shall see in the final chapter. But there was more to be done before the dawn of this new era.

Freiburg, Federal Republic Of Germany: theories of liquid crystals

Theory had been much despised by the pre-war liquid crystal pioneers Vorländer and Friedel. In the immediate post-war period there were important new advances in the theory of liquid crystals and, like the identification of new varieties of smectic phase, they came from Germany. In Halle, the Professor of Physics, Wilhelm Kast, had introduced a new research student, Wilhelm Maier, to liquid crystal studies for his Habilitation thesis, i.e. the thesis necessary to become a university professor. After the occupation of Halle by the American forces, many staff from the university were sent to other parts of Germany, to what was to become West Germany. In 1945 the Western allies had some idea of what would happen to central Europe under the Russians. In due course Kast and his former research student Maier, now a qualified university teacher, ended up in Freiburg-im-Breisgau close to the Rhine, about as far west as one could go and still remain in Germany.

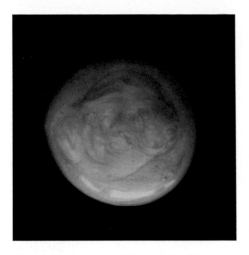

Plate 1 A droplet of cholesteric liquid crystal exhibiting the hues of a changeable silk. (See Figure 1.1)

Plate 2 Liquid crystal sample in a test-tube, warmed from room temperature. Samples go left to right. The sample starts cloudy, develops a region in which it is clear, with the interface between the two regions advancing until the whole sample is clear. On cooling, the process is reversed. (See Figure 2.2)

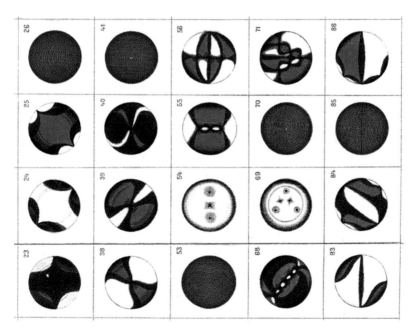

Plate 3 Coloured pictures of liquid crystal droplets. Reproduced from Lehmann's 1904 book. These are a few from an enormous collection of similar images. (See Figure 2.6)

Plate 4 The schlieren texture. This is an optical pattern seen in a polarizing microscope, characteristic of Lehmann's liquid crystals. (See Figure 2.4)

Plate 5 The fan texture of a smectic liquid crystal © I. Dierking. (See Figure 4.4)

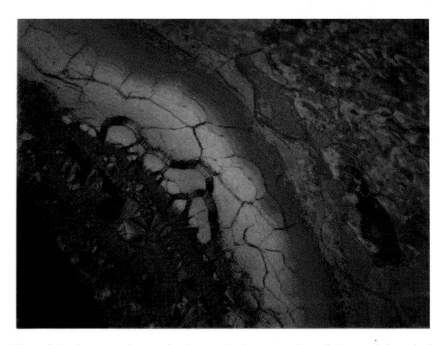

Plate 6 A microscope image of a cholesteric phase. A variety of characteristic optical textures is shown, including fans (left), Grandjean planes (centre) and schlieren (right). © D. Dunmur. (See Figure 4.5)

Plate 7 The collection of cigar boxes in which Daniel Vorländer stored his liquid crystal samples. Courtesy of the University of Halle. (See Figure 3.3)

Plate 8 Dark threads, appearing as coloured lines, marking fluid defects (known as disclination lines) in a nematic liquid crystal. The black areas emanating from points are known as brushes. They are areas of optical interference which connect through points on the surfaces marking the end of defect lines. © D. Dunmur. (See Technical Box 4.2)

Plate 9 A characteristic optical texture of a smectic B phase forming from a nematic phase. The low temperature smectic B phase appears as butterfly shapes in the nematic fluid. There is no way of knowing if Sackmann observed this particular texture. (See Figure 8.3)

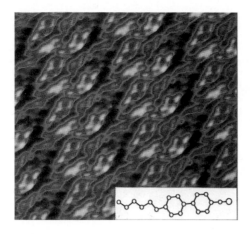

Plate 10 An image of a thin film of liquid crystal molecules obtained using scanning tunnelling microscopy. The alignment of the molecules is similar to that adopted in a nematic liquid crystal. Inset is a representation of the molecular structure of the molecules being viewed. Courtesy of Dr Jane Frommer, IBM Almaden Research Centre. (See Technical Box 4.3)

Plate 11 One of the first twisted nematic liquid crystal displays (Courtesy of Martin Schadt). (See Figure 10.4)

Plate 12 A modern flat-screen liquid crystal display television set. These are currently manufactured up to a size in excess of 1 m diagonal. For scale, an image one of the first twisted nematic displays is included. Main image courtesy of Sharp, first twisted nematic display image courtesy of Martin Schadt. (See Figure 11.2)

Figure 8.4 Alfred Saupe (1925–2008).
© Peter Palffy-Muhoray.

In the post-war era Freiburg became one of the great European centres for liquid crystal research, the roots of which are directly traceable to the school of Vorländer in Halle. In the years immediately following the collapse of Germany, universities in Western Europe began to function again. Released prisoners of war began to return to their homes, and one young Luftwaffe pilot and parachutist, Alfred Saupe (see Figure 8.4), came back to Freiburg, having been captured in The Netherlands and imprisoned in England for three years. Saupe joined the university in Halle and after completing his degree studied for a Ph.D. on liquid crystals under Wilhelm Maier. Maier was an experimental physicist, and so not unnaturally his new student started his research career measuring various properties of liquid crystals. However, the interpretation of the measurements was not so easy, and Saupe tried to make sense of his results using a new theory.

What emerged was a mathematical description of something like the swarm theory, except that the model avoided some of the contentious concepts that had exercised Friedel, Zocher and others by simply ignoring them. The great thing about mathematical models is that they are, or should be, self-consistent. One might take exception to the model, but the outcome is certain. In fact we now recognize that Saupe's theory is actually rather good, so good in fact that it endures under the description of the Maier–Saupe theory,[9] not bad for an aspiring Ph.D. student. Perhaps even more remarkable, however, is the fact that the Maier–Saupe theory already existed in the literature, having been formulated by Friedel's assistant François Grandjean in 1917. This important work was noted in Chapter 4, but its significance has only become apparent through the success of the theory developed by Maier and Saupe. The mathematics used by Maier and Saupe is essentially the same as that used by Grandjean, although the problem is tackled with different initial assumptions. Notwithstanding the resemblance between the two approaches, there is no question of plagiarism. Maier and Saupe would not have struggled manfully toward a resolution of their problems had they known the answer already. Their papers would have been phrased entirely differently. Grandjean's work really was forgotten. The question is why?

The immediate reason is that the paper was published during World War I in France, while much of the relevant work was going on in Germany, so the obvious people (Born, Zocher, etc.) would not have read it. By the time they were in a position to read it, time had passed by, and there was no database or abstracting service, as there is today, to draw it to their attention. Secondly, Georges Friedel's review paper in 1922 had been so influential that it tended to relegate prior scientific work to the status of history. We know that Georges Friedel held Grandjean in very high regard, and thus Friedel would not have downgraded Grandjean's work gratuitously, therefore Friedel must have been aware of this paper, but Friedel's long 1922 paper does not cite it. Copious literature tells us that Georges Friedel was at the very least impatient with the theoretical arguments between swarmists and distortionists. His plea was for more experimental work: the theory could sort itself out later. His omission of the Grandjean theory from the 1922 paper was therefore no accident: he felt that the time was not right.

Grandjean survived until 1975, when he died at the ripe old age of 93. One might have expected that at the time of the rediscovery of his theory by Maier and Saupe in 1958 he would have made some comment. By this time, however, having failed to obtain funding for his liquid crystal work in the 1920s, Grandjean had abandoned this field and had become a biologist. His studies of mites and ticks were to make him famous and probably consumed all his passions. If the liquid crystal people were not reading his paper, Grandjean was unconcerned;* more fools them!

None of this should detract from the achievement of Maier and more particularly his student Saupe. The latter was to have a distinguished career in liquid crystals in the USA, making seminal contributions to many aspects of the subject. Saupe's supervisor, however, met an untimely end in 1964 at the young age of 50. While in Italy to give a lecture, presumably on liquid crystals, Wilhelm Maier suffered a tragic

---

* Even in 2010, few liquid crystal scientists know of Grandjean's work, and it continues to be hidden from view. The first comprehensive volume detailing the physics of liquid crystals appeared in 1974, written by Nobel Prize winner Pierre-Gilles de Gennes. He deals with the Maier–Saupe theory, but does not mention Grandjean. This omission is rectified in the second printing of the book, which appeared in 1975, but then the reference to Grandjean disappears again in the second edition of 1993. Grandjean's theory really should be remembered.

drowning accident on the coast near Pisa. The circumstances of his death have never been revealed, but his contribution to liquid crystals will always be recognized in his eponymous theory.

The entreaties, noted in Chapter 5, of Sir William Bragg in 1934 to scientists of the developed nations that they should research liquid crystals had fallen on deaf ears. The scientific progeny of Vorländer from Halle continued to work, but for the rest of the world it took some time before the pre-war interest in liquid crystals was rekindled. By the late 1950s a renaissance of interest in mesomorphic phenomena began to flower. Two indicators of the quickening of the pace of liquid crystal research were the appearance in 1957 of a major review article by the American chemist Glenn Brown and his student Wilfrid Shaw in *Chemical Reviews*[7] and the organization of another liquid crystal conference.

Leeds, UK: Faraday Discussion on liquid crystals II

In Great Britain, the rebirth of liquid crystal research was stimulated in an unexpected way by the polymath 'Sage', J.D. Bernal. As we have described, he was active in various aspects of the planning and execution of military initiatives in World War II, while at the same time strengthening his links, scientific and political, with the USSR. Others perceived that there was a deep-rooted conflict between Bernal's political sympathies and his scientific work, at least in the Western world. A speech made by Bernal at a peace conference in Moscow in August 1949 in which he described the wickedness of capitalist science provoked wide condemnation in Britain and the USA. An example of the opprobrium sparked by Bernal's outburst in Moscow was an after-dinner speech made shortly after by Nobel Laureate Sir Edward Appleton, Principal of Edinburgh University, which was fully reported by *The Times*:[10]

> *It would appear that . . . there are two Professor Bernals: one is the brilliant natural philosopher of world-wide renown, the other a fervid convert to extreme political theory. Such a perverse interpretation of the situation as regards research in Britain would never have been made by the former.*

Despite the unease provoked by some of Bernal's activities, he was still highly regarded amongst his fellow scientists. Indeed the 1950s were exciting times scientifically, when many of Bernal's ideas about molecular structure and biological function were beginning to bear fruit.

Foremost amongst the advances in molecular biology was, of course, the discovery of the structure of DNA. This Nobel Prize-winning achievement, which received the award for medicine in 1962, is attributed to Francis Crick, James Watson, and Maurice Wilkins. However, it was based on developments in the application of X-rays to the determination of structures of biologically important molecules. These developments had been inspired to a considerable extent by the work of Bernal on tobacco mosaic virus (TMV),[11] and applied by others such as Rosalind Franklin at King's College London. Indeed Franklin had taken the X-ray pictures of DNA that were to form the basis for the structure revealed by Crick, Watson, and Wilkins in 1953. In that same year, Rosalind Franklin moved from King's College to Bernal's department at Birkbeck College, 21/22 Torrington Square,* London. There has been controversy over the lack of recognition that Rosalind Franklin received for her part in unravelling the structure of DNA, but now Franklin is established as a heroine of science for her careful work.[12]

The double-helix structure of DNA is now recognized as one of the most important scientific discoveries of the twentieth century, and perhaps of all time. For Bernal it showed the importance of what he described as the 'secondary structure' of macromolecules. It was not enough just to know the chemical formula for a molecule, i.e. its atomic content and how the atoms are connected. Much more important in determining the function of molecules was how they arranged themselves. Most molecules, and almost all macromolecules, are flexible and adopt a variety of different structures; some of these might do something—react or stimulate other molecules to react or change—but others may not be effective. The way to find out about this was through X-ray studies, the technique that Bernal had made his own. It was clear that a conference was needed.

Amidst what we would now describe as his 'jet-setting' around the world, Bernal arranged a General Discussion of the Faraday Society on Configurations and Interactions of Macromolecules and Liquid Crystals. This took place in the University of Leeds on the 15–17 April 1958. There are no original contributions from Bernal, but he gives the introduction in which it is clear that his is the guiding spirit for the meeting.[13]

> *The printed title of this discussion is necessarily an abbreviated one and does not give a sharp enough idea of the range and limitations of the subject to be covered. A more correct title would be*

---

* Bernal's flat at Torrington Square was above his department at Birkbeck College, and was the scene of many quiet and not-so-quiet social encounters. Famously, one evening in November 1950, Bernal was hosting a party for delegates to a World Peace Congress in Sheffield. Amongst the guests was the artist Pablo Picasso. Fuelled by red wine and encouraged by other guests, Picasso drew a mural on the wall of the flat. Bernal presented the mural to the Institute of Contemporary Arts in 1969, and it has recently been acquired by the Wellcome Trust (*The Times*, 2 April 2007).

'The Configurations and Close Interactions of Macromolecules and their Expressions: in solutions, in liquid crystals and in the solid state'.

He goes on:

This is a deliberate restriction for the purpose of this discussion. A general discussion on the configurations and interactions of macromolecules should go beyond the close interactions which we discuss here to consider those between colloidal particles in solution and a more condensed gel state...

We get a better idea what the meeting was about from the submitted papers. There are 17 papers on macromolecules, of which more than half deal with biological macromolecules. Poignantly, in the printed record of the meeting there are two papers from (the late) Rosalind E. Franklin: she died aged 37 in Cambridge on the second day of the conference. This must have been a terrible blow for Bernal, who had encouraged her so enthusiastically, especially since her contributions to the unravelling of the structure of DNA work had been side-lined by Watson and Crick.

In addition to the papers on macromolecules there were just four papers on liquid crystals, and three of those were on lyotropic liquid crystals. Why had liquid crystals been included at all? One can only surmise that Bernal, remembering his work on TMV, had a vague notion that the liquid crystalline state was going to be important for biological function. If he had thought it was possible to detect changes in molecular configurations in liquid crystals that might favour particular biochemical activity, then, not for the first time, he was way ahead of the game. As it turned out, the liquid crystal papers are not earth-shattering and there were not many liquid crystal scientists at the meeting. The interest of the meeting for our story is twofold. Firstly, F. Charles Frank from Bristol presented a much-quoted paper on the elastic theory of liquid crystals,[14] which had absolutely nothing to do with the subject of the conference. The second point of interest is the roll call of those who did attend and contribute to the general discussion on the liquid crystal papers.

The paper by Frank was nothing at all to do with molecular structure, but was a reworking of the distortion theory of liquid crystals of Oseen and Zocher. Despite this, the paper is often cited as the origin of the so-called 'Frank' elastic constants, and has achieved a prominence denied to the earlier original work, not unlike papers from the 1933 General Discussion on Liquid Crystals. Frank himself was always quick to deny any elevated status in the development of the theory of liquid crystals.

Apart from Bernal, who was largely responsible for both the General Discussion on Liquid Crystals held in London in 1933 and the General

Discussion in Leeds in 1958, there was only one person who contributed to both meetings, A.S.C. Lawrence, the soap expert.* At the 1933 London meeting, Lawrence had made the connection between the varieties of soap preparations and liquid crystals. At the 1958 Leeds meeting Lawrence presented part 10(!) of an on-going series of papers on soap solutions and their liquid crystallinity.

The preamble to the written proceedings of the Leeds meeting records the attendance of 38 foreign visitors, of whom we note in particular Professor Dr W. Maier of Freiburg-im-Breisgau. In the General Discussion, Maier contrasted his molecular theory of liquid crystals to Frank's continuum approach, Maier's goal being to promote his new approach to the scientific community. As papers describing the new Maier–Saupe theory had thus far only appeared in German, a personal appearance by Maier at a major international meeting would justifiably be thought to be important for the acceptance (as we would now say) of the Maier–Saupe paradigm.

Another contributor to the liquid crystal discussion was Dr G.W. Gray of the University of Hull, who in a few years would assume the mantle of Vorländer as the liquid crystal chemist *par excellence*. We shall return in Chapter 10 to Gray's work, which had begun in the late 1940s when few researchers felt that liquid crystals had any further surprises in store. Gray's researches were to be extremely influential, not only for the development of the subject as a whole, but also for the commercialization of the science of liquid crystals which was to take place in the 1970s. In 1958, however, this was all in the future.

Globalization

In the early 1960s the first signs of what we now call the 'globalization' of liquid crystals were emerging. One of these signs is the adoption of English as a common scientific standard, and another is the involvement of the USA.

Earlier in this chapter we referred to the review of liquid crystals by Brown and Shaw. As is appropriate for an article in *Chemical Reviews*, Brown and Shaw were chemists, from the University of Cincinnati in Ohio, USA. However, far from being an expert critical review by scientists working in the field, it is a literature review by those beginning research in liquid crystals. Despite this, the paper contains an

---

* At the time of the Leeds General Discussion, Lawrence had left the Navy (see Chapter 6) to take up a position in the Department of Chemistry at the University of Sheffield. Stuart Lawrence's friendly disposition and deep knowledge of colloid science had earned him the nickname 'Soapy' Lawrence.

extraordinarily rich bibliography, going right back to the early days of liquid crystals. Also included is a comprehensive survey of the chemistry and physics, not only of conventional (thermotropic) liquid crystals, but also of lyotropic phases. Furthermore, and perhaps controversially for the time, the possible importance of liquid crystals to biological activity is explored.

The article still contains elements of scientific interest, but some parts are now of more obvious historical interest, most notably the theoretical section entitled 'Hypotheses as to the Nature of the Nematic Structure'. We quote some brief extracts, which show the rather surprising conclusions of the authors.

> *A theoretical interpretation of the mesomorphic state has intrigued many physicists and chemists since its discovery in 1888. There can be found in the literature exchanges of arguments for and against two hypotheses that evolved out of the data that were interpreted. Even today there is no one theoretical interpretation of the mesomorphic structure that completely explains all of the experimental data. However, of the two hypotheses that have been proposed, namely the swarm hypothesis and the distortion hypothesis, the swarm hypothesis is the more widely accepted.*

> *First, brief mention will be made of the distortion hypothesis as it was proposed by Zocher...the hypothesis has its limitations when one attempts to interpret the properties of light extinction and wall effects. The reader is referred to (an article by) Zocher in which he argues that the two hypotheses do not have the same physical significance.*

The London Faraday General Discussion, in which Zocher had argued so forcefully against the swarm hypothesis, had been 24 years earlier than this. The Leningrad school, as we have seen from Tsvetkov's work, had long since abandoned swarms. It was a perhaps a mark of the paucity of recent research in this area, and the lack of real debate, that Brown and Shaw were able to write off the distortion point of view in just one paragraph.

After completing his doctoral thesis, Shaw's subsequent career was spent in the chemical industry. His supervisor, Professor Glenn Brown (see Figure 8.5), continued as an academic, later moving from Cincinnati to Kent State University. Kent is an unprepossessing middle-American small town (census returns for 1970 indicate a population of around 28,000) in the State of Ohio some 50 miles south of Cleveland. A teacher training college had been established in the town in 1910, which acquired university status in 1935.[15] It was at Kent State University that Glenn Brown founded the Liquid Crystal Institute, which has become famous as a centre for liquid crystal

Figure 8.5 Glenn H Brown (1915–1995).

research. In August 1965, the first *official* International Liquid Crystal Conference* was organized at Kent State University by Glenn Brown, who is regarded as a founding father of liquid crystal research in the USA. This conference also marked the opening of the Liquid Crystal Institute at Kent, which was to play an important organizational role in the transformation of liquid crystal science from a niche interest to a major contributor to modern science and technology.

Among those invited to the conference in Kent was Horst Sackmann from the German Democratic Republic. Unlike many from the eastern bloc, Sackmann was able to travel to this international conference. In his lecture, Sackmann presented for the first time an English version of his ground-breaking work on the identification of different varieties of smectic phase,[16] but there is no still no mention of the possible structure of smectic B, and still less of smectic C. All that Sackmann can bring himself to say is:

> *Because of the general connection between double refraction and structure of matter our present texture system points to a structural connection between the phases with the same characteristics of texture, especially because there is a coincidence of texture characteristics of the liquid crystal phases with their miscibility characteristics. But an explanation of the given texture characteristics on structural backgrounds is needed urgently. Today no explanation between texture and structure of phases is known beyond that of Friedel.*

* In July 2010, the venue for the 23rd International Liquid Crystal Conference is Krakow, Poland.

Following the publication of the conference proceedings in 1966, Sackmann's classification of smectic phases eventually did excite a response, most specifically from Alfred Saupe.

In 1968, Brown organized the second international conference, and invited Saupe, who was spending the year as a visiting professor at the Liquid Crystal Institute, to give a plenary address. In an overarching review, entitled 'On molecular structure and physical properties of thermotropic liquid crystals', Saupe discusses textures in nematic, cholesteric, and what he now specifically calls smectic A liquid crystals, before passing onto what he calls 'tilted smectic liquids'.

> *The tilted smectic liquid has some interesting similarities with a*
> *nematic liquid which are not obvious at first sight.*

Interestingly enough, at the same conference there were papers by I.G. Chistyakov and colleagues from Ivanovo in the Soviet Union, investigating different forms of smectic liquid crystals.[17,18] Chistyakov, who used both X-ray diffraction and thermodynamic methods, does identify the smectic C phase with a tilted smectic phase (Figure 8.6).

The ghost of Georges Friedel is finally laid to rest. His scepticism about the existence of more than one smectic modification was misplaced. His opponents (as Lehmann would have put it) were not fooling themselves: the different smectic modifications they had seen were not simply flights of imagination brought on by wishful thinking, as Friedel had thought. Just as there are many crystalline phases, so there are also many mesomorphic (liquid crystalline) phases. Nevertheless, to estab-

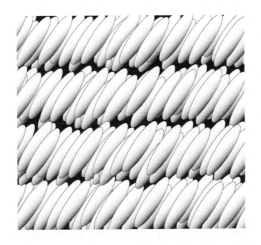

Figure 8.6 A representation of the molecular structure of a tilted smectic C phase. Compare this figure with the untilted structure of a smectic A phase in Technical Box 4.3.

lish this fact required some considerable time and several different strands of evidence. There was X-ray evidence, combined with thermodynamic studies, combined with a better understanding and more experience of what was meant by a phase transition. The primary signature, however, was that the optical textures themselves were subtly differentiated. Friedel, with all his observational care, either had not noticed, or more likely had not been exposed to, the errant smectics.

In 1969 Alfred Saupe moved permanently to the Liquid Crystal Institute at Kent State University. His presence there greatly enhanced its international reputation, and the Institute could now claim a direct link with Vorländer through Saupe's supervisor Maier and his supervisor Kast. The Kent State University Liquid Crystal Institute will figure strongly as the liquid crystal story draws to its climax and has established its place in the history of liquid crystals.

There is another history in which Kent State University has a lasting place, that of human conflict. The war in Vietnam was at its height in 1970. In April of that year, President Richard Nixon announced that US Forces had entered Cambodia, an escalation which rapidly sparked anti-war protests throughout the USA. Many university campuses erupted into noisy demonstrations, and Kent State University was one of these. On Friday 1 May 1970, there were disturbances across the campus of Kent State University, which also spread into downtown Kent and the headquarters of the university Officers Training Corps was set alight. The local mayor was sufficiently alarmed to declare a state of emergency, which came into effect on 2 May. The National Guard was summoned to restore order. The scene was set for confrontation, and sure enough with the resumption of classes on Monday 4 May a detachment of National Guardsmen faced about 2000 students on The Commons, an open area of the campus.

What prompted the shootings, no one knows, but after a few seconds around noon on 4 May 1970, four students were dead and nine injured. This was tragedy enough, as commented later by Kent State University President, Michael Schwartz,[19]

> *People were killed here, people who hadn't really done anything.*
> *They were killed by the authorities of their own government.*

The greater tragedy was the failure after decades of enquiries and litigation to resolve the conflict between the right to peaceful protest and the authority of the state.

The infamy of Kent State University is undeserved. There have been many student protests around the world, often resulting in the deaths of innocents. Perhaps the story of Kent State University endures because of the extreme contrasts it represents. On the one hand there is the tranquillity of Kent, Ohio, a small, almost sleepy, mid-western town with a dominating Ohio State University campus. On the other hand, there are the grotesquely violent scenes witnessed there on a few days in May 1970.

It is surprising that Kent State University should have become a centre for liquid crystal research. The university scarcely had an international reputation. Still less did it possess any tradition of liquid crystal research. Nonetheless, Glenn Brown's enthusiasm and commitment enabled him to establish a centre which would rapidly acquire fame. It was to attract the world's leading liquid crystal scientists and produce world-class research over the next half century.

It might be argued that the world's leading liquid crystal scientists had nowhere else to go, but this in no sense diminishes the stature of the Liquid Crystal Institute. In the 1960s the subject of liquid crystals was a minority sport, not at all part of conventional science.

This was soon to change. It had been a Frenchman, Georges Friedel, who first put liquid crystals onto a rational basis in the 1920s. Two generations later it was to be another Frenchman whose theoretical insights would transform the study of liquid crystals from a minor footnote into a central part of the scientific mainstream. It is that Frenchman, the flamboyant, attractive, and mercurially brilliant Pierre-Gilles de Gennes, to whom we turn next.

Pierre-Gilles de Gennes

Born in 1932 in Paris, de Gennes studied physics at the *École Normale Supérieure*. He impressed at the Ecole from an early age, and went on to study for his doctoral thesis on magnetism at the Centre d' Nucléaire at Saclay on the southern outskirts of Paris. After completing military service and a brief post-doctoral visit to Berkeley, USA, de Gennes was recruited in 1961 to join the staff of the newly instituted *Laboratoire de Physique des Solides* at the *Université de Paris-Sud*, Orsay. One of the driving forces for this new laboratory, which would rapidly achieve an outstanding international reputation, was Professor Jacques Friedel, son of Edmond Friedel and the grandson of Georges Friedel. Jacques had continued the family tradition of an academic life in science, and had also inherited his father's and grandfather's interest in liquid crystals.

It was not long before the Friedel inheritance emerged in the laboratory of solid state physics at Orsay and, no doubt strongly encouraged by Jacques Friedel, a number of groups, including that of de Gennes, turned their attention to liquid crystals. The year was 1968, which has achieved historical significance, especially in France, for student unrest and a lasting change to the political legacy of World War II. In Paris there had been several student protests against the Vietnam War during the early months of 1968. However, these became more locally focused in May when one of the new suburban campuses of the University of Paris at Nanterre banned male students from visiting women in their rooms.[20] Such a ban in France was unthinkable and rapidly led to violent clashes between students and police in many parts of Paris: the barricades were back. There was a tough response from the Gaullist regime, and students were arrested and campuses closed. The situation rapidly escalated and the President, General de Gaulle, even felt moved to take refuge at a French air force base in Germany for a few days. With thousands of students fighting on the streets and millions of workers on strike throughout France, de Gaulle refused to resign, but threatened to call a state of emergency. He did call for new elections in June, and was returned by a landslide: a backlash to the wide-spread disruption. However, his new power was short-lived, and within a year de Gaulle had been ousted.[21]

Amidst this political turmoil, the liquid crystal groups, inspired by de Gennes, at the University of Paris South were creating a revolution of their own in the understanding of the physics of liquid crystals. The fruits of these labours began to be revealed in 1968, under the anonymous communal authorship of the Orsay Liquid Crystal Group. While it would be presumptuous to ascribe any political tint to the liquid crystal group at Orsay, this authorship does pay homage to the Paris Commune of an earlier age.*[22]

The charismatic Pierre-Gilles de Gennes (see Figure 8.7) possessed all that people love and hate, or at least envy, in a certain type of person. He was confident and accomplished: a maestro in his chosen field of science, but also appreciative and knowledgeable about the arts, and a

Figure 8.7 Pierre-Gilles de Gennes (1932–2007).

* The Paris Commune of 1871 followed the defeat of France in the Franco–Prussian War at the beginning of 1871. Disenchanted by their government and army, left-wing groups in Paris came together to form a self-governing commune in Paris. The commune lasted only from March to May 1871 before the rebellion was put down bloodily by the National Army.

talented graphic artist in his own right. He was also a competent sports-man, in mountaineering, kayaking, windsurfing and skiing. At winter conferences in ski resorts he would often give his lecture in ski clothes, ready to challenge the slopes at the earliest opportunity. Above all, de Gennes was a Frenchman, tall, a Gaulois between his fingers, and some would say handsome. Certainly he was popular with women; he stayed married to Anne-Marie, with whom he had three children, but he also had a further four children by his long-time partner and scientific col-league Françoise Brochard.

De Gennes opened the eyes of many to the remarkable phenomena of liquid crystals through his lectures, articles, and his classic textbook *The physics of liquid crystals*, first published in 1974.[23] In the scientific litera-ture, the eponymous Landau–de Gennes theory of liquid crystals will perpetuate the name of de Gennes, linked to the great Russian physicist and mathematician Lev Landau (1908–1968). It was Landau who, in the 1930s, revolutionized the theoretical understanding of phase transitions. The details of his theory need not concern us, but it is interesting that as long ago as 1937 Landau was already thinking about liquid crystals.[24]

> *One often finds the opinion that liquid crystals represent bodies in which the molecules are arranged in 'chains', oriented in one direction…*
>
> *…we can imagine liquid crystals as bodies in which the molecules, or more precisely their centres of mass, are distributed completely randomly, as in ordinary liquids. Anisotropy of the liquid crystal is caused by the equal orientation of its molecules. For instance, if the molecules have an elongated shape, then they can all be arranged with their axes in one direction.*

The essence of Landau's theory was a mathematical expression of how the properties of a fluid change in the vicinity of a phase transi-tion.[25] Different types of phase transition could be described by Landau's theory depending on the details of the mathematics, but it was de Gennes who formulated the connection to liquid crystals.

The physics of liquid crystals I: properties that do not change with time

So what did de Gennes contribute to our understanding of liquid crys-tals through his insight and the experiments that he inspired? We recall from Chapter 5 the wars of words between the distortion theorists and the swarm theorists, voiced in the 1933 Faraday Discussion, but which were never properly resolved. With hindsight these two groups can be identified as coming, respectively, from the camps of continuum mechanics and molecular statistical mechanics. The former—the continuum mechanics people—regarded matter as continuous to be described in terms that Newton would have recognized, but the latter—molecular mechanics lobby—preferred to think in terms of the microscopic particles of matter, atoms and molecules.

Thanks in part to de Gennes, we now realize that there is no conflict. It is all a matter of length-scales. The world according to an ant, or a hiker, or a jumbo jet pilot looks very different, but in reality they all exist in the same space and are subject to the same environmental influences. The description we apply depends partly on who we are, but more importantly for what purpose we want to describe our environment.

Whatever theory is used to describe the liquid crystal environment, one inescapable experimental fact is that at a certain temperature the liquid crystal 'melts' to an ordinary liquid. This is the phase transition that separates the liquid from the liquid crystal, and is what defines liquid crystals as the fourth state of matter. An understanding of this phase transition would provide the evidence to support liquid crystals as a new state of matter, and it was de Gennes that provided the necessary physical insight. He recognized that one property which could be used to represent the liquid crystal phase close to a phase transition was an order parameter. This had been introduced for the nematic liquid crystal phase by Tsvetkov,[26] and had by implication at least been hinted at by Grandjean as early as 1917,[27] but for de Gennes' theory it was not necessary even to define the order parameter, all that was necessary was that it existed. The order parameter identified by de Gennes is a bit like a colour: we all know what is meant by it, but we don't have to define or explain it. Most importantly, we can talk of shades of colour—more or less, or none, and within the context of a mathematical theory we can quantify it.

The order parameter was non-zero in the liquid crystal phase, but changed to zero in the isotropic phase, which is for all intents and purposes an ordinary liquid. How the order parameter varied around the change from liquid crystal to ordinary liquid reflected the nature of the phase transition. If the order parameter slowly decreased with temperature and reached zero at the temperature of the phase change, then this was a 'second order phase transition', whereas if the order parameter dropped suddenly to zero at the temperature of the change from liquid crystal to liquid, then this was a 'first order phase transition'.* These were

---

* The classification of phase transitions as *first order* or *second order* was introduced by Paul Ehrenfest (1880–1933) a physicist working in Leiden in The Netherlands in 1933. He noticed that the measured changes in the heat capacity of liquid helium close to a phase transition were very different from the behaviour in normal liquids. Accordingly, the transition in liquid helium at –272°C was designated as a second-order transition; subsequently, second-order transitions were discovered in many materials less exotic than liquid helium.

terms instantly recognizable to all physicists and no longer were the mysteries of liquid crystals hidden within a jungle of jargon. De Gennes had brought liquid crystals into the mainstream of physics and an immediate consequence was that many more scientists started to take an interest.

Not only did de Gennes bring respectability to liquid crystals, he introduced new challenges to experimentalists, and perhaps unintentionally provided a bridge between the distortionists and the swarmists. The distortionists are happy that their liquid crystals can be described by a macroscopic order parameter and the swarmists can, at least amongst themselves, pretend that they can calculate it. Not only that, continuum theories and molecular theories can be developed to interpret a wide range of experiments and everyone is happy, thanks to de Gennes. Actually de Gennes went even further and suggested that the order parameter was not just a number, but had directional qualities as well; like the wind, which has a strength, but also comes from a particular direction. The wind is a vector and the order parameter is similar but more complicated, a second-rank tensor,* and so even more exciting to mathematicians and theorists, who love complexity.† Even experimentalists were excited because the treatment of de Gennes opened up endless possibilities for new experiments to test the theory. A whole industry was started devoted to understanding and verifying the consequences of the Landau–de Gennes theory.[28]

The physics of liquid crystals II: properties that change with time

The Landau–de Gennes theory and all its manifestations explain equilibrium systems, that is, systems that do not change with time, what physicists would describe in terms of static properties, and perversely chemists describe by thermodynamics (no wonder physicists and

---

* The order parameter due to de Gennes is a second-rank tensor. A vector, like the wind or any velocity or force, has magnitude and direction, and is an example of a first-rank tensor. A vector can be represented by a line, the orientation of which gives the direction and the length of which is the magnitude. However, there are properties that cannot be simply represented as a single magnitude and direction. They have to be described in two, three, or even more dimensions. Having broken out of a one-dimensional world, there is no limit. Einstein was one who used the mathematical power of tensors to describe the space–time continuum in general relativity in terms of fourth-rank tensors.

† The second-rank tensor order parameter of liquid crystals has a fascinating twin in the context of cosmology and the Big Bang Theory (or standard model as it is known by scientists!). Within the next year (2010) it is confidently expected that the Large Hadron Collider will identify for the first time the origin of mass: the Higgs Boson. The Higgs field, like the order parameter, is also a second-rank tensor. Remarkably, liquid crystals have been used as experimental models for the creation of the universe, an application beyond the wildest dreams of most liquid crystal scientists [see Trebin (1998) and references therein].

chemists have difficulty communicating with each other). However, de Gennes was not content to leave liquid crystals in a sedentary state. He tackled their dynamics, that is, how things change with time. Others had also done and were doing this, but de Gennes did it in a way that made connections with experiments: new experiments on the scattering of light, periodic flow, and the effects of charges and electric and magnetic fields. All of these would be formulated in detail by talented theoreticians and mathematicians, but it was de Gennes that brought the theories alive by suggesting and explaining new experiments.

Fortunately there were experimentalists around to follow up the ideas. An impromptu gathering with de Gennes in one of the laboratories at Orsay would inevitably end with the blackboard covered with scribblings and equations, and with an all-important message written at the bottom of the board: Please Leave, Do Not Erase.[29] The scribblings would be pondered over by others for days, and more often than not resulted in a new experiment and a new scientific publication, which always closed with the catch-phrase,

> ... as predicted by de Gennes.

De Gennes was able to explain a number of phenomena that were a consequence of motion in liquid crystals. One of the visual characteristics of liquid crystals, especially nematic liquid crystals, is that they shimmer. This optical effect results from a constant fluctuation of the refractive index of the liquid crystal—the director is constantly in motion. The liquid crystal group at Orsay were able to carry out experiments to investigate this and explain their results, thanks to de Gennes.[30] Optical shimmer or scattering of light is the visual manifestation of fluctuations in the local orientation of liquid crystals and it is strongly influenced by something called rotational viscosity. One normally associates viscosity with the ease with which a fluid flows, but rotational viscosity is a measure of how easily a liquid can be stirred.

The long history of observations of liquid crystalline viscosity, going all the way back to Schenck in 1905, demonstrated clearly that liquid crystals were non-Newtonian fluids. This does not mean, as Oseen had implied, that the laws of Isaac Newton are violated, but simply that the flow varies in an unpredictable way with the head of pressure causing the flow. For liquid crystals the situation was even more complicated. As Mięsowicz and others showed in the early years of liquid crystal research, more than one, though *exactly* how many was unknown, viscosities would be required to describe the motion of a liquid crystal.[31] It was a problem that challenged even the most talented physicists, and only the attentions of an even rarer breed of scientists, applied mathematicians, could unravel the complexities.

The theory has come to be known as the Ericksen–Leslie theory, despite the fact that Ericksen's name does not appear as a co-author in the ground-breaking paper.* Jerry Ericksen (b.1924), a mathematician

at John Hopkins University, Baltimore, Maryland, USA, became interested in the dynamics of liquid crystals in the late 1950s. His hunch was
that the non-Newtonian behaviour of liquid crystals was the result of
competition between the elastic effect of orienting surfaces and the viscous effect of orienting flows. Another suggestion at the time was that
the non-Newtonian behaviour was due to so-called 'fractional orientation', in which stronger flows would orient the molecules more perfectly, which turned out not to be the case here, but is in fact true for
lyotropic liquid crystals in strong flow fields.

The primary source for the distortion or continuum theory is, as we
have seen, the extensive body of work developed by Oseen. In the paper
Oseen presented at the 1933 Faraday Society Meeting on liquid crystals
he reports some preparatory work by his graduate student Adolf Anzelius
on the dynamics of nematic liquids.[32] Anzelius never published his thesis work; one of his examiners drily commented that:[33]

> *the dissertation should have technical importance for lubrication*
> *problems for semi-dry fabrication.*

During the early part of the 1960s, Ericksen worked on the problem
of flow in liquid crystals, but the breakthrough only occurred when a
Scottish mathematician working at the University of Newcastle,
England, spent a sabbatical year at John Hopkins University.[34] The
young mathematician was Frank Leslie (1935–2000) (see Figure 8.8),
who had already worked on complex fluids and was aware of the
outstanding problems that nematic liquid crystals posed. After some

Figure 8.8 Frank Leslie (1935–2000).

* It is with good reason that the theory has taken the names of both Ericksen and
Leslie. Ericksen's strict policy was to permit his postdoctoral workers and graduate
students to publish under their own names alone. He reports surprise, shared by few
others in the field, at finding his name appended to the theory. The theory is sometimes
known as *nematodynamics* or the *ELP* theory (the *P* is Parodi; see note 38).

discussions with Ericksen about the form the full theory of liquid crystals should take, Leslie spent the whole of the next month in the library. Ericksen gave him a copy of Anzelius's doctoral thesis on the dynamical theory of nematics. The library owned a comprehensive German dictionary. With thesis in one hand and dictionary in the other, progress was slow. After two years of work, first at John Hopkins University and then back at Newcastle, the final form of the theory was published in 1968. However, an experimental test for the theory was still awaited.

The test was soon to come. At Ericksen's suggestion, Ray Atkin, a postdoctoral worker from England, examined the properties of Poiseuille flow—flow in a capillary with circular cross-section. There is a nineteenth century formula that relates the fluid flow of a standard Newtonian fluid to the capillary radius, the pressure gradient across the capillary and the fluid viscosity. However, if the fluid is non-Newtonian, of the type described by the Ericksen–Leslie theory, then Atkin determined that there is a different relationship between flow, pressure, and capillary radius. The experimental test would be unambiguous.

The result, although published only in 1970, was ready in 1968 before the 2nd International Liquid Crystal Conference at Kent State University. To this conference, unbeknownst to Atkin, Ericksen, or Leslie, came the experimentalists Frederickson and his student Fisher,[35] who had performed just the flow experiment envisaged by Ericksen, in which they changed the size of the capillary and also the nature of the orienting condition at the boundaries. Their results were unexplained, but, encouraged by Atkin, Fisher and Frederickson rapidly re-plotted their data. They immediately found that the results were in agreement with the predictions of Atkin and the Ericksen–Leslie theory. In the years that followed, it became clear that it was possible to interpret essentially all macroscopic flow data on nematic liquid crystals in terms of this theory. Indeed using similar principles, Leslie was later able to extend the theory not only to discuss cholesteric materials,[36] a relatively straightforward matter, but also smectic materials.[37]

The story was completed by one of the Orsay liquid crystal group, O. Parodi,[38] who used a thermodynamic argument to reduce the number of independent viscosities from the six needed by the Ericksen-Leslie theory to five. One way or another, de Gennes' influence had pervaded all aspects of liquid crystals.

Soft matter and the Nobel Prize

De Gennes only devoted about seven years of his scientific life to liquid crystals before he moved on to bring understanding and inspiration to other fields. In 1971, aged 39, he moved from the solid state physics laboratory at Orsay to a prestigious professorship at the Collège de

France in Paris. Here he created a world-class group of experimental and theoretical scientists devoted to the study of complex systems, mostly fluids, an area which came to be called soft condensed matter.* The problems tackled by the group, inspired as ever by de Gennes, included the behaviour of polymers, the wetting of surfaces, an understanding of adhesion, and the formation of bubbles in liquids, the latter betraying an ever-present interest in champagne.[39]

In 1991 de Gennes was awarded the Nobel Prize in physics for his work on liquid crystals, polymers, and other complex fluids. He is the only scientist so far to have been awarded a Nobel Prize for work on liquid crystals.* As the citation reads:[29]

> *De Gennes has discovered relations between different, seemingly quite unrelated, fields of physics—connections which nobody has seen before.*

The area of study embraced by de Gennes now has its own identity and indeed has spawned learned journals devoted exclusively to research in the subject of soft matter. However, de Gennes was not one for categorization and labelling, rather he looked for similarities and unity in systems. The wide applicability of his ideas and concepts to many areas of science prompted one of the Nobel Prize judges to label de Gennes as the 'Isaac Newton of our time', but de Gennes distanced himself from such a remark, claiming that the stature of Newton was much higher than contemporary researchers.

Following his Nobel Prize, de Gennes embarked on a crusade to attract young people to science, and his charisma and dramatic lecturing style thrilled many audiences. He was invited by one of the authors to give his Nobel lecture, entitled 'Soft Matter' at a university in the North of England. He agreed, provided he could lecture to school children. Eight hundred young people aged from 12 to 15 from around the region filled the assembly hall of Sheffield University. This modern space, more like a football stadium than a lecture theatre, was full of critical youngsters waiting with interest. Despite the environment, de Gennes captivated his audience. His visual aids were four hand-drawn overhead transparencies, crumpled and travel weary, bearing signs of previous rainstorms and spilt coffee. The connection between lecturer and lectured was total, and de Gennes' performance even inspired questions

---

* The significance of soft matter is that it embraces many types of material encountered in daily life, such as toothpaste, vinaigrette, rubbers, jelly, most living cells, even spaghetti. In science, solids and gases were first to be explained, then liquids, but the understanding of other sloshy-types of material was much harder until de Gennes came along.

* As we noted in Chapter 3, despite a number of nominations, the Nobel Prize was denied to Lehmann. The scientific community at that time simply did not understand the importance or significance of liquid crystals as a new state of matter.

from the young audience. After the lecture, the Nobel Laureate was led away to lunch with the top officials of the university, yet such had been his connection with the audience that a young member of the audience rushed up to de Gennes to ask a further question.

After retirement from the College de France in 2003 and from his Directorship of the Ecole Spéciale de Physique et Chimie Industrielles in 2002 (a position he had held from 1976), de Gennes turned his attention to biological systems. At the Institut Curie he became involved with groups researching cell function and artificial muscle, and de Gennes brought his insight to a new field, which he had previously criticized. In an interview, he said,[40]

> *I have been very critical about biophysics…Biophysics is doing only the details, not addressing the big question.*

In the same interview, he was challenged about his constant shifts between different areas of research, but rather than bringing new perspectives to new areas, de Gennes claimed, characteristically,

> *I tend to see it more as a process of learning. For instance when we entered the field of polymers we were like students…we made many mistakes. Our lives have been a cascade of student lives.*

Sadly, de Gennes died on 18 May 2007 aged 75, and did not live long enough to unravel the mysteries of his new-found subject of interest, biophysics. He had impacted on many areas of science, but it is for his contributions to liquid crystals that he will mostly be remembered. He left a subject transformed, in which his paradigm, building on that of Landau, was the first port of call for the puzzled experimentalist. From the point of view of the fundamental science of liquid crystals, after de Gennes the paradigm was set, and the practice of the science was normal rather than extraordinary. This is a suitable point at which to move from the foundations of liquid crystal science to applications, which we shall address in the coming chapters.

9
An unlikely story

I have been asked if the crystalline-liquid substances may be technically exploited? I do not see any such possibility.

D. Vorländer, *Chemische Kristallographie der Flüssigkeiten*,
Leipzig 1924

Who better to know than Herr Professor Dr Daniel Vorländer, Vice
President of the Deutsche Academie der Naturforscher Leopoldina,
the oldest scientific academy in the world, and Director of the Chemical
Institute in Halle? During his career at Halle, spanning the years 1890
to 1935, Vorländer and his large team of organic chemists synthesized
more than 1000 new liquid crystalline materials. The compounds had
been characterized, purified, and examined in Vorländer's laboratories, and many samples resulting from this work still exist in the collection of chemicals in the University of Halle. Despite the fact that
most of the basic physics required for liquid crystal displays had
already been discovered, Vorländer had no inkling that, 50 years on,
liquid crystals would form the basis of a multi-billion dollar display
industry.

The confidence of Vorländer's dismissal of possible applications
for liquid crystals is perhaps understandable, given the nature of technical exploitation of new materials in the first part of the twentieth
century. The great German chemical industry had been founded to
produce chemicals for processing raw materials, fuelling and oiling
machines, dyeing fabrics, cleaning, painting, curing simple medical
conditions, and weapons. In any case, the German chemical company
E. Merck, at Lehmann's behest, had included liquid crystals in their
catalogue of chemicals since 1905, but without any commercial
success. The idea that the properties of apparently inert (i.e. unreac-
tive) chemicals could be exploited as electrical or optical components
was unheard of, and really had to await developments in solid-state
physics.

Television comes to town: the birth of electro-optics

Until World War II, popular entertainment was based around the cinema. To display moving images in a cinematograph it was necessary for it to have a shutter. Without a shutter the succession of frames of a film would just appear as a continuous blur. Until the advent of the electronic age, these shutters were almost always mechanical, although their speed greatly improved from the flickering silent movies that made Charlie Chaplin's name. At the dawn of the electronic age, television and the transmission of moving images by wire or radio waves became a challenge for the emerging electronics industry, and there developed a need for optical switches or shutters that could be activated by electrical signals.

Electrical current is a flow of electrons that can be manipulated using vacuum tubes or valves. With these devices, which were invented towards the end of the nineteenth century, it is possible to modulate electron beams to carry information, and to amplify and control the strength of the electronic signal. In order to convert an image into an electrical signal, it is still necessary to use some sort of shutter. The concept of transmitting images by electrical signals had been around for a long time, but it was not until the 1920s that recognizable television sets began to appear. The invention of the television cannot be attributed to a single person. Most of the early systems used some sort of mechanical shutter, as with the rotating disc of John Logie Baird.* However, the idea of using the electrical and optical properties of chemicals in the form of liquids or solids as shutters was beginning to catch on and the electro-optics business came into being, if only in a small way.

Electro-optical materials are those in which application of an electrical voltage will change the optical properties. The changes in optical properties may be dramatic, as in an opaque material becoming transparent, or vice versa, or the material may become a source of light. More subtle changes may also occur, such that a beam of light travelling through the material may change colour, be slowed down or speeded up, or even sent around a corner. There is not just one phenomenon responsible for electro-optical effects, and numerous scientists have contributed to our understanding of the many processes involved. Perhaps the most famous is Einstein, who in 1905 explained the photoelectric effect.† This is when a beam of light of a particular colour at a

---

* John Logie Baird (1888–1946) is acknowledged as the UK's pioneer of television. He studied electrical engineering at Glasgow and West of Scotland Technical College, and formed a company, Baird Television Limited, in 1925.

† Albert Einstein (1879–1955) was awarded the Nobel Prize for physics in 1921 for his attainments in mathematical physics and especially for his discovery of the photoelectric effect. Today Einstein is remembered for many things, most especially his theory of general relativity. His contribution to electro-optics no longer ranks as his key discovery.

frequency greater than some critical threshold value, depending on the metal, impinges on a metal surface and generates an electrical current. The phenomenon is undoubtedly an electro-optic or opto-electronic effect, but its importance is due less to the applications it spawned than the deep understanding it provided of the quantum nature of light and the electronic structure of metals.

Despite starting his scientific career in a patent office, Einstein was not one for applications. Nor indeed was the Scottish cleric the Reverend John Kerr (1824–1907),[‡] who discovered an effect that can in some sense be regarded as the basis for the liquid crystal optical displays of today. Following a degree in divinity, Kerr became a Minister of the Free Church of Scotland.[1] He was a student and life-long friend of Lord Kelvin, and his consuming passions were religion and science. With remarkable good luck, and perhaps helped by his influential friend, John Kerr managed to combine his passions. In 1857 he was appointed as a lecturer in physical science and mathematics at a training college of the Free Church of Scotland in Glasgow. This position provided him with laboratory facilities, and in 1875 he discovered the *Kerr electro-optic effect*, which nearly 60 years later was to figure in the first liquid crystal display patent.

TECHNICAL BOX 9.1 The Kerr electro-optic effect

The Kerr effect, as observed by John Kerr, was a change in the optical properties of a piece of glass when an electrical voltage was connected across the glass. Since glass is a good insulator, no current flowed, but the effect observed was a change in the optical transmission of the glass when viewed between crossed polarizers. For the effect to be seen, a particular arrangement for the direction of the light, the position of the electrodes, and the orientation of the polarizers was essential. What Kerr deduced was that the refractive index of the glass changed along the direction connecting the positive and negative connections of the applied voltage (the electric field direction). Thus the glass acquired two refractive indices, one along the direction of the field, and a different one perpendicular to this direction. In the language of the geologist or mineralogist, the glass had become doubly refracting or birefringent: just like certain natural crystals, such as calcite. The optical signature for such crystals was always that they could be viewed between crossed polarizers in a microscope, unlike non-birefringent crystals, which did not show up in the dark image plane of the microscope. It was not long before the Kerr effect was detected in many other materials, especially electrically insulating organic liquids, and related optical effects due to magnetic fields (e.g. the Cotton–Mouton effect) were also discovered.

[‡] The Kerr effect was discovered by Kerr in 1875, when he was 51. For this and other contributions to optics, Kerr was elected as a Fellow of the Royal Society in 1890 and was awarded the Royal Medal of the Society in 1898.

A Kerr cell consists of a sealed glass vessel containing an insulating transparent liquid or solid and a pair of electrodes. Windows are attached to the cell and a parallel beam of light is passed through the cell between the electrodes. The cell is placed between two polarizers, the polarization admittance directions of which are at 90° to each other and ±45° to the direction of the electric field. The amount of light transmitted depends on the voltage applied to the electrodes and the material inside the Kerr cell. For almost all materials, the electric field increases the refractive index along the direction of the field so that component of the light polarized along direction of the field is slowed down with respect to the component of the light polarized in a perpendicular direction.

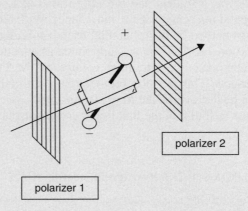

polarizer 2

polarizer 1

The figure shows the arrangement of polarizers and Kerr cell for an electro-optic shutter. The electric field connecting the electrodes is at an angle of 45° to the axes of the polarizers. The arrow indicates the direction of incident light. In the absence of an voltage applied to the electrodes, no light is transmitted by the second polarizer. Application of a voltage causes some transmission of light.

If the amount by which the component of the light is slowed (the retardation) is equivalent to one whole wavelength of light, then it is as though nothing has happened (see Technical Box 1.1) and the light impinging on the second polarizer will be blocked. For retardations between one whole wavelength and zero, light is transmitted by the second polarizer and is a maximum intensity when the retardation is half a wavelength. The voltage necessary to produce a retardation of half a wavelength is called the 'half-wave voltage' and for the materials used in Kerr cell shutters is typically thousands of volts applied across a pair of electrodes separated by a few millimetres.

Application of a half-wave voltage to a Kerr cell then switches between the light being blocked and the light being transmitted: hence the description 'Kerr cell shutter'. Its advantage was the speed with which it could be switched—in millionths of a second, compared to a mechanical shutter which was only thousandths of a second at best.

By the first decade of the twentieth century the Kerr effect had been measured for a number of materials and there was a reasonable understanding of how the change in refractive index is caused by an electric field. However, the Kerr effect of liquid crystals had not been investigated. The liquid crystals available at the time only became liquid crystalline at high temperatures, and their turbidity and tendency to conduct electric charge prevented their Kerr effects from being measured. Nonetheless, there were electro-optical experiments carried out on liquid crystals and the first comprehensive study was that by the Swedish physicist Y. Björnståhl (1888–1942), working in Uppsala, which was published in 1918.[2] We do not know much about Björnståhl, but he probably discovered the electro-optic effect that was to be exploited in the first liquid crystal displays manufactured in the late 1960s.

Liquid crystals, when melted into their nematic or smectic states, are optically opaque and appear to be inhomogeneous, rather like a badly stirred pot of paint. They are not attractive materials for optical studies. However, Björnståhl made a cell filled with nematic PAA,* applied an electric voltage, and observed what happened. Initially nothing, but on increasing the voltage there was a critical voltage at which the sample turned from partial transparency for light to a completely opaque state: he had demonstrated an electro-optic shutter using a liquid crystal. The origin of the effect is likely to be electrically induced turbulence in the liquid crystal, which is now reasonably well understood, but in fact is not the best electro-optic effect to exploit for liquid crystal displays.

The period between Björnståhl's work on the electro-optics of liquid crystals and the beginning of another war in Europe was one of great activity in the physics of liquid crystals. As has been explored in earlier chapters, the work of Friedel, Frederiks, Oseen, and Zocher during the 1920s and 1930s laid the foundation for the description and understanding of the properties of liquid crystals, but still the idea of using them was beyond the imagination of most, as Vorländer had succinctly indicated. Electro-optic technology was advancing, however, especially in the entertainment industry. By the 1930s the Kerr cell shutters or light valves commonly used in cinematography were potentially of use in nascent research into television and facsimile transmission. In such cells the material of choice was liquid nitrobenzene (a close chemical relative of trinitrotoluene, or TNT, the explosive), and in order to operate the device it was necessary to apply voltages of typically 4000 volts. Under these circumstances electrical breakdown followed by fire and explosion was not uncommon!

In the search for better materials to use in Kerr cells, the Marconi Wireless Telegraph Company, London, was investigating the use of

---

* PAA = *para*-azoxy-anisole, see Chapter 3.

nematic liquid crystals. In 1934 the Marconi Company applied for a patent entitled 'Improvements in or relating to Light Valves'. This patent,[3] prepared by brothers Barnett Levin and Nyman Levin,[†] research scientists at Marconi, demonstrated the use of liquid crystals in an electro-optic shutter. The patent is generally accepted as a landmark in the development of liquid crystals for display applications. It came only 10 years after Vorländer's crushing remark, which he might well have come to regret.

In reality, the application described in the Marconi patent was a false start on the road to modern display technology. The liquid crystal light valve was an improvement on existing Kerr cells, to the extent that the voltages required were about 20 times smaller than necessary for nitrobenzene cells. However, the device did not exploit the special physical characteristics of nematic liquid crystals, and the materials used would almost certainly have decomposed within weeks (if not days) of continuous use. Liquid crystal light valves did not catch on, and nitrobenzene-filled optical shutters were still in use by the Rank Organization during the 1950s and 1960s. Dr Nyman Levin was appointed in 1955 as Chief of Research and Development at the Rank Organization, but he had either forgotten, or discounted, his earlier work on liquid crystals.

The search for new displays

Alongside the discovery of nuclear fission and fusion, the discovery of the transistor in 1947 must rank as one of the most significant developments in physics of the twentieth century. Solid-state electronic devices rapidly replaced cumbersome vacuum-tube technology, and the instruments containing transistors and solid state diodes were much smaller, more reliable and consumed far less electrical power. Solid-state semiconductor devices were able to perform almost all the functions of glass electronic valves (e.g. diodes, triodes, tetrodes, etc.) and also exhibited new phenomena that could be exploited in new applications.

One new development arising from solid-state electronic technology was envisaged by David Sarnoff (1891–1971), Chairman of the Board of the Radio Corporation of America (RCA), in a speech given in September 1956:[4]

[†] Nyman Levin (1906–1965) was the senior author on the patent and subsequently took his scientific expertise to the Royal Naval Scientific Service and Rank Precision Industries (UK). In 1958 he moved to the Atomic Weapons Research Establishment at Aldermaston (Berkshire, UK), where he was Assistant Director and then Director until shortly before his death.

It [semi-conductor technology] will also give us brighter and bigger
TV pictures, and ultimately replace the TV tube altogether with a
thin, flat surface screen that will be hung like a picture on a wall.

Sarnoff was not talking about a liquid crystal display, but a device based on the solid-state crystal phenomenon known as electro-luminescence: light emitted from certain semiconducting materials when an electric voltage is applied to them. It was natural that RCA research teams should be looking for new developments in television technology. The company was a major player in the industry, holding many patents and wished to keep ahead of any potential rival technologies. Indeed it was in RCA that research into liquid crystal displays began around 1962. However, despite its early entry into the field, RCA closed down its liquid crystal research activities in 1972 and never capitalized on its early successes.

With hindsight, the 10 years from 1962 to 1972 can be recognized as the formative years for the liquid crystal display industry. Two unconnected events, on different continents, occurred in 1962, which introduced two key liquid crystal scientists to the world. In the UK the organic chemist George Gray authored a monograph entitled *Molecular Structure and Properties of Liquid Crystals*.[5] This book resulted from Gray's 16 years of research into liquid crystals at the University of Hull, and its comprehensive coverage of the field apprised many other scientists of the fascinating properties of this new state of matter. In America, Richard Williams, a lone physical chemist working in the RCA research laboratories, discovered the phenomenon[6] that would become known as 'dynamic scattering' from nematic liquid crystals. Only a short time later, William's discovery would become the basis for the first commercial liquid crystal displays.*

Neither George Gray nor Richard Williams were research managers, and it would be others who realized the possibilities of liquid crystals for commercial exploitation. Within RCA it was an electrical engineer, George Heilmeier, who persuaded his laboratory director in 1964 to embark on a secret programme of liquid crystal display research. The objective would be to develop a liquid crystal display, which could then be used as a flat-panel television. This programme was intended to keep RCA at the forefront of TV technology. But while RCA wanted to maintain its monopoly on some aspects of TV design, there were others who were equally keen to break this monopoly and stop paying RCA large royalties and licence fees.

In March 1967 the Minister of Technology, the Right Honourable John Stonehouse MP (1926–1988) visited the Royal Radar Establishment

---

* It is now recognized that Williams' observations were a re-discovery of the effect noted by Björnståhl in 1918 and referred to earlier. However, the conditions of the experiment were much more controlled and it become possible to develop ideas to explain the phenomenon.

(RRE), Malvern, to be informed about new developments in defence-oriented research. Malvern is a Victorian spa town close to the border between England and Wales, and also home to a major defence research facility. During his visit, Minister Stonehouse learned of developments in colour TV displays. He was appalled to hear that the cost of licensing 'shadow mask' technology from RCA was greater than the development costs of the supersonic passenger jet Concord. Surely there was an alternative technology that could save the UK millions of dollars in royalties and licence fees? The Minister proposed that there should be a UK programme of research to find a solid-state alternative to the vacuum TV tube, just as David Sarnoff of RCA had envisaged nearly 10 years earlier.

The narrative now shifts to 1976 and a pile of clothes on a beach near Miami, Florida, apparently the last memorial to John Stonehouse, former Postmaster General and Minister of State for Technology in Her Britannic Majesty's Government. His political career had come to an end with the fall of the Labour Government in 1970, and Stonehouse had then set up a chain of finance companies connected with Bangladesh. This business got into difficulties, as had his marriage, and Stonehouse sought a way out of his problems. Unfortunately his attempt to abandon his former life on a Florida beach was unsuccessful.* In due course he was caught in Australia, prosecuted, and jailed on various charges of fraud, deception, and impersonation. Despite trying, he did not escape the painful memories of his former life. However, in the development of liquid crystal displays he can be remembered as an unlikely and unsuspecting hero.

Following the visit of Stonehouse to RRE in 1967, there was a policy directive to find a replacement for the cathode ray tube.† Government works by committees, and the directive was referred to the Committee for Components, Valves and Devices, which was to be chaired by the distinguished scientist Dr Cyril Hilsum. A career Government scientist

---

* John Stonehouse was born in 1926 in the English port of Southampton. He came from a family of politicians and achieved high office in the Labour Government under Prime Minister Harold Wilson. Two years after his apparent suicide in 1976, Stonehouse was discovered in Australia living in domestic bliss with his former secretary, Sheila Buckley. He was arrested and sentenced to a prison term for various offences. Having served his sentence, Stonehouse had a short career as a novelist and died in 1988.

† The cathode ray tube was the development of a vacuum tube with a fluorescent screen which lit up when activated by a beam of electrons. By arranging for the beam to scan across the screen inside the vacuum tube, pictures could be depicted on the screen and viewed from outside the tube. The electron beam acted as drawing device, activated by electrical signals applied to the electron beam source (the electron gun). To make the image coloured was complicated. Three different electron beams were required to activate different materials (phosphors) deposited on the screen that would light up with different colours (red, blue, and green) when suitably activated by an electron beam. To make a picture, the beams had to be kept apart, which was difficult. The technology that RCA had was certainly smart and had cost a lot to develop. RCA was in a strong position to charge heavily for its use by other companies.

based at RRE, Hilsum had pioneered important developments in solid-state electronic devices, but knew nothing of liquid crystals. Just at this time, information about liquid crystals was beginning to filter through to the physics community. There had been an International Conference on Liquid Crystals held at Kent State University chaired by Glenn Brown in 1965, as recorded in the last chapter. Articles were appearing about possible applications of liquid crystals in the popular scientific press, such as that published in *Scientific American* in 1964,[7] written by James Fergason. He was a liquid crystal scientist from Kent State University and one of the first to recognize the potential of liquid crystals in technology. Despite the burgeoning liquid crystal literature, the secret programme of research into displays initiated by RCA remained hidden from other scientists.

A meeting was convened early in 1968 in London at which the Great and the Good of the UK defence staff and research scientists reviewed the possibilities for liquid crystal applications. At the meeting self-styled authorities propounded; admirals, generals and air vice-marshals discussed. By all accounts the occasion almost collapsed in farce. It was saved by the intervention of the youthful and unknown chemist George Gray, who, uniquely amongst the gathering, seemed to know what he was talking about. The determined chairman of the meeting, Cyril Hilsum, did not suffer fools gladly. He quickly realized that George Gray was perhaps the only person who could help the Ministry of Defence get a liquid crystal research programme off the ground.

Meanwhile, in the USA, research on liquid crystal displays was up and running. Although inspired by scientist Richard Williams, the team was led by the bullish and ambitious George Heilmeier, who was determined to bring liquid crystal displays to the American public as soon as possible. Preparing to enter the fray was the UK Ministry of Defence liquid crystal research group, inspired by scientist George Gray, directed by the equally bullish and determined Cyril Hilsum, and with military applications for liquid crystal displays as their primary objective.* In time both groups would have a significant involvement in the final triumph for liquid crystal displays, but in totally unexpected ways. There were a few others in universities and industrial research groups working in the area of liquid crystal applications at this time, and their contributions were also invaluable to the development of liquid crystal science. Key amongst these was James Fergason, who had been researching applications of liquid crystals since 1965 at the newly formed Liquid Crystal Institute of Kent State University.

---

* Although RRE was a military establishment, in the post-war years opportunities for commercially exploitable 'peaceful' technologies were increasingly sought. Thus, funding for the liquid crystals research programme came from both the Ministry of Defence and the civilian Department of Trade and Industry.

James Fergason (1934–2008), like George Gray, had a history in liquid crystals. Unlike the academic Gray, however, Fergason had been working in industry at Westinghouse Corporation Pittsburgh on applications. In particular, he was developing cholesteric liquid crystals to be used as coloured temperature sensors, using the phenomenon of selective reflection of light that had been discovered many years earlier. These sensors have found various markets and are discussed in more detail in Chapter 12. Fergason was more of an inventor than a scientist, always looking for new ideas or effects that might be turned into applications. In 1998 he was admitted to the US Hall of Fame for Inventors. However, as we shall see, Fergason's involvement with liquid crystal displays, although important, is surrounded by controversy.

In 1965 Fergason moved from Westinghouse to the Liquid Crystal Institute at Kent State University, a move which coincided with the first International Liquid Crystal Conference held on the Kent campus in August of that year. At the meeting, there were three papers presented with Fergason as a co-author, all on cholesteric liquid crystals. Members of the recently-formed RCA liquid crystal display group also made a number of presentations at the conference, but not about applications. Indeed, according to Joe Castellano, who was one of the RCA team attending the conference:[8]

> The important conclusion that I reached from attending this conference was that while many brilliant scientists were working in the field of liquid crystals, applications to displays in general and television in particular had not yet become apparent.

Just like Vorländer then, except that the scientists at RCA working on their secret research programme on liquid crystal displays thought they knew better.

The first liquid crystal display

So what was going on at RCA?* There is no doubt that the driving force for liquid crystal displays at RCA was George Heilmeier. He had been supported by RCA while completing his Master's degree and Ph.D. in electrical engineering at Princeton. Then, moving to permanent employment with RCA, he quickly established himself as a key figure at the

---

* The events are well documented in the book *Liquid Gold*, written by Joe Castellano, who was one of the liquid crystal research team at RCA during the period 1965–1972. This account illustrates the conflicts that arise in translating scientific knowledge into technical applications which might make money. Not only are there commercial considerations, but the ambitions and personal characteristics of individuals are often as important in determining the outcome, successful or otherwise, of a development project.

Sarnoff laboratories. His interest was caught by liquid crystals and he was well aware of the work of Richard Williams on the phenomenon that was to become known as dynamic scattering.

However, Heilmeier, with co-worker Louis Zanoni, first explored the 'guest–host effect' in which a thin film of a nematic liquid crystal (the host), doped with a small amount of a dye molecule (the guest), could be induced to change colour when activated by an electric voltage. This excited Heilmeier to imagine that it would be possible to use the guest–host effect as the basis for a display and ultimately the wall-mounted TV anticipated by David Sarnoff. Heilmeier then used his forceful personality to persuade laboratory managers to increase the resources available to his group, with the objective of producing a commercial liquid crystal display.

With an expanded team it became possible to explore a number of configurations and variants on cell designs, and search for better liquid crystal materials. It was found that the guest–host display worked better if polarizers were used, but these brought the disadvantage of reduced overall transmission and so less brightness. Heilmeier then returned to the effect discovered two years earlier by Richard Williams. It did not use polarizers and so perhaps could be used as the basis for a brighter and better display. Indeed it could, and in due course a display was designed, built, and patented. Technical details of the device were submitted for publication in the technical press early in 1968, nearly two years after its invention.[9] Finally, in May of 1968, RCA went public. They held a press conference where they announced the first liquid crystal display, which would have 'a major effect on the electronics industry'.[10]

This was the dynamic scattering liquid crystal display, the first liquid crystal display to be commercially developed. In its relatively short life this display was sold in millions, in wrist-watches and calculators, but not by RCA.

Despite the public acclaim for the dynamic scattering mode liquid crystal display, at RCA things were not going smoothly. The liquid crystal research group headed by Heilmeier was keen to get the company to take up the commercial challenge of manufacturing liquid crystal displays. Eventually they did, with a short-lived 'point-of-purchase screen' displaying the price of an item, which could be changed electronically from a central control. It was not a commercial success. The research team, however, were producing new and better materials and also novel configurations for future displays. In the commercial world outside RCA many other groups and companies jumped onto the liquid crystal display bandwagon, developing improved technology, new materials, and new products, but it was to be a rough ride for some, perhaps most of all for the RCA team.

As has already been mentioned, the success of the 'shadow mask' technology for colour TVs was reaping vast rewards for RCA. It was clear that senior management was not going to do anything to upset this by developing a rival technology, the commercial protection of which was not firmly established. Furthermore, the company had over-extended itself in acquisitions and was headed for financial trouble.[11] Huge investment in a new and untried liquid crystal display business, for which they lacked the production capabilities and expertise, was definitely not on the agenda.

Despite the company's lack of commitment to liquid crystal displays, Heilmeier's team at RCA had been responsible for creating a new technology. Their efforts were recognized by the company in 1969 through the David Sarnoff Outstanding Team in Science Award. Included in this award was Richard Williams, who, although not part of Heilmeier's team, received the recognition of his company bosses for the original discovery that led to the dynamic scattering liquid crystal display.

At this early stage in the development of liquid crystal display technology it was not clear if the dynamic scattering mode or guest–host effect displays would become the preferred technology or if one of the other possible configurations that were being explored would turn out to be more successful. As it happened, the configuration that was to become the great success in liquid crystal displays had indeed been anticipated within the RCA research group, but it had been spectacularly ignored.

In the great expansion of the RCA liquid crystal research activities, group leader George Heilmeier had recruited a number of talented scientists. Amongst these was a German physicist named Wolfgang Helfrich. He was primarily a theoretical physicist, but had worked on organic semiconductors at the National Research Council of Canada from 1964 to 1966. After a short spell back in Germany, Helfrich had joined RCA in 1967. During his time at the Sarnoff laboratories, Helfrich's task at RCA was to help the group understand the physics of their electro-optic effects. He formulated a mechanism, now known as the Carr–Helfrich effect, for the formation of William's domains.* These are the liquid crystal instabilities, identified earlier by Richard Williams, which mark the onset of dynamic scattering.

Helfrich had also apparently speculated on other configurations for liquid crystal displays.[12] On two occasions, first in 1969 and again in 1970, he had proposed to Heilmeier an idea for a display based on a twisted liquid crystal structure. The twisting of liquid crystal films had been demonstrated by Charles Mauguin many years before in 1911 (see Chapter 4), but Helfrich's idea was not pursued. The display would require polarizers, which reduced the brightness. More important perhaps, the idea would challenge the favoured RCA display, which had

---

* These domains are more properly referred to as Kapustin–Williams domains, since they were first observed by the Russian physicist A P Kapustin.

already been announced to the public. The boss, Heilmeier, was in no mood to receive any rivals to 'his' dynamic scattering mode display.

RCA and Heilmeier had missed the opportunity to patent and develop the display which ultimately would not only realize Sarnoff's ambition of a wall-mounted flat screen TV, but would also generate billions of dollars in revenue. Meanwhile in 1970 Heilmeier left RCA to become a successful US Government Director of Defense Research and subsequently a much-honoured business leader in technology. Helfrich, too, left RCA. He joined a laboratory in Switzerland where he would be able to exploit his idea, later returning to academic life in Germany.

The technology moves on

The febrile days of 1969 and 1970 are remembered by all who participated in the early development of liquid crystal displays. A new technology had been born, but precisely how and where it would mature was uncertain. Rival liquid crystal display systems were being investigated and the issue of improved materials was becoming of supreme importance. It was all very well to demonstrate a display in the laboratory, but commercial devices would have to work under unpredictable conditions, such as high and low temperatures, subject to vibration, sunlight, water, etc, and the public would expect a reasonable working life for any devices for which they had paid good money. The first requirement for a liquid crystal to be used in a device is that it remains liquid crystalline under use. This is not a trivial requirement. Early prototype liquid crystal devices needed internal heaters to maintain the material of the display liquid crystalline.

It did not take long for chemists to devise mixtures of compounds that would remain liquid crystalline at room temperature. A key development was published in 1969,[13] with the preparation of a compound designated MBBA that was liquid crystalline at room temperature.[†] Mixtures could then be prepared with a sufficient range of operating temperatures for commercial devices.

Thus, the materials available to the device engineers were improving, but still the vital issue of what device configuration to use had not been settled. Since the liquid crystal properties required for different display configurations were different, the chemists involved in the preparation of new compounds had to develop a variety of design

---

[†] MBBA (methoxybenzylidenebutylaniline) was first prepared in 1969 by H Kelker and B Scheurle working in the laboratories of the German chemical company Farbwerke Hoechst. It turned out to be less than ideal as a display material because of its toxicity and lack of chemical stability under natural light.

strategies. It has to be said that this situation still applies now, with different companies using different display configurations requiring liquid crystals with different properties. However, the compounds used now are quite different from, and much better than, those available to the display-makers of the late 1960s and early 1970s. At that time the configurations being investigated for commercial displays were the dynamic scattering display and the guest–host display, neither of which needed polarizers for their operation.

In the early years of research on liquid crystals, their remarkable optical properties had attracted much attention. Most dramatic were the reflected colours observed from cholesteric liquid crystals, which changed as the temperature changed, and which James Fergason was investigating for possible applications; these are described in Chapter 12. The physics of the effect was well understood, and was due to a twisted or helical structure developed by the optically active liquid crystal molecules of the cholesteric phase. Left-handed or right-handed versions of the liquid crystal molecules developed helical structures of opposite twist senses.

As long ago as 1911, Charles Mauguin had reported remarkable optical effects with twisted structures of non-optically active liquid crystal molecules (Figure 9.1). For these materials the twist had been

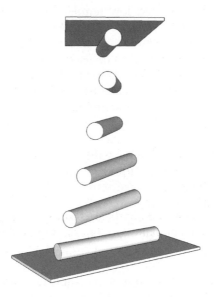

Figure 9.1 Mauguin's twisted structure. The top and bottom plates represent microscope slides, which Mauguin twisted with respect to each other. The alignment of the molecules, or more accurately the optic axes, was fixed at the surfaces, but in between the surfaces the optic axes twisted by the same amount as the angle between the microscope slides. Mauguin did not treat the surfaces of the slides, so he selected regions (domains) where the optic axes were parallel to the surfaces.

applied externally. A film of liquid crystal was confined between two glass plates which had been rubbed. It was then twisted so that the rubbing directions were no longer parallel to each other. Under these circumstances, polarized light passing through the films would suffer a rotation of polarization equal to the twist angle between the rubbing directions of the containing glass slides. There were certain conditions of refractive index and film thickness that had to be satisfied, but Mauguin understood these and had worked out the optics of the effect. Mauguin also discovered that the twisted structure could be destroyed by a magnetic field, causing the liquid crystal film to behave optically as a normal isotropic film of liquid.

These days all scientists concerned with the optical effects of liquid crystals are aware of the original work of Mauguin. Those scientists investigating optical effects for liquid crystal displays in the 1960s and 1970s also claimed, when questioned, to have read the early literature. This may be the case, but not all cited the earlier work; Mauguin had after all published in French. Wolfgang Helfrich, working at RCA, would certainly have known about Mauguin's experiments, so anecdotal reports of Helfrich's proposal in the summer of 1969 for an electric-field addressed twisted nematic cell using polarizers seem entirely believable; it is a pity the details were not written down.

According to non-public reports,[14] Mauguin's twisted structure was also 'discovered' by James Fergason in December of 1969 and subsequently incorporated into a working device in April 1970. Details of a twisted nematic device were published by Fergason and colleagues in the January 1970 edition of *Electro-technology*.[15] This article does indeed mention a twisted nematic structure being switched by an electric field from an opaque to a transmitting state. However, the context of the paragraph strongly implies that this is a guest–host effect and that no polarizers were used.

Looking back, Fergason always insisted that he was familiar with Mauguin's work, but there was no mention of it in the *Electro-technology* article. Nor apparently did Fergason think that his idea was sufficiently novel to patent, going ahead with open publication. It can only be assumed that Fergason was indeed describing a guest–host effect, admittedly in a twisted nematic cell, but relying on the dye molecules to generate the optical switching from opaque to transmitting. The addition of polarizers to this configuration would not have enhanced the effect, rather the reverse, so the all-important optical principle of Mauguin's twisted nematic was not being exploited.

There is one clearly documented event that took place in 1970: the Third International Liquid Crystal Conference held in West Berlin, 24–28 August. One of the papers presented there gave a full mathematical description of the effect of a magnetic field on the twisted nematic structure.[16] The author was the Scot Frank Leslie, who will be remembered as the co-author with Jerry Ericksen of the full theory for the hydrodynamical behaviour of nematic liquid crystals. At the time of the 1970 Berlin conference Leslie had no interest in, or knowledge of, potential display applications for liquid crystals. Nevertheless, assuming they understood it, a few members of his audience must have been electrified by his paper. Leslie's magical skill with equations was sadly not matched by an illuminating lecture style, and if you missed a partial differential or a subscript or two, separated by a comma, then you were likely to be in trouble. However, Frank Leslie was eminently approachable and he would have explained his key results to any that asked. Given the commercial sensitivity of the topic, it is unlikely that any of the individuals 'in the know' at the conference would have explained to Leslie the deep significance of his results for the liquid crystal display industry.

What Leslie had unknowingly developed in the guise of a solution to an academic problem was a full and accurate description of the switching behaviour of a twisted nematic cell. Admittedly he had used a magnetic field, but the transposition to using an electric field was mathematically trivial, even for the most applied physicist. The optical response was not given, although this was standard optical physics, and Mauguin had already explained the principles.

Who might have picked up on this highly significant result? The liquid crystal group from RCA were certainly represented at the conference, although they had no interest in twisted nematic devices, apart from the frustrated Helfrich, who had already proposed such a device at RCA. There was now a sound piece of theory which provided additional understanding of his twisted nematic effect. Indeed Helfrich was alarmed by Leslie's talk, but his idea for a display having been rejected, Helfrich was leaving RCA and would take his idea and the new information from Leslie with him.

James Fergason might have picked up the significance of Leslie's result, but his subsequent patent does not suggest it.[17] A few months after the Berlin conference, on 8 December 1970, the key paper was submitted on the twisted nematic display[18] and simultaneously a patent was filed in Switzerland from a previously unknown liquid crystal display group attached to a company that made cosmetics and pharmaceuticals.[19] What was going on?

10
The light dawns in the West

The twisted nematic was a classic invention, putting together two old ideas....to produce a new idea that was so simple everyone called it obvious—that is once they had actually heard about it!

E. P. Raynes, Internal Report, Defence Research Agency,
Malvern 1992

For the victors of World War II, the years immediately following the cessation of hostilities were a time when applied science burgeoned. The secret successes of scientists working towards military objectives could be exploited for the benefit of the all. However, the need for military research continued, as the Cold War and the arms race became central themes of international politics. Much of the scientific expertise was concentrated in military and government-controlled research establishments, and it was here that many of the peaceful applications of war-time research were developed. These establishments continued to attract some of the most talented post-war young scientists; indeed a period researching in a government laboratory was often considered more prestigious than joining a university.

In the USA there were centres of excellence such as the National Bureau of Standards and the National Institutes of Health as well as a number of military research laboratories. Amongst these, the US Army Research and Development Laboratory contributed to liquid crystal research during the 1960s.

Likewise, the UK had its National Physical Laboratory. Here, the Basic Physics Division, led by John Pople (1925–2004), a future Nobel Prize winner in chemistry, contributed to the development of new types of radio, microwave, and far infra-red spectroscopy.* The UK Ministry

---

* Spectroscopy is a generic term that encompasses many techniques that harness the properties of electro-magnetic radiation (the spectrum) for studying the properties of all types of materials, for the imaging of objects, including parts of the body, and for the chemical and structural analysis of materials.

of Defence research centre, the Royal Radar Establishment (RRE), Malvern, originally established as a telecommunications research centre during World War II, had initiated a programme of research into liquid crystal displays in 1968.

During the 1950s and 1960s, one of the leading national research institutes in the world was the National Research Council of Canada (NRC), based in Ottawa. This had a record of attracting some outstanding researchers, particularly from Europe, and indeed fostered its own Nobel Prize winner in chemistry, Gerhard Herzberg (1904–1999), famous for developing many types of optical spectroscopy useful for studying the properties of atoms and molecules. Although the NRC did not have a significant liquid crystal research activity, it played a key role in the progress towards the modern liquid display industry.

A chance encounter and a new display

Amongst many, two particular young scientists from Europe spent their postdoctoral years at the laboratories of the NRC in Ottawa. One we have already met was Wolfgang Helfrich (b. 1932). In 1964, fresh from completing his Ph.D. at the Technical University of Munich, Helfrich took up a position with NRC researching the structure and properties of organic crystals. In 1966 Helfrich returned to Germany to complete his Habilitation degree (D.Sc.), which would qualify him to become a professor of physics in Germany, but in fact he had already taken on a new position at the RCA laboratories in Princeton, New Jersey, USA. Here, as we have seen, he became a member of the new liquid crystal display group.

By some coincidence, just as Helfrich moved to the USA in 1967 to take up his new job, Martin Schadt (b. 1938), a young Swiss physicist from Basle, began postdoctoral research at the NRC. Schadt spent two years at NRC researching, like Helfrich, the properties of organic crystals. The two scientists did not know each other at that stage. However, Schadt would have learnt of Helfrich's work, since during his period at NRC Helfrich had published a paper recording the first observation of electroluminescence in the organic crystal anthracene.[1]

At NRC, Martin Schadt, a talented experimental physicist, was researching similar materials to those studied by Helfrich. The work was very successful and culminated in the invention and patenting[2] of the first primitive display using an organic light emitting diode.*

---

* A diode is the basis of all semiconductor devices. It consists of a wafer of two different materials, the structures of which are such that electrons travel more easily from one of the materials to the other, rather than in reverse. Depending on the choice of materials, it is possible to generate light from diodes by application of an electric voltage. The light is only generated if the bias (positive or negative voltage) is connected the right way round to the different materials of the diode.

Schadt's interest in applications had been fired. In 1969 Schadt moved back to Switzerland to find a new job, and took a position with the Laboratoire Suisse de Recherches Horlogères in Neuchâtel.

The preoccupation of the Swiss with clocks and watches is legendary. Not surprisingly, the clock and watch industry of Switzerland viewed with concern the emergence of electronic time-pieces. Watches using arrays of light-emitting diodes to display the time were being investigated by a number of companies in Europe and America. To keep abreast of these developments in the time-keeping business, the Swiss National Research Centre for clocks and watches recruited an expert in the field, Martin Schadt, fresh from his research successes at NRC.

It is not clear whether Schadt's employers wished to develop their own rival electronic watches or simply wanted to be informed of the competition. Whatever the intentions, Schadt did not stay long and soon joined the Central Research Department of Hoffmann-La Roche, a chemical and pharmaceutical company based in his home town of Basle.

On the face of it, this was a strange move: there was no overlap of interests or expertise between Schadt and Hoffmann-La Roche, a fine chemicals company. One attraction may have been an agreement between Hoffmann-La Roche and the Swiss electrical equipment manufacturer, Brown Boveri et Cie (BBC), to jointly develop electronic medical equipment. However, a new line of research was clear for Hoffmann-La Roche: they wanted to become involved with liquid crystals. A possible reason for this was the search for improved displays for medical devices, but it was also known that liquid crystals are of importance to the life sciences. The liquid crystalline nature of cell membranes, lipids, and other biological molecules was attracting attention amongst researchers.[3] As a manufacturer of cosmetics and pharmaceuticals it was reasonable that Hoffmann-La Roche should have established a liquid crystal research group. They did, and with unexpected consequences.

Another unlikely recruit for the new liquid crystal group at Hoffman-La Roche was Wolfgang Helfrich from RCA (see Figure 10.1). He certainly knew about liquid crystals, but as a theoretical physicist he was not, at the very least, an obvious choice of researcher for a cosmetics and pharmaceutical company. There is no doubt that Helfrich wished to return to his native Germany and in due course seek an academic position, but he was also nursing his thwarted idea for a new liquid crystal display. The significance of his idea must have been strengthened by the papers presented at the Third International Liquid Crystal Conference, which he attended in Berlin in August 1970.

Early in 1970, RCA had a visit from Dr Conradin von Planta, Head of the physics division of Hoffmann-La Roche. Since the liquid crystal group at RCA was being run down, Helfrich was looking for a job and was invited to join Hoffmann-La Roche by von Planta. The meeting between Schadt and his prospective colleague Helfrich at the 1970

Figure 10.1 Wolfgang Helfrich in 1976.

Berlin conference was highly significant, although Helfrich made no mention of his idea for a display until he joined Hoffmann-La Roche later in the year. The professional partnership between Schadt and Helfrich was destined to launch a new display technology.

It was in the autumn of 1970 that Helfrich joined the liquid crystal group at Hoffmann-La Roche. Within a very short time he and Schadt had invented and built the first twisted nematic liquid crystal display. This development had been carried out under the cooperative research agreement with BBC. Later, in the autumn of 1970, the BBC laboratories received a visit from a member of the Liquid Crystal Institute, Kent State University.[4] Somewhat naively, a member of the BBC research team told the visitor about the twisted nematic display of Schadt and Helfrich. This information was immediately communicated by the visitor to his colleague James Fergason back in Kent.[4] It will be recalled that earlier in 1970 Fergason had published an account of a 'guest–host' device incorporating a twisted nematic structure. One can imagine Fergason's anguish and anger on realizing that Schadt and Helfrich had stolen a march on him.

There may have been anguish and anger in Kent, but it is likely that sheer panic prevailed in Basle when Schadt and Helfrich realized the significance of the information leak about their device. To forestall any prior claims from Fergason, a patent application was rapidly prepared on behalf of Hoffmann-La Roche and within two weeks filed at the Swiss Patent Office in Berne on 4 December 1970. It is an important principle of patent law that prior public disclosure invalidates any patent claim. Thus, shortly *after* the twisted nematic display patent was filed, an article describing the twisted nematic device was submitted to *Applied Physics Letters*:[5] the article was received by the journal on 8 December 1970. This journal, published in the USA, exists for the rapid communication of important advances. The discovery had been disclosed to the scientific community and this was to be the real beginning of the modern liquid crystal display industry.

The twisted nematic liquid crystal display

To understand the scientific significance of the invention, and to follow the subsequent legal battles about the ownership of the idea, we need to describe the elements of the twisted nematic liquid crystal display. First and foremost, the device is an optical shutter: on being activated by a voltage the shutter can change from clear to opaque, or vice versa. Numbers, letters, and complex pictures are formed from arrays of these shutters, and the visualization of the images is by transmitted or reflected light. None of these features are particular to a liquid crystal display. The importance of the liquid crystal is that it is a fluid whose optical

TECHNICAL BOX 10.1 The twisted nematic display

The twisted nematic cell, shutter element, or pixel consists of a thin film of nematic liquid crystal, just a few thousandths of a millimetre thick, sandwiched between two glass plates that have transparent electrodes deposited on them. The transparent electrodes can be of a very thin metal coating, but normally they are made from a transparent electrically conducting coating of indium-tin oxide. These electrodes are connected to a source of voltage, which activates the liquid crystal film and hence changes the optical transmission of the cell. The particular feature of the twisted nematic liquid crystal cell is that the surfaces of the glass plates are treated, usually by rubbing in one direction, so that the molecules of the liquid crystal align along the direction of rubbing.

The cell is assembled so that the rubbing directions of the two plates are almost at right angles to each other: hence the 'twisted' description of the cell. The final elements of the device are two sheets of polarizing film (Polaroid) placed each side of the liquid crystal cell. The polarizing film will only transmit light with a vibration direction along a defined direction of the film. Thus light incident on the cell will be 'plane-polarized' by the initial polarizer; this means that the vibration direction of the light is confined to a plane defined by the polarization axis of the film and the direction of travel of the light beam. Provided that the optical properties of the liquid crystal and the thickness of the cell are between certain limits, the thin film has the effect of twisting the plane of polarization of the light as it passes through the cell. The amount by which it twists is equal to the angle between the rubbing directions of the glass plates of the cell. Having passed through the liquid crystal cell, the light beam emerges with its plane of polarization rotated by 90°. This was demonstrated by Mauguin in 1911 (see Technical Box 4.1).

If the exit Polaroid film is arranged so that its transmittance direction is parallel to the plane of polarization of the emerging light, then the cell will appear clear. On the other hand, if the transmittance direction of the exit Polaroid film is at right angles to the plane of polarization of the emerging light, then the cell will appear dark. The twisting of the plane of polarization is known as guiding.

Top glass plates of cells. The alignment direction for the liquid crystal is perpendicular (90°) to the alignment direction on the bottom plates. This is indicated by a twist of the top plate with respect to the bottom plate. The inside surfaces of the plates are electrically conducting to allow application of a voltage. The outside surfaces of the plates have thin polarizers attached so that their optical transmission direction is perpendicular to the transmission direction of the bottom plates.

Bottom glass plates of cell. The alignment direction for the liquid crystal is perpendicular to the alignment direction on the top plates. The inside surfaces of the plates are electrically conducting to allow application of a voltage. The outside surfaces of the plates have thin polarizers attached so that their transmission direction is perpendicular to the transmission direction of the top plates.

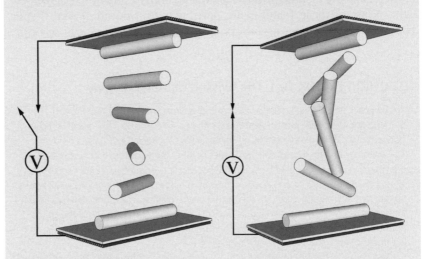

Schematic representation of the operation of a twisted nematic cell: left = OFF STATE in which light is transmitted; right = ON STATE in which no light is transmitted.

The source of voltage is indicated by **V**: it is unconnected in the OFF STATE (*left*) and is connected in the ON STATE (*right*). The liquid crystal alignment through the cell is indicated by cylinders, which can be thought of as representing the average alignment of liquid crystal molecules through the cell. In reality the volume between the glass plates will be occupied by more than a billion billion molecules.

Application of an electrical voltage to the liquid crystal film will cause the alignment of the liquid crystal to change from a twisted structure to a structure where the liquid crystal molecules are aligned along the electric field direction, i.e. perpendicular to the glass plates. This destroys the guiding of the plane of polarization of the light and so the transmission of the cell changes from transmitting to opaque or vice versa, depending on the arrangement of the initial and exit polarizers. There is a further, as yet unstated, requirement for the nematic liquid crystal used in the twisted nematic device. The liquid crystal must have a positive dielectric anisotropy, which means that the molecules of the liquid crystal prefer to align in the same direction as an electric field. This property requirement contrasts with that for the 'dynamic scattering mode' liquid crystal display presented by Heilmeier of RCA, which needs a nematic liquid crystal having a negative dielectric anisotropy, i.e. the liquid crystal molecules prefer to align perpendicular to an applied electric field.

In summary, the key features of the twisted nematic cell revealed to the world in December 1970 were:

1. a twisted configuration of the liquid crystal structure defined by glass plates rubbed at 90° to each other
2. polarizing films placed each side of the liquid crystal cell
3. a liquid crystal material of positive dielectric anisotropy.

properties can be made to change when a voltage is applied. As we have seen in the previous chapter, this effect can be achieved in a display simply by using the light scattered by certain liquid crystals, as successfully demonstrated by Heilmeier. However, the change from opaque to clear in a twisted nematic display is based on the optical principle of guiding, already demonstrated by Mauguin in 1911.

The reaction to the Schadt and Helfrich invention was mixed. In Basle, the press ridiculed the idea of an electronic replacement for the Swiss watch (see Figure 10.2). Perhaps Schadt's former employers in Neuchâtel (The Swiss Clock Research Laboratory) thought he had sold out to the opposition and used the influence of the national watch industry to mount a campaign against alternative time-pieces. They needn't have worried: even in the twenty-first century the clockwork watch retains its cachet and we even have clockwork radios, torches, and laptop computers!

At this time, early in 1971, there was already a nascent liquid crystal display industry in the USA based on the technology of RCA's dynamic scattering device. The scientists from the RCA group were distinctly unimpressed by the Schadt–Helfrich device. After all, if verbal accounts are to be believed, Helfrich had already proposed the twisted nematic display at the RCA laboratories and it had been discounted.

Furthermore, as noted in the previous chapter, Fergason, at Kent State University in the USA, had already published a description of a twisted nematic device earlier in 1970.[6] This device did not rely on 'guiding' nor did it utilize polarizers, and Fergason had not felt it necessary to patent his device. Despite this, and the existence of the Schadt–Helfrich patent, Fergason rapidly went ahead with his US patent application for a twisted nematic device, which was filed early in 1971 and eventually granted in 1973.[7]

The reaction of the UK research group based at RRE Malvern was more perspicacious. The group had been set up in 1970 and in April

Figure 10.2 Timely cartoon of the week: the latest novelty from Basle published by the *Swiss Watch Journal* in 1972. Reproduced in M. Schadt, *Milestone in the history of field-effect liquid crystal displays and materials*, Japanese Journal of Applied Physics 48, 03B001 (2009).

1971 was joined by a physicist from Cambridge, Peter Raynes, who had just completed his Ph.D. In his words, the reaction of the UK group to the Schadt–Helfrich invention was:

> *The TN [twisted nematic] was a classic invention, putting together two old ideas [optical guiding by a twisted nematic liquid crystal—Mauguin 1911 and reorientation of a nematic by an electric field—Frederiks 1933] to produce a new idea that was so simple everyone called it obvious—that is once they had actually heard about it! The TN was the first LCD [liquid crystal display] with a right feel to it... We therefore quickly decided to concentrate our efforts at RRE into investigating TN LCDs.*

The wall-mounted flat-screen liquid crystal TV was off the drawing board, but it would be nearly 20 more years before a working model was produced.

Despite interest from some of the scientific community in the twisted nematic device, the management of Hoffman-La Roche research were not convinced of its commercial significance and halted liquid crystal research.[8] Having seen his idea for a liquid crystal display demonstrated, Wolfgang Helfrich turned his interest to the physics of liquid crystalline cell membranes. The disbanding of the liquid crystal group at Hoffmann-La Roche prompted Helfrich to follow his academic aspirations, and in 1973 he left Basle to become a physics professor at the Free University of Berlin, where he remained until his retirement in 1997.*

The battle for priority

In Basle there was yet another change of business strategy at Hoffmann-La Roche. An approach from the Seiko watch company of Japan to buy the twisted nematic liquid crystal display patent finally alerted senior management to the fact that the twisted nematic liquid crystal device might have some commercial value. The research team under Martin Schadt (see Figure 10.3) was hastily reassembled and charged with licensing the twisted nematic display technology. The scene was set for a series of scientific and legal battles that would drag on for many years and not end until 1992.[†]

The development of new science for applications is nowadays strongly moderated by patents or, as the jargon has it, intellectual property rights (IPR). The search for profits from science is a powerful motivation, indeed these days perhaps the only motivation for companies, but the patent system can also have a constraining effect on new science. Companies do not want to get involved with patent battles if they can be avoided.

Figure 10.3 Martin Schadt (b.1938).

---

* Professor Wolfgang Helfrich made important contributions to the understanding of the behaviour of membranes and his work has been recognized by a number of awards. In 1993 he received the Hennessy Prize for Innovation and the prestigious Wolfgang Ostwald Prize for his fundamental work on complex fluids and colloids.

† Detailed accounts of the battle over patent rights can be found in Buntz (2005) and Castellano (2005).

The twisted nematic patent based on the invention of Schadt and Helfrich had been filed by Hoffmann-La Roche in December 1970, but filing a patent and being granted patent rights are very different things. A patent pending will be examined by patent examiners to see if the new invention is already covered by existing patents, or is an obvious application of an existing patent, or has already been described in the previous literature, *prior art,* as it is called. Furthermore, a patent is always open to challenge from others who may have competing intellectual or commercial claims. It would take years of legal argument before the issue of ownership of patent rights over the twisted nematic display was resolved. Certainly there was no competition from the original 1936 Marconi patent for a liquid crystal light valve, described in the last chapter, but there was strong opposition to the Schadt–Helfrich patent from other quarters.

The twisted nematic patent was originally filed in Switzerland and shortly afterwards a second patent was filed in the UK, but to be effective, it was essential that the new display device was protected right across the industrialized world. Of paramount importance was the establishment of patent protection in the USA. However, here there were problems.

Fergason had filed a US patent describing a twisted nematic device in February 1971, two months later than the original Swiss Schadt–Helfrich patent. Under normal circumstances, the prior date of the Swiss patent would have invalidated Fergason's claims. However, he maintained, supported by witnessed laboratory note books, that he had indeed demonstrated the twisted nematic device in April 1970. This pre-disclosure claim is allowable under US law, but not elsewhere.

The evidence presented was sufficient to persuade the US Patent Office, and indeed the German Patent Office, that Fergason should be granted rights to the twisted nematic liquid crystal display. In Japan attempts by Hoffman-La Roche to patent the twisted nematic device initially failed, but this failure was not due to competition from Fergason, but rather because of strong arguments produced by Japanese scientists that the invention was already in the public domain as *prior art*. So things were looking distinctly shaky for the Schadt-Helfrich patent.

The conflict between Hoffmann-La Roche and Fergason over the US rights to the invention was bitter, and extended over a number of years. Hoffmann-La Roche received legal advice that an action in the USA against the Fergason patent would fail. As a result, the application for a US patent for the Schadt–Helfrich invention was withdrawn. Despite this, Fergason's private company, International Liquid Crystal Company (ILIXCO), owner of his US patent, decided to sell the twisted nematic patent. Hoffmann-La Roche was a ready buyer, and now had patent protection for the twisted nematic display in the USA.

Nevertheless there was still a dispute. At the time of Fergason's invention he had been a faculty member of Kent State University Liquid Crystal Institute, so the university thought they deserved a slice of the action. There was also a court case between Fergason and Hoffman-La

Roche over payment of the buy-out fees for the patents. Only after months of expensive court discussions was the action settled out of court in 1976. Royalties generated by the twisted nematic patent would henceforth be divided according to a formula: 30% to Hoffmann-La Roche, 30% to Brown Boveri et Cie, 30% to Fergason, and 10% to Kent State University,[9] so almost everyone made money (millions of dollars over the lifetime of the patent). Hoffmann-La Roche made money. Brown Boveri et Cie made money. Kent State University made money. Fergason made money. The lawyers especially made money. Only RCA, the company which had launched the liquid crystal display, made no money. If Wolfgang Helfrich had recorded his proposals for a twisted nematic device in 1969, then events would have turned out very differently.

There was still the small matter of patent rights in Japan, which was rapidly becoming the centre of the liquid crystal display industry. Eventually, in May 1986, the Japanese authorities also accepted the Schadt–Helfrich twisted nematic patent. It took even longer, until 1992, for this patent to be granted in Germany. Apparently, on examination of the original application, the Chief Judge of the German Federal Patent Court had declared:[10]

> ... if I can understand the underlying physics, it could not possibly be an invention.

Hoffmann-La Roche and their Swiss partners BBC had successfully acquired sufficient world-wide patent protection to consider the commercial exploitation of the twisted nematic display. But there was a problem: if the twisted nematic liquid crystal display was the right technology, who was going to manufacture the devices?

Neither Hoffmann-La Roche, a drugs company, nor BBC had facilities or expertise to develop liquid crystal displays for the market. In any case there were still some obstacles to overcome before full commercialization could be envisaged. The main stumbling block was the choice of liquid crystal material. It had to remain liquid crystalline over a reasonable range of temperatures and most importantly it had to have the correct electrical and optical properties. Schadt and Helfrich's first device had used a mixture of chemicals that were similar to those developed in Helfrich's old group at RCA. The mixture was tailored to have the right properties, except that it was not liquid crystalline below 20°C. If the display were to be used at lower temperatures, it would have to incorporate heaters. An early prototype device from Hoffmann-La Roche is illustrated in Figure 10.4.

One of the driving forces to replace existing displays was the desire to reduce power consumption so that portable devices could be made which did not constantly demand replacement batteries. If the new generation of liquid crystal displays were going to need heaters, then the objective of portability was destroyed. The battery life would be measured in hours, not even days: so much for a portable low-power display. Hoffmann-La Roche therefore began a programme of liquid crystal

Figure 10.4 One of the first twisted nematic liquid crystal displays (Courtesy of Martin Schadt). (See colour plate 11.)

materials development for twisted nematic displays that was well within its capabilities as a chemical company.

Hoffmann-La Roche should have had a head start developing the best materials for its protected twisted nematic liquid crystal display, even if the company had no chance of actually making the displays. However, there were other players in the field. Foremost amongst these was the German chemical company E. Merck of Darmstadt, who had been supplying liquid crystals for research since the days of Lehmann. Furthermore, as related in the previous chapter, Hans Kelker and Bruno Scheurle, researchers at the Hoechst Chemical Company, Frankfurt, had discovered a room temperature nematic liquid crystal, MBBA, in 1969.

MBBA was increasingly used in display mixtures of various types and, anxious to promote their new product, Hoechst placed an advertisement in the German press (see Figure 10.5). This challenged display manufacturers to produce a wall-mounted TV. However, the advertisement was withdrawn after a successful court action by TV tube manufacturers against Hoechst on the grounds that the advertisement undermined their products and gave a false message. As we shall see the message was true, but it took a long time to be delivered.

MBBA as a liquid crystal was by no means ideal. It tended to decompose readily, and would react with other materials in liquid crystal displays, moreover it was to prove to be strongly toxic and carcinogenic. The search was on for room temperature liquid crystals which were stable and harmless, and had the right properties for a twisted nematic device. What followed was a triumph and a reward for years of painstaking and largely unrecognized academic research in the 'small is beautiful' English University of Hull.[11] The organic chemist George

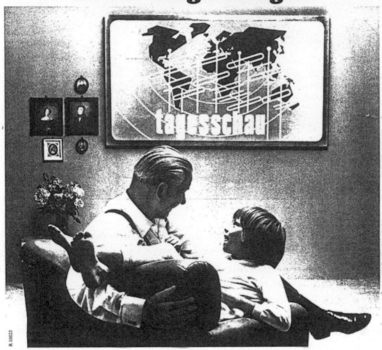

Bald können Sie Ihren Fernseher an den Nagel hängen

Figure 10.5 'Soon you will be able to hang your television on the wall' (der Nagel = a nail). A concept advertisement for a wall-mounted television that appeared in 1971.

Gray returns to the story, and he will be a key figure in the development of materials for liquid crystal displays for more than a decade.

Born in Edinburgh, Scotland in 1926, George Gray narrowly escaped being enlisted for military service during World War II. Instead he went to the University of Glasgow to study chemistry, at which he excelled. On graduation in 1947 he was offered a faculty position at the University College of Hull as an Assistant Lecturer in Chemistry at the young age of 20, even before he had started his Ph.D. There began a successful association between Gray and the University of Hull which lasted for more than 40 years and brought fame and good fortune to both. At Hull, George Gray was introduced to liquid crystals by his senior colleague Brynmor Jones,* and Gray's early research into liquid crystals was to

---

* Sir Brynmor Jones (1904–1989), head of the Department of Chemistry and subsequently Vice-Chancellor at the University of Hull, had been an early researcher into liquid crystals. He was responsible for encouraging the organic chemist George Gray to study liquid crystals.

Figure 10.6 George Gray (b.1926).

lead in 1953 to his Ph.D. At that time Hull University was a relatively new institution, not yet empowered to award its own degrees, so qualifications obtained from Hull were external degrees bestowed by the University of London.

As a Scot with strong connections to both Edinburgh and Glasgow, George Gray combines the characteristics of both those great cities (Figure 10.6). The quiet, mild but confident manner of a native of Edinburgh could sometimes give way to the tough determined demeanour of a Glaswegian. In either guise, George Gray attracted great loyalty from his friends and respect from his colleagues. His professional respect was in no small way a consequence of his attention to detail. In particular, it was his detailed researches into the relationships between molecular structure and liquid crystal properties that would provide the initial boost to the twisted nematic display technology.

The original objective for the UK Ministry of Defence liquid crystal research team at RRE Malvern had been formulated in April 1970 when Gray was offered a contract to work on 'Substances Exhibiting Liquid Crystal States at Room Temperature'. A few months later, in December, came the publication of the patent on the twisted nematic liquid crystal display from Schadt and Helfrich. The UK research programme went into overdrive to find suitable materials for the twisted nematic liquid crystal display. There was not long to wait. By the summer of 1972 a class of liquid crystalline materials known as alkyl- and alkoxy-cyanobiphenyls had been discovered by George Gray's small team at Hull.[12] The compounds, or at least some of them, were liquid crystalline at room temperature, stable, colourless, and had the essential property for a twisted nematic device of having positive dielectric anisotropy.

The years of solitary, painstaking research by George Gray had paid off and his experience had enabled a revolutionary set of materials to be synthesized with suitable properties for commercializing the twisted nematic display. However, it was not the end of quest, rather the beginning, because rights to the two components of the display, the device and the material to make it work, were held by different organizations in different countries. It was as though George Stephenson had invented the railway engine, but the invention of the rails was down to someone else, and they weren't talking to each other.

There was another complication: a small liquid crystal display industry was already growing, but it was not using twisted nematic technology, rather the dynamic scattering mode promoted by Heilmeier at RCA. This had been taken up by a number of companies in the USA and Japan, and a few products were becoming commercially available. Digital wrist watches using liquid crystals were being sold, but according to one of the engineers manufacturing them,[13]

> ...most watches became inoperable after a while.

Liquid crystal displays had entered the market, but as small novelty items. They were no challenge to the large-scale industrial display sector. However, behind the scenes there was a battle of technologies raging in the board rooms of companies. For the most part these battles were private, but occasionally glimpses of controversy emerged.

Having left RCA, Heilmeier, as a senior member of the US Defense Administration, was in a strong position to influence technological developments. Surprisingly, he was equivocal about the potential for liquid crystal displays, and in a keynote address to the American Institute of Electronic and Electrical Engineers, delivered in October 1972, he said:[14]

> How many realistic scenarios are there in which we win because we have a flat-panel, matrix-addressed display in the cockpit, while the other guy has a CRT? And how many would we buy if they did exist?

These were not exactly encouraging words for a new liquid crystal technology, which he had helped to launch, struggling to make an impact on the existing display industry.*

Not all agreed with Heilmeier, and from a similar senior position in defence research Cyril Hilsum countered, as robustly as ever, in a lecture:[15]

---

* The early scepticism of Heilmeier was misplaced and liquid crystal flat-screen displays are now everywhere in military and non-military environments. In later life, George Heilmeier has been much-honoured as a pioneer of liquid crystal displays, receiving many awards, including in 1990 the NEC Corporation of Japan Foundation Prize for the discovery and practical applications of liquid crystal displays.

So, my friends, I am here because I believe in flat panel displays, and in research on these displays...

This was a thinly veiled attack by Hilsum on those senior figures in defence research management, Heilmeier included, who were opponents of new display developments.

Japan was not interested in electronic display innovations for military applications, but was looking hard at commercial opportunities in domestic and business markets. A significant development for liquid crystal displays was to come from the Sharp Corporation in Japan. In May 1973 they announced the introduction of the world's first portable liquid crystal calculator using the dynamic scattering mode. Initially it was a great success, but, as Sharp found to its cost, its portability was limited. A lumberjack working in a timber yard in northern Japan found that his Sharp liquid crystal calculator would not function in the low winter temperatures. He complained to Sharp's sales division, the manager of which immediately halted production of the calculator.[16] He demanded from the scientists and engineers a device that would work inside or outside, winter or summer. That device was not long in coming, but it would be a twisted nematic liquid crystal display using Gray's cyanobiphenyls.

While the patent disputes over displays were being fought in the court room, there was conflict developing in the market place concerning the choice of liquid crystal display technology and the materials they used. Those companies that had started up production of displays using dynamic scattering were reluctant to change their product, which would entail revised production methods, but the quality and lifetime of these dynamic scattering mode displays were proving to be a problem. If Japanese loggers and display engineers were losing confidence, then the rest of the public would not be far behind. However, the materials for twisted nematic displays were improving rapidly and some of the early problems, such as patchy images,* had also been solved by the UK research group at Malvern.[17]

In 1974, just one year after the announcement of the cyanobiphenyl liquid crystals, the Malvern group formulated a mixture of four components that met all the material requirements for a twisted nematic

---

* The origin of patches in the display was traced to different regions of left or right twist in the liquid crystal cell. In a simple 90° twist cell, the liquid crystal film could untwist either to the right or to the left, and the areas appeared as patches in the display. This could easily be corrected by adding a small amount of a chiral additive, which would favour one of the twist directions. However, this was not enough. Not only were there two possible directions of twist, it turned out that there was a very small tilt of liquid crystal optic axis at the surfaces. Putting two surfaces together meant that the tilt directions on opposite surfaces might be parallel or anti-parallel. The surface tilt could couple with twist and produce optical inhomogeneities in the switched cell. Once identified it was possible to design cells that avoided areas of *reverse twist* and *reverse tilt*, with much improved optical characteristics.

display.[†] Most importantly for the timber men of Japan, the mixture would operate over a temperature range from –10 °C to +60 °C. The UK chemical company BDH was licensed to produce the new liquid crystal materials, and they were then offered to the liquid crystal display industry as the best for purpose. What followed is summarized in the words of George Gray:[18]

> *On a remarkably short time scale, thanks to the efforts of Research Director Ben Sturgeon at the UK company BDH Ltd, the materials were quickly produced in bulk and sold worldwide. It is interesting to note that around this time (1973) BDH Ltd came under the ownership of E. Merck of Darmstadt, Germany, but the UK company was left to operate its successful liquid crystal business independently, although there was close collaboration between the research teams in Darmstadt and at Hull University and RRE. By the end of the 1970s, mixtures based on cyano-biphenyls and -terphenyls were the industry standard for nematic displays. Their great contribution was that at a time when user confidence in twisted nematic displays was faltering because of life-time problems with materials such as cyano Schiff's bases, they made possible for the first time the manufacture and sale of TN devices that had long lifetimes and were attractive to users, so providing the emerging liquid crystal display industry with the secure base that it needed in order to consolidate before developing into the burgeoning multi-billion dollar display device industry that we have today.*

It is certainly true that by the end of the 1970s BDH was the leading manufacturer of liquid crystals in the world and cyanobiphenyls were their largest selling product. As well as the reward of commercial success, there was recognition for the scientists involved. In 1979 the liquid crystal groups at RRE Malvern,[19] the University of Hull, and BDH were awarded the Queen's Award for Industry and in 1995 George Gray was awarded the prestigious Kyoto International Gold Medal for his contributions to liquid crystal display technology.*

[†] The mixture devised became known as Eutectic 7 (E7). The science behind the mixture was the same that allows the sprinkling of salt to melt ice. Salt solutions freeze at a lower temperature than pure water. This is a manifestation of a general thermodynamic principle that states that mixing compounds lowers the melting point, with respect to the melting points of the pure materials. The group at Malvern realized that the lower temperature limit of a nematic liquid crystal could be reduced by making a mixture. Thermodynamics allows the prediction of the magnitude of the effect, and the mixture with the lowest melting point is known as a eutectic.

* In 2005 Heilmeier was also awarded the Kyoto Prize in Advanced Technology for his efforts to create a foundation for liquid crystal display technology. Following Gray's award in 1995, this was the second Kyoto Prize to be awarded for liquid crystal display technology.

New liquid crystal materials for displays

In May 1973 the privately owned German chemical company Merck purchased BDH (formerly known as *British* Drug Houses) from the pharmaceutical conglomerate Glaxo. The Merck directors who negotiated this deal were totally unaware of the small-scale liquid crystal activities at BDH, but had been attracted by the very successful Analar range of fine chemicals produced by BDH. Such chemicals are the feedstock for the research business of chemistry in general, and research for new drugs in particular. As we shall see, Merck had developed a new interest in liquid crystals in the late 1960s, but this also was on a very small scale. No one could have foreseen the consequences of this link-up between the German and British liquid crystal materials groups of Merck and BDH.

Founded in 1668 as a pharmacy, Merck claims to be the oldest chemical company in the world. Over the centuries it has supplied chemicals for many purposes. Even before the nature of liquid crystals had been firmly established, Otto Lehmann had approached the company to ask them to supply liquid crystals for research. In response, in a letter dated 17 June 1905 they replied:[20]

> I must point out that the requirements to be met regarding the purity of your substances are extremely difficult. Consequently, I can only entrust very experienced people with their synthesis. I must reserve a considerable margin concerning the time for delivery and price. I agree that you may announce the willingness of my company to prepare the compounds as far as possible. Please emphasise that the compounds are not in stock, but will be delivered if possible by special request.

Not long after, E. Merck listed liquid crystals in its catalogue, as an advertisement from 1907 indicates (Figure 10.7).

The involvement of Merck with the supply of liquid crystals had been reactivated in the 1960s by publications from Fergason and subsequently Heilmeier on possible applications. An internal report by Dr. Ludwig Pohl from the research department at Merck at the end of 1967 stated:[21]

> The increasing amount of work with liquid crystals in chemistry and medicine may offer a new market for this substance class. Since so far there is, to my knowledge, still no supplier of such compounds it would be desirable to become the first company offering a range of liquid crystals on the market.

As a consequence, Pohl was sent by the company to the Second International Liquid Crystal Conference held at Kent State University, USA, in 1968. He returned full of enthusiasm for these new materials:[21]

Chemische Fabrik · Darmstadt

empfiehlt alle gangbareren
Präparate zur Demonstration der Eigenschaften

flüssiger und fließender Kristalle,

ferner

sämtliche Chemikalien für analytische Zwecke,

insbesondere:

Garantiert reine Reagenzien.

Volumetrische Lösungen.

Indikatoren und Farbstoffe
für analytische und mikroskopische Zwecke.

Chemikalien und Lösungen
zur Trennung von Mineralgemischen.

Reagens- und Filtrierpapiere.

Mineralien-Sammlungen.

Zu beziehen durch die Großdrogenhandlungen.

Figure 10.7 Merck catalogue 1907.[20] The first five lines read: 'E. Merck, Chemical Factory, Darmstadt, recommends all available [current] preparations for the demonstration of the properties of liquid and flowing crystals.' The advertisement goes on to list other fine products produced by the company.

The area of liquid crystals looks like neither a scientific nor technical red herring.

In 1969 the company launched a new range of liquid crystal products, labelled Licrystal. Initially these were for rather esoteric applications such as cholesteric mixtures for thermal imaging and as solvents for specialized spectroscopic experiments. The market for liquid crystals for displays had not yet developed, but as the first commercial suppliers of liquid crystals, Merck were well-placed to make the most of new opportunities.

The purchase by Merck of BDH meant that there were now two liquid crystal research and production groups within Merck. The former BDH manager Dr Martin Pellatt recalls:[22]

> That was not a great situation... in addition, there was inevitably a competitive situation between two existing liquid crystal groups within the same company.

These two groups soon found themselves developing, manufacturing and selling liquid crystal materials for displays as competitors, but the materials they were selling were different. The cyanobiphenyls described above had revolutionized the prospects for twisted nematic displays, and licences to manufacture were sold by the patent holders (the UK Ministry of Defence and the University of Hull) to BDH (as part of Merck) and to Hoffmann-La Roche, Merck's principal competitor.*

Merck scientists at Darmstadt were also developing their own room temperature liquid crystal materials with properties suitable for use in twisted nematic displays, and these also joined the market alongside the cyanobiphenyls. Clearly there were opportunities for collaboration and joint developments, but relationships between Merck and BDH were complicated. A confidential memorandum by the BDH Research Director Dr Ben Sturgeon to the Merck Board of Directors, dated November 1974, explains the situation:[23]

> Active interest in liquid crystals dates from November 1972 when the MOD (UK Ministry of Defence) offered to extend our 12 year old materials contract with RRE (Royal Radar Establishment)...
>
> A general clause in any British government contract prohibits disclosure of details of the work to foreign nationals, so the full range of the work at BDH... may not be discussed with Merck.

Yet the company was totally owned by Merck!

It was not long before new ranges of liquid crystals suitable for displays were discovered[24] at Merck by a researcher, Dr Rudolf Eidenschink (b.1938), who had joined the company in 1975. These new materials, known as phenylcyclohexanes and bicyclohexanes, shared some of the beneficial characteristics of the cyanobiphenyls, though had different properties. They were also eminently suitable for use in twisted nematic displays. Thus the marketing teams of BDH and Merck, both of the

---

* The major UK chemical company during the 1970s was ICI. According to Hilsum (personal communication) ICI was approached to produce, under licence, liquid crystal materials for displays, but nothing came of the negotiations.

same company, were promoting different ranges of liquid crystals to display manufacturers around the world, apparently as competitors. This bizarre situation lasted for some years, until eventually a joint marketing strategy was agreed.

TECHNICAL BOX 10.2 Liquid crystal materials for displays

The details of chemical design and synthesis of compounds for liquid crystal applications is complicated and specialized. As has been mentioned, the first requirement is for a compound that remains liquid crystalline over the range of operating temperatures of a display device. This had been achieved by Hans Kelker from Hoescht, but his compound was very susceptible to decomposition, with a corresponding deterioration of the display. The origin of this decomposition was a 'weak-link' in the structure, illustrated below, which could be disrupted by light and or moisture.

$$CH_3O - \bigcirc - CH=N - \bigcirc - C_4H_9$$

The molecular structure of MBBA; the weak link is the unit CH=N, which is readily broken.

Gray's great achievement was to design and synthesize a molecule without the weak link. The two 'rings' of the cyano-biphenyl were directly connected, thus avoiding any possibility of decomposition, but the received wisdom was that such a strong molecule would only form crystal phases, not liquid crystal phases. The addition of a substituent, the cyano-group (**CN**), to the central structure promoted the formation of liquid crystal phases, and by introducing a flexible alkyl chain (**C_5H_{11}**) also, the temperature range of the liquid crystal phase was stabilized around room temperature. Making a mixture of similar compounds enabled the temperature span of the liquid crystal phase to be extended to the required operating range of $-10\,°C$ to $+60\,°C$.

$$C_5H_{11} - \bigcirc - \bigcirc - CN$$

The molecular structure of a room-temperature liquid crystal compound, pentyl-cyanobiphenyl is illustrated above.

The discovery of liquid crystal compounds with directly linked benzene rings (known to chemists as phenyl rings) opened up many new possibilities. Given the extreme interest at E. Merck in liquid crystals for applications, they quickly turned their attention to the new possibilities, and Eidenschink discovered two new series of liquid crystals suitable for displays: cyano-phenylcyclohexanes, structure (a) below, and cyano-bicyclohexanes, structure (b) below.

(a)

C_5H_{11} —⬡— CN

(b)

C_5H_{11} —⬡—⬡— CN

(c)

C_5H_{11} —⬡—⬡— CN

Liquid crystal materials prepared by Merck (a, b) and BDH (c) are shown above. In structure (a) a cyclohexane ring is joined to a phenyl ring, while in (b) two cyclohexane rings are joined together.

Together these compounds invented by George Gray and Rudolf Eidenschink dominated the materials used in twisted nematic displays for more than a decade. However materials development never stands still, and new and better compounds were discovered for the twisted nematic and other types of liquid crystal display that became available. The chemical stability of the cyano-biphenyls, phenylcyclohexanes and bicyclohexanes meant that their core structures were ideally suited to liquid crystal displays. Chemical modification of the basic structures was possible to produce a wide variety of new compounds, and these were marketed by BDH in the UK, E. Merck in Germany and increasingly by chemical companies in Japan.

Following the adoption of twisted nematic technology by the liquid crystal display industry, E. Merck expanded and consolidated its position as the major supplier of liquid crystal materials by buying liquid crystal interests from Hoffmann-La Roche, and Brown Boveri et Cie in 1985.

In contrast to the liquid crystal materials suppliers, the situation for device manufactures in Europe and the US was not so rosy, and increasingly the manufacturing centre for liquid crystal displays was Japan. As early as 1973, the Seiko Watch Company had been the first to produce a commercial product using twisted nematic technology. Following the debacle with the Japanese lumberjack, Sharp Corporation abandoned its dynamic scattering mode technology. For its portable calculator, it now switched to the twisted nematic display, using the cyanobiphenyls invented in the UK by Gray. US display companies such as Fairchild Corporation and Motorola and many others also changed to the twisted nematic technology, but they were no match commercially for the liquid crystal displays flooding the market from Japan. By 1981 there were no longer any major US companies manufacturing liquid crystal displays.

Apart from Japan and the US, the other great economic power was that of the Soviet Union. Its scientists had made important contributions to the liquid crystal story. It was, after all, Frederiks who had done the first experiments on the switching of liquid crystal films, which would eventually be exploited in displays. There had even been research activity in Tsvetkov's liquid crystal laboratory in Leningrad during the war years, but had there been research into liquid crystal displays?

The key meeting in the development of liquid crystal displays was the third International Liquid Crystal Conference, held in West Berlin in August 1970. But there had been few participants from the Soviet bloc. The political sensitivity of West Berlin put it 'out of bounds' for all but a few liquid crystal scientists from behind the Iron Curtain. Research on liquid crystal displays began in Russia and its satellite countries during the early 1970s. It seems likely that a number of key discoveries of new materials and electro-optic effects were first made in the USSR, but the absence of a recognized patenting system and late or inadequate publication meant that they were ignored by the West.[25] However, in 1976 Horst Sackmann organized in Halle the first of a series of conferences 'Liquid Crystal Conferences of Socialist Countries'. This provided a forum for exchange of information and also limited opportunities for invited western scientists to learn of liquid crystal research from behind the Iron Curtain.

New centres for liquid crystal research began to be established in Soviet countries, the most important of which were in the Institute of Crystallography of the Russian Academy of Science in Moscow and in the Research Institute for Organic Intermediates and Dyes (NIOPIK) also in Moscow. It was in the latter that a thirty five year old organic chemist, Victor Titov, was appointed head of the liquid crystal group in 1970. As a talented organic chemist, a jovial colleague, and a member of the Communist Party, Titov quickly became established as a leader in Soviet liquid crystal science. In 1981 he was appointed Director General of the Institute, and prestigious state awards followed: The State Prize for Science and Technology in 1983, and The Council of Ministers Prize of the USSR in 1986.

As a respected scientist and reliable party member, Victor Titov was able to travel outside the USSR, and became an international figure. However, in the often troubled times of the cold war, westerners were cautious in their dealings with him. In 1984 Titov travelled to the UK to attend the 10th International Liquid Crystal Conference held in York. At this time much of liquid crystal research in the UK was funded by the Ministry of Defence, and from time to time the UK Security Services would take an interest in related liquid crystal activities. A 'secret' meeting between Titov and a mysterious well-dressed lady contact at the York Conference Reception held in the National Railway Museum might have caused consternation, but it betrayed a level of naivety that he thought such a meeting could go undetected. Much later it was Titov's naivety that would have tragic results, not as a result of the Cold War, but a perverse consequence of perestroika.

The process of political and economic reform in the Soviet Union, known as perestroika or reconstruction, was first proposed in 1979 by Prime Minister Aleksei Kosygin (1904–1980). However this attempt to modernize the economy was thwarted by President Leonid Brezhnev. Kosygin was dismissed in September 1980, and died two months later. It was only after Mikhail Gorbachev (b.1931) took over in 1985 that real progress was made, leading in due course to the break-up of the Soviet Union.

As a committed Communist Party member, Victor Titov was not always in tune with the reforms, but his professional success continued. In May 1991 he was appointed as Director of the Research Institute for Complex Problems of Printing (Institute of Graphic Art), while remaining close to his liquid crystal interests and former colleagues in NIOPIK. In August of 1991, events in Russia began to move fast. Gorbachev had prepared a protocol to give more autonomy to the member states of the Soviet Union, but there was considerable opposition from his Ministers and the Russian Parliament. In the event, Gorbachev was deposed by anti-reformers in a 'palace coup', but this had the effect of mobilising the much stronger pro-reformers led by Boris Yeltsin. After a brief stand-off with the army, with tanks again in Red Square, Yeltsin triumphed and the Soviet Union collapsed.

Following this new Russian Revolution, the consequences of the changes in the political and economic structures were catastrophic, and suddenly there was virtually no financial support for State Institutions. 'Business' thrived, though much of it was run by organized crime, and fortunes were there to be made and lost. Unfortunately Victor Titov, now President of the Liquid Crystal Society of the Commonwealth of former Socialist States, got caught up in this maelstrom, and paid a heavy price.*[26]

As Director of a State Research Institute, Victor Titov (see Figure 10.8) was faced with huge financial problems to keep the Institute running. But he was resourceful, and leased the top floor of the Institute to a private company based in the Netherlands which imported food from Western Europe. In exchange for occupying part of Titov's Institute and parking its trucks in the yard, the private company paid for the maintenance and utilities of the building and even provided meals for the Institute's staff. The import company's owner, a young entrepreneurial

---

* The following is taken from an account by one of his closest colleagues and friends Lev Blinov, who had worked with Titov at NIOPIK. Professor Blinov was a pioneering liquid crystal physicist in the NIOPIK research institute and made many significant advances. He continues to research in Russia and Italy.

Figure 10.8 Victor Titov (1935–1996).

Russian, invited Titov to become the company's President and Titov accepted. For a few years the food importing business and the Institute of Printing prospered. Finally, the latter was privatized, yielding Titov a substantial stake in the new graphic art company. There were now two prosperous businesses occupying a substantial property in an attractive part of Moscow, with Titov president of one and managing director and major shareholder of the other. At that time, the real estate market in Moscow was booming and Titov received a very attractive offer to sell the Institute. This offer was of the type that 'could not be refused'. Unwisely, he did.

In September 1996, the father of the young Russian owner of the food import business, who was also a director of the company, was found dead at his dacha with a knife in his forehead, an apparently motiveless murder. Titov was not unnaturally disconcerted by this, but did not connect the killing with any company business. The following month, on October 24th and 25th, there was a meeting of the Liquid Crystal Society of the Commonwealth at Titov's Institute, at which he was re-elected as President of the Society. Following the traditional round of toasts, Victor Titov left for home accompanied by one of the foreign guests at the meeting. Lev Blinov takes up the story:

> *The car, driven by the Institute's driver, stopped at Victor's house. Sonya Torgova, Scientific Secretary to the Society and Alfredo Strigazzi, Professor of Physics from the Politecnico of Turin, were also in the car to be taken to their residences. A not very sober Alfredo wanted to go up with Victor to his apartment and present Victor's wife Valentina with a gift from Italy, but Victor suggested that he should do this the following day and thereby saved Alfredo's life. In a few seconds Victor closed the external door behind him to meet the killer standing in the elevator hall. Three bullets were*

found in Victor's body. Victor was the only person to have seen the killer because after the first bullet in his back he turned round to receive the second one in his face.

Murders of this type were daily occurrences in Moscow at the time, and it was typical of a 'contract killing'. Probably the killer himself shortly met the same fate to destroy the evidence. Two years later the Russian owner of the food import company was himself murdered in Eindhoven, Holland. Neither his nor Titov's murder was ever solved.

Today a huge new building stands on the site of Titov's Research Institute, but the liquid crystal research group of NIOPIK has vanished. The Liquid Crystal Society of the Commonwealth of former Soviet States founded by Titov continues. He is remembered as a successful scientist and businessman, but his fatal weakness was an almost child-like naivety. He was often influenced by unscrupulous operators, and this led him into situations that wiser counsel would have warned against. A warm-hearted friend and colleague, Titov was a good pianist and wrote poetry in his spare time. He was no match for the ruthless Russian mafia, which ended his life. He did not live to see the fruition of the research on liquid crystals that would flood the world with flat-panel displays.

11

The sun rises in the East

It is a suitable kind of technology for the Japanese mentality. It was a manufacturing puzzle. I am not sure American management would wait a quarter of a century to solve it.

Shinichi Hirano, Japanese engineer with
IBM, *New York Times*, 16 December 1990

In 1971 the liquid crystal television was a pipe-dream of scientists working in the field, yet a decade later small-screen portable TVs began to appear. Not only was the realization of the pipe-dream a triumph, it was also a success that depended on contributions from many working in different institutions, universities, and companies around the world. No longer was there a battle between individuals over whose was the idea; the implementation of the idea was down to companies and their research and development groups working together. Of course there was still plenty of scope for commercial competition, but there were mutual benefits to be had from collaborative agreements and exchange of information. The final part of the liquid crystal display story is a tale of companies rather than individuals.

The companies involved were mostly Japanese, and their eventual success is a tribute to their focused research and long-sighted business plans. Research into liquid crystal displays in Japan dates from the public launch of Heilmeier's dynamic scattering display. In 1968, scientists from Sharp Japan visited the RCA laboratories, and were impressed by what they saw.[1] Most importantly, key figures in the Japanese electronics industry believed liquid crystal devices had the potential for large-area high-resolution displays. However, they probably did not anticipate the long period of time necessary to develop the successful TV-on-the-wall technology. The big companies that would take up the challenge were Seiko, Sharp, Hitachi, Matsushita and Toshiba. However, at the beginning, there were many technical problems still to be resolved. Eventual success was far from guaranteed.

Bigger and better displays

With the introduction of liquid crystal watches, calculators, and other simple displays the representation of numbers and characters using a seven-segment image became recognized as the futuristic script, or at least a type-face characteristic of the late twentieth century. This display was designed so that the individual segments of the display could be separately switched. The display in Figure 11.1 has nine connections: seven for the segments, another for the back (earth) electrode, and a spare. The liquid crystal is sandwiched between the front and back glass plates, which carry the electrode pattern shown. With a row or two of such individual display elements, the number of separate connections will quickly increase. Such a simplified scheme is not able to display more complicated images such as Chinese or Japanese characters, so there was a strong demand for a more flexible format that would allow the display of more complex patterns and eventually high-resolution graphic images.

To understand the problem and the solution, it is necessary to delve into the technical aspect of displays known as 'addressing'. The basic element of any liquid crystal display is a shutter, or pixel, which when activated can change from black to white. To create a character or picture requires a number of these pixels to be juxtaposed and individually switched to produce the desired image. Clearly, the more pixels and the smaller they can be made the greater the resolution of the final image.

The techniques of matrix addressing and multiplexing enabled the seven-segment display format to be replaced by a matrix of pixels, and so a device could now display several lines of text or numbers. To do better than this required the imagination of liquid crystal scientists from different companies working in different countries.

By the beginning of the 1980s, the twisted nematic configuration was the design of choice for liquid crystal displays. In essence this

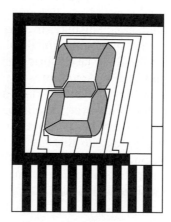

Figure 11.1 The familiar seven-segment display.

display works by application of a voltage to a film of liquid crystal which has been twisted through a quarter turn, 90°. The voltage untwists the liquid crystal and changes its optical properties, hence the switch from light to dark, or vice versa. However, the response of the liquid crystal to the voltage is only gradual and depends on the material properties of the liquid crystal mixture used. It is this behaviour that limits the complexity of images that can be displayed, as discussed in Technical box 11.1.

TECHNICAL BOX 11.1 Matrix addressing and multiplexing

The problem with displays made up of separately switched (addressed) pixels (picture elements) is that for complex images, the number of electrical connections becomes very large. One way of dealing with this is to use a technique known as **matrix addressing**. This is where the elements of the image (pixels), now small square dots, are arranged as a matrix, each row and column of which has a separate electrical connection. Since the display is a sandwich with electrical connections on both sides of the liquid crystal film, the row connections can be on one side of the film, and the column connections on the other. If, say, half the required voltage to activate a particular pixel is applied to a row, but then the other half applied to a single column, then only that pixel identified by the specific row and column numbers will receive the full voltage and be switched on. This is the principle of matrix addressing. If the number of rows is n and the number of columns is m, then the total number of connections is $(n + m)$ instead of $n \times m$, which would be required for individual addressing of each pixel.

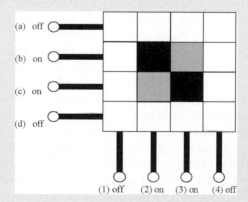

But it is not quite as simple as that, since an image will require a number of individual pixels in the matrix to be activated. For example, in the figure above pixels 2b and 3c are to be activated by switching rows b and c and columns 2 and 3, but unwanted pixels 3b and 2c are also activated. To deal with this, another concept is introduced: **multiplexing** or time-sharing. If the pixels required for our image are activated individually and sequentially, and the activated pixels stay illuminated long enough after activation for the whole matrix to be addressed, then the eye will register a true representation of the image. This is what is done, with complex sequences of electrical pulses applied to the rows and columns of a matrix display. It is similar to the way a cathode ray television picture is built up by scanning in lines, but for a liquid crystal display both rows and columns are scanned.

Matrix addressing and multiplexing go some way towards solving the problem of complex images, but there are still limitations imposed by the liquid crystal itself. When a voltage is applied to a liquid crystal display element, the light transmission changes progressively, the more voltage the greater the change. The brightness of a display element will, within limits, depend on how much voltage is applied. The optical quality of the display is measured in terms of a **contrast ratio**, which is the ratio of the brightness of the on-state to that of the off-state. In the process of multiplexing, pixels may have a residual voltage even in the off-state, and the ratio of on to off voltages for a particular contrast ratio depends on the choice of liquid crystal material. More specifically, it depends on the elastic properties of the liquid crystal mixture used. It is possible to calculate the maximum number of rows of pixels that can be addressed for a desired contrast ratio. For typical liquid crystal materials in a twisted nematic cell the number of rows is about 20 for a visually acceptable image. This is not sufficient for a complex image, but better than the seven-segment display, and sufficient to display a few lines of text or a simple graphic.

The more you twist a piece of elastic, the more it twangs on release: its response sharpens. Rather remarkably, this works for liquid crystals as well. In 1978 the Americans Cole and Aftergut, working at the General Electric Company Schenectady, in upstate New York,[2] discovered that increasing the twist of a nematic film from a quarter turn increased the speed of response of the film. This idea was picked up by Peter Raynes, at RSRE Malvern, UK.* Raynes and his colleague Colin Waters constructed a new display based on this principle. They found that not only did it switch faster, it also responded to the applied voltage much better, avoiding the gradual change seen with the twisted nematic device. For a multiplexed matrix display this meant that more rows could be addressed, up to 100, for a given level of contrast, and as a result more complex characters and images could be shown. Raynes and Waters called this a super-twisted nematic display. A patent was quickly filed in June 1982, followed by publication in the technical literature.[3,4]

The device described in the patent was a display without polarizers, but one in which the nematic liquid crystal film had been coloured by adding a suitable soluble dye. On switching, the film changed colour

* This was formerly RRE, but in 1976 became the Royal Signals and Radar Establishment (RSRE) after merging with other Ministry of Defence research laboratories.

rapidly and a new display mode had been discovered. The authors of the patent also noted that the device could work without a dye if polarizers were used, as in the normal twisted nematic device. The super-twist technology was offered by RSRE to a number of UK electronics companies, but the tough demands made by the technology, particularly the narrow limits on cell spacing, made it hard for UK industry to manufacture super-twisted nematic displays competitively.

The Swiss electronics company Brown Boveri et Cie was also working on possible new configurations for liquid crystal displays, and had also come up with a super-twisted variant of the twisted nematic display. A year after the Waters-Raynes UK patent had been filed, the team from BBC, headed by scientist Terry Scheffer, applied for a patent for a super-twisted birefringence effect liquid crystal display.[5] The patent was similar in many respects to the earlier UK patent, but it described the device operating without added dye, but with polarizers. However, the resulting image was still coloured, switching between an off-state (blank screen) of blue and on-state switched characters and images of yellow. The source of the colour was the optical interference effect generated in the liquid crystal film. Such a display could be, and indeed was, used for images where colour was not important, diagrams, and of course text, but it was not going to be suitable for TV or computer images. This did not hinder the development and production of super-twisted or STN displays as they became labelled, and various Japanese companies, most importantly Sharp Japan, started selling displays for devices such as mobile phones and portable computers using these displays. This significant development generated substantial royalties for both BBC and RSRE (i.e. the UK Ministry of Defence).

Coloured text or graphic STN displays having yellow images on a blue background or bright blue images on a light green background were attractive to look at, but would not do for TV screens. To eliminate the colour required 'sleight of hand' to remove the optical interference that came from the strongly twisted structure, and amazingly the liquid crystal scientists at Sharp, Japanese manufacturers of STN displays, already knew how to do it.

A few years before the STN displays became commercially available, Sharp scientists had been experimenting to remove interference colours from the simple 90° twisted nematic displays. They had found that if two as far as possible identical displays were placed back-to-back, so that the twists were compensated, that is they cancelled out, optical interference colours were eliminated.[6] More importantly, if only one of the back-to-back displays was activated by a voltage, then the images created were also free of any discolouration from optical interference and it was possible to produce black images on a true white background. The extension to STN displays was very successful, and the result was the perfect visual display unit for a portable word processor. With the addition of red–blue–green filters to the display it was possible to make a full-colour liquid crystal screen capable of displaying complex coloured graphics and images.

The new compensated or 'double' STN displays were very successful as portable word processors, but they were complicated to make and expensive. However, the principle of optical compensation was actually very simple. A double, back-to-back display was unnecessary. All that was needed was to compensate for the optical properties of the supertwisted nematic film, and this could be achieved with a stretched film of polymer. Not any old stretched film of course, but a carefully engineered film having exactly the right characteristics of thickness, refractive indices, and birefringence.[7] This was still an expensive component, but nothing like as complicated as making two identical liquid crystal panels for each display.

By the late 1980s the world had portable full-colour word processors incorporating STN liquid crystal displays, and the demand was great. However, these were not going to be suitable for the elusive liquid crystal television. Even if the resolution of the multiplexed STN matrix displays was good enough, the response times were far too slow to permit moving images to be displayed. This was partly due to the liquid crystal materials being used, which were progressively improving, but more important was an inherent inadequacy of the matrix addressing scheme used. Another breakthough was needed.

A prototype portable television

As the story of liquid crystals approaches its climax, the plot gets much more complicated. More characters are swept into the narrative and the level of technical understanding required may strain even the most committed reader.* The focus of the story shifts to Japan, which by the mid-1980s had established dominance in the display manufacturing business. Despite this, the final breakthrough that allowed wall-mounted flat-screen liquid crystal televisions still owes much to post-war developments in electronics in Europe and the USA.

The essential problem still to be solved in the development of a liquid crystal display for high resolution moving images was that of addressing. How could the electronic information be passed effectively to the display in as short a time as possible? The concepts of matrix

---

* An excellent technical account of the history of liquid crystal displays by Hiro Kawamoto was published in 2002 in the Proceedings of the Institution of Electrical and Electronic Engineers.[1] The author is an electrical engineer who worked with Sharp Japan, one of the key liquid crystal companies involved in the development of the liquid crystal television.

addressing and multiplexing helped, but a limit had been reached, partly due to the technique and partly due to the less-than-perfect liquid crystal materials that were available.

After World War II, there was a revolution in electronic devices prompted by the invention of the transistor. Innovation followed innovation, and techniques for building complex circuits on wafers of silicon or other semiconducting materials became established. These circuits could be activated and controlled much more easily than jumbles of individual resistors, transistors, and capacitors soldered together in complex arrays, and they were much more reliable. The possibility of putting together an array of transistors with a flat-panel display to make a TV had been proposed as early as 1971 by an RCA scientist in the context of electroluminescent devices.[8] These transistors, known as thin film transistors (TFTs), would be deposited as a matrix directly onto one of the inner surfaces of the display cell. The individual TFTs act as separate switches for the voltages applied to the pixels of the display. They can be switched on and off by electrical pulses applied to the elements of the matrix, and the problem of unwanted pixels being activated disappears. Such a technique could readily be applied to liquid crystal displays.

This idea was put into practice a few years later by a team of scientists at Westinghouse Company in the USA led by Peter Brody.[9] In 1973 they described a prototype TFT liquid crystal display 15 cm × 15 cm consisting of 14,000 pixels. A few years later a black and white video liquid crystal display was revealed to the public.[10] This development was labelled as an *active matrix display*, and the first such displays used cadmium selenide semiconductors rather than silicon. It was, however, a much smaller active matrix liquid crystal video display that captured the imagination of the public.

At the beginning of the 1980s, the interest in developing and manufacturing complex liquid crystal displays was fast catching on in Japanese electronic companies. Seiko produced a wristwatch-sized video display using active matrix addressing. These displays, although much smaller than the original display of Brody and colleagues, were manufactured using crystalline silicon transistor arrays, thus avoiding the problems associated with cadmium selenide. They attracted considerable public attention and featured in dramatic fashion in the 1983 James Bond film 'Octopussy'.

We cannot leave Dr Brody and his Westinghouse team without a postscript from *The New York Times* of January 1991. A few weeks earlier *The New York Times* had published an item about an award by a Japanese company to 'George Heilmeier, the inventor of liquid crystal displays', under the headline 'Invented in the US, Spurned in the US, A Technology flourishes in Japan'. This article prompted correspondence, especially from Peter Brody, who reminded readers that:

These displays [Heilmeier's dynamic scattering displays], how-ever, are very distant relatives of the flat-panel televisions being developed in Japan today....

The technology that turns otherwise sluggish and low performance liquid crystal layers into color television panels is the active matrix technology that was born and developed, not at the RCA laboratories, but in my department at Westinghouse Laboratories.

But even this is not quite the whole story.

As always the devil is in the detail. The important detail that Brody skates over in his claim to fame is the material used for the active matrix array of his TFT liquid crystal display. The first active matrix liquid crystal displays, including Brody's, used cadmium selenide as the semiconductor. This is a highly toxic material and devices manufactured from cadmium selenide tended to suffer from electronic drift, which rendered them unreliable. There were alternatives, indeed the harmless silicon had been the semiconductor of choice for integrated circuits for decades, but the crystalline silicon used by Seiko for their wristwatch TV could only be used for small area displays (approximately 2 cm square).

Another alternative was non-crystalline or amorphous silicon. In 1976, Cyril Hilsum, the liquid crystal group leader at RSRE in the UK, had encouraged Walter Spear and Peter LeComber at the University of Dundee to look into the use of amorphous silicon for TFT active matrix liquid crystal displays. This research was very successful and the results were published in 1981.[11] In 1982 Spear took a trip to Japan, where he gave a series of lectures on the use of amorphous silicon (a-silicon) technology, principally for solar cells. However, he did mention the work with LeComber and RSRE on applications to displays. The liquid crystal scientists at Sharp and other Japanese companies were quick to realize the significance of a-silicon for liquid crystal displays. There began a race in Japan to introduce liquid crystal TVs to the public using TFT technology.

The significance of this race for the world at large can be appreciated by recording some of the competitors: Canon, Fujitsu, Toshiba, Hitachi, Matsushita, Sharp; all giants in the electronics business. They were certainly taking the possibilities offered by liquid crystal televisions seriously, more seriously perhaps than the buying public, who were only mildly interested in the small displays, typically 8 cm × 8 cm, available in 1987.

There were still commercial conflicts with other display technologies; the traditional cathode-ray tube TV continued to develop with bigger and bigger screens and higher definition, so the visual standards were constantly increasing. Also within those companies involved with cathode-ray tube production there were conflicts of commercial interest between the traditional technology, which was making money, and the new liquid crystal technology, which was costing the company huge amounts in research and development. It had been just these considerations that resulted in RCA pulling out of liquid crystal display development in 1972.[12] However, Japanese liquid crystal scientists were more persistent and persuaded their managers to keep faith with flat-screen liquid crystal television.

Flat-screen liquid crystal television arrives

Despite the commercial and technical issues, the liquid crystal television had captured the imagination of Japanese business, and they were determined to succeed. One company that was particularly sensitive to the competition between cathode-ray tube and liquid crystal televisions was Hitachi, at the time one of the largest producers of cathode-ray tubes in the world. Hitachi and some other companies solved the problem of internal competition between technologies by transferring cathode-ray tube production to China. However, these companies realized that the days of the cathode-ray tube were numbered.

On a visit to Hitachi Research Laboratories in the 1980s by one of the authors, the question was posed: 'What would be the weight of a television tube having screen dimensions of one metre by one metre?' The cathode-ray tube is a glass vessel, inside which is a vacuum. The larger the tube the larger its external area, which is subject to forces of atmospheric pressure. The area of the tube subject to atmospheric pressure, and hence the force, increases as the square of the linear dimensions of the tube. In order to have sufficiently thick glass to withstand the pressure, the estimated weight of a tube with a display area of 1 m square would be approaching 1 metric tonne. Not a household item that would be in great demand!

In 1987 the liquid crystal television display was a long way from an acceptable size for home viewing, and once again there was a danger that the technology would become stranded. It was saved by the imagination of one man, not a research scientist but a manager, Mr Isamu Washizuka, General Manager of the Liquid Crystal Division of Sharp Japan.[13] In August 1987, he demanded a 14-in (35-cm) diagonal display that would be suitable for a domestic TV, and incidentally for a computer. This was a real challenge for the engineers, who had set up a production line for 3 × 3 inch display panels. As it happened, the panels were processed nine at a time and so it was in principle possible to deal with a larger area display. However, the production line was fully committed, and only by occasionally sneaking in their larger panels could the research engineers try out the larger design displays. There were many problems, but eventually by February 1988 the researchers had processed some of the larger displays, and,[14]

> *To the surprise of all the people involved, the first 14 in panels looked okay when examined by the naked eye. They secured four good panels.*

There were other electronic problems to be resolved, but Sharp were well-placed to deal with these and eventually Manager Washizuka showed the liquid crystal display to the President of Sharp.[13]

> *The panel was too good to be true. His [the President's] first reaction was that Washizuka was fooling him with a phoney set-up: it was only some time later that he realized to his surprise it was the real thing.*

At last the technology had produced a display suitable for the elusive wall-mounted TV.

Despite the focus of this chapter on events in Japan, it would be unreasonable to ignore the considerable activity in liquid crystal displays that was going on in Europe and the USA during the 1980s. Amongst the electronics companies in the West there was a realization that active matrix-addressed liquid crystal displays could be the visual interface of the future. These companies tried to establish manufacturing facilities and open up profitable markets,[15] but for the most part these attempts were unsuccessful. The novelty of the new liquid crystal displays did not catch on with the public, and they were expensive. The quality of the images was improving all the time, but in terms of value for money liquid crystal displays did not represent a good buy. This conservatism for new devices in Western markets was in sharp contrast to the enthusiasm for new electronic gadgets shown in south-east Asia. However, in the West, one consumer always on the lookout for new technology, where cost was not an issue, was the military. They were prepared to pay almost any price for the highest specification displays, and a number of companies both in Europe and the USA focused on this apparently lucrative sector. Not for the first time, however, the attractiveness of military markets for displays was to prove to be false.

In 1987 the US company Optical Imaging Systems (OIS) was contracted by the US Air Force to develop 6 × 6 in liquid crystal displays for military portable computers. While this technology was successful, OIS, like others, could not make a profit from military applications for their displays. However, there was a strange development, perhaps a death wish in the company management, who had entered into a joint product development agreement with Samsung Electronic Devices of South Korea.[16] Samsung employees were trained in the new technology by OIS. In due course the Korean company began to manufacture active matrix liquid crystal displays for non-military applications. Back in the USA, OIS was developing its military and avionics markets, but it still could not make a profit. The company ceased to exist in 1998, while

Samsung is now one of the largest manufacturers of high-quality liquid crystal displays in the world.

The story is similar for many other companies in Europe and the USA that tried their hand at exploiting the inventiveness of their compatriot scientists. A few Western companies, notably Philips of Eindhoven in The Netherlands, remained in the liquid crystal display business until relatively recently. Philips had entered liquid crystal displays through a joint venture with the French company Thomson CSF, one of the pioneering companies in liquid crystal technology research. Eventually, in 1999, Philips formed a joint company, LG Philips, with LG Electronics of Seoul, South Korea, which went on to become the world's largest manufacturer of liquid crystal displays in 2003.[17]

As liquid crystal display technology developed in the 1980s, the basic ingredients for the displays, nematic liquid crystals, were also being constantly improved. Each new stage in the technology required materials with different properties, and each new demand from the technology resulted in new liquid crystal compounds and new mixture formulations. Whenever possible, Merck acquired the rights to new and interesting liquid crystals, but also realized that the real action in liquid crystal displays was now in Japan and neighbouring countries. Even as early as 1980, Merck established a liquid crystal applications laboratory in Atsugi, a town just outside Tokyo, the first of a number sites in the Far East. Partnerships and collaborations were now important to the development of the technology, and one more technical innovation in displays is worth a mention.

In the 1990s, liquid crystal displays became familiar in portable devices such as computer games, hand-held televisions, mobile phones, and even in a new generation of satellite navigation systems. These were devices to be viewed by one person, but they were personal in another sense. The viewers had to position themselves in such a way that the contrast and colour representation of the image was optimized. This is the problem known as 'angle-of-view' and it is inherent in the optical physics behind liquid crystal displays. It is not a problem with fluorescent screens in which the picture elements send out light uniformly in all directions. Anything that could be done to improve the angle of view would enhance the quality of liquid crystal screens and improve their competitiveness. This would be an important development, since other flat-screen technologies, such as plasma displays, were coming to the market. These used small gas discharge elements as pixels and gave bright, high-contrast images. They were emissive devices generating light, unlike the back-lit liquid crystal screens. As a result plasma displays could be viewed from any angle without any distortion of the image.

There are a number of optical tricks that can be applied to improve the angle of view of liquid crystal displays. One is to use a variant of the

TECHNICAL BOX 11.2 Improving the view-ability of liquid crystal displays

The optical basis for the second generation of liquid crystal displays, after dynamic scattering, was a twisted configuration of the liquid crystal between the confining plates of the display cell. Activation by a voltage applied across the plates caused the liquid crystal twist to unwind and change the optical properties so that a pixel could be activated. Imagine viewing a twisted structure, left or right, down the axis of the twist and begin to untwist it. Moving your head from side to side the partially twisted structure will look different as you move to the left or to the right. The optical pathways are different, and this translates in a liquid crystal cell as differences in the image when viewed from the left or the right. This is the problem of the angle of view and is so extreme that in early liquid crystal displays the contrast could completely reverse, black becoming white, or the colours could change into their complementary colours.

An imaginative idea which partially solved this problem was proposed by Professor G Baur and colleagues at a German workshop on liquid crystals held in Freiburg in 1993. Instead of applying the activating voltage across the film, causing the molecules of the liquid crystal film to lift from the surface plates, the orientation of the molecules on just one surface of the film could be changed by applying a voltage between two sets of electrodes deposited on one plate only. This was to become known as the *in-plane switching mode*.

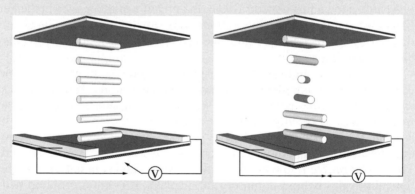

IPS off state – no light transmitted   IPS on state – light transmitted

The optical characteristics of the film still changed from untwisted to twisted, but the molecules of the liquid crystal remained in the plane of the containing plates and the electrical signals were connected to just one of the plates. The upside of this arrangement was that the angle of view of the cells was very much improved; the downside was that the active plate was more complicated to manufacture and the electrical connections all on one plate reduced the light transmittance through the display.

stretched film compensation layer that removed the colouration in STN displays. Clever design and manufacture of polymer films can provide an optical compensation that goes some way to correcting the limited angle of view. Another method was to try to remove the source of the problem by changing the configuration of the liquid crystal within the display. One source of the angle of view problem is that on switching the twisted nematic film, the untwisting of the liquid crystal introduces an optical asymmetry into the film. This shows itself as different amounts of optical transmission when the film is viewed from the left or right, or from the top or bottom. By changing the way in which the film was untwisted by the applied voltage, it was possible to reduce the optical asymmetry and hence improve the appearance of the display.

This solution to the angle-of-view problem was proposed by Professor Gerhardt Baur from the Frauenhofer Institute of Solid State Physics in Freiburg and patented.[18] In 1993 Merck acquired the patent, and in collaboration with Hitachi Japan started to manufacture in-plane switching mode displays. These had significantly better angle-of-view characteristics than their rivals, and were soon in the marketplace.

The success of this innovation illustrates generally that new developments are always possible, and in particular it stimulated other research groups to experiment with new liquid crystal configurations. In fact it was not long before the need for a twisted structure was overcome and so the angle-of-view problem virtually disappeared. Most modern liquid crystal televisions use an old concept of variable birefringence, but in a vertically aligned configuration. That is not to say that all the problems have been solved and the outstanding problem now is how to portray fast-moving images on liquid crystal screens. This is partly a problem of electronics and addressing, but more importantly a problem of finding suitable liquid crystal materials.

The triumph of liquid crystal display technology is on view everywhere for all to see. Modern liquid crystal display televisions (see, for example, Figure 11.2) have now largely replaced the old cathode-ray tube technology, but new technologies are emerging and may in time replace the liquid crystal display. However, the development of the liquid crystal display is far from complete, and new innovations are being researched and implemented all the time.

In this history of liquid crystals, especially with regard to the development of displays, we have adopted a particular perspective. For us, it was the invention of the twisted nematic display that opened the door to modern display technology. Indeed the twisted nematic display still has a significant market share of small displays, but it is no longer the configuration of choice for the largest and most sophisticated liquid crystal displays.

Figure 11.2 A modern flat-screen liquid crystal display television set. These are currently manufactured up to a size in excess of 1 m diagonal. For scale, an image of one of the first twisted nematic displays is included. Main image courtesy of Sharp; first twisted nematic display image courtesy of Martin Schadt. (See colour plate 12.)

TECHNICAL BOX 11.3 Hang-on-the-wall large-area flat-screen liquid crystal televisions

There are many new configurations that are being exploited for different applications, but in coming to the end of the story of liquid crystal displays, we will outline one of the configurations used for modern liquid crystal display televisions.

The liquid crystal remains at the heart of the display as the optically active element. That is to say, changes in the optical characteristics of a thin aligned film of liquid crystal are used to provide an optical switch. Polarizers are still needed to make the optical switch work. The essence of the switch is still to change the orientation of the liquid crystal, but this does not have to use a twisted configuration in the film. One of the problems of using twisted films was optical non-uniformity when viewed from different angles. Remarkably, the modern television screen illustrated in Figure 11.2 uses the optical effect first reported by Charles Mauguin in 1911 and then exploited by Frederiks in his fundamental studies of liquid crystals, published in 1929 and 1933. The twist is no longer used, and the configuration is known as vertical alignment of nematic (VAN) liquid crystals. The switching process is illustrated below.

In a VAN cell the initial alignment is for the liquid crystal molecules to be vertically aligned at the electrode surfaces, top and bottom as in the figures opposite. When a voltage is applied across the cell, above a certain threshold voltage, the liquid crystal molecules and hence the director and optic axis, tilt out of the field direction. This requires the molecules of the liquid crystal to have a structure such that the dielectric anisotropy is negative.

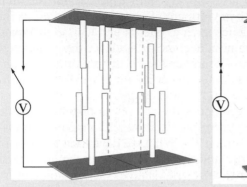 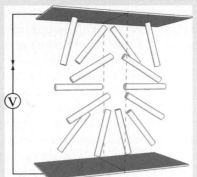

VAN cell off, no light transmitted   VAN cell on, light transmitted

The VAN cell illustrated above is divided into two and treated in such a way that one half of the cell tilts to the right, while the other half tilts to the left. This is one way of dealing with the problem of the angle of view. By having half the pixels tilting one way and the other half tilting the other way, the symmetry of the image is maintained: it looks exactly the same from the left as from the right. The images above are only schematic, but illustrate that the response of the liquid crystal to the applied voltage is greatest in the centre of the cell, away from the aligning surfaces of the electrodes.

One final, and ongoing, issue concerning liquid crystal nematic displays of all types is the speed of response. This is influenced by many things, including the cell configuration chosen for the display. However, above all, the most important factor that determines the speed with which an image can be represented, and changed, is the liquid crystal material. Many properties of the liquid crystal have to be optimized to produce a visually acceptable display, but the property that is most difficult to optimize is the switching time, which in turn depends on the rotational viscosity. Indeed it is possible that modern materials have reached the limit of what is possible for response times in nematic displays. In modern TV displays the problem of image blur when displaying moving pictures is dealt with electronically by adjusting the pixels as they change. It is an optical trick, but the eye sees what it wants to see.

While the electronics companies of Europe sold their expertise in liquid crystal displays to companies in the Far East, the European chemical companies kept a firm grip on their materials. One company in particular, Merck, had been involved with liquid crystals since their discovery by Reinitzer and Lehmann, and they remain at the forefront of liquid crystal materials development. As we have noted in earlier chapters, Merck benefitted from their acquisition in 1972 of BDH Chemicals, the source of the new generation of stable liquid crystal materials for displays. They also quickly built a successful research team of their own at Darmstadt in Germany. The search for materials for displays goes on, with companies such as Merck and Chisso in Japan, but the details of this research now become difficult for the non-expert reader to comprehend. What can be understood is the importance of ongoing research into new liquid crystal materials for other display applications, which will be examined in the next chapter.

In some sense we have now reached the end of our history. Our title notwithstanding, no modern philosopher's stone can magically transform a bar of soap into a flat-screen TV. What we have tried to show in these pages is that the soap bar and the flat display screen are linked through the science of the fourth terrestrial state of matter: liquid crystals. In so doing, we have also tried to communicate a view of the scientific process. We want to dispel a widespread view that science consists of boring data collection, mind-numbing theories, unintelligible equations, and inhuman conclusions.

Science is driven by scientists, some good and some bad. They are subject to the same pressures and emotions that affect everyone. Scientists do not always come up with the right conclusions, but if they are good scientists, then there is no going back to the beginning. The data remain to be re-examined and retested, and new hypotheses can be proposed. The heroic image of a scientist in the bath having his or her eureka moment is not one that we recognize. Certainly there are eureka moments, but, more often than not, these are retrospective. They record particular observations or moments of understanding that with hindsight mark the beginning, rarely the end, of a new discovery or invention. The liquid crystal example serves as a parable for all scientific endeavours. Neither Reinitzer nor Frederiks shouted that they had made the discovery of a lifetime, but looking back they surely had.

But don't stop reading just yet. There is another chapter; the liquid crystal story continues. Carrots may have been the beginning, but flat screens are not the end. We now look forward, and a bit sideways, at other aspects of liquid crystal science that may prove to be just as revolutionary as the flat-screen liquid crystal display TV.

12

The new world of liquid crystal materials

The story we have told of liquid crystals has mostly followed the time-line of history. We have tried to show how the advancement of science and the people involved have been influenced by the events of history. While scientists are inevitably people of their time, their special talent (shared with creative artists) is that they are not conditioned by the prejudices of accepted fashion and knowledge. This was clear at the outset of our story, but, in pursuing particular themes, we have by-passed many important developments in liquid crystal science. To have lingered over every new piece of science would have delayed the nar-rative to the extent that we would have lost the trail. In this final chap-ter we will look again at the liquid crystal story and examine some of the liquid crystal discoveries and applications that we have so far neglected to explain.

Cholesteric liquid crystals and temperature sensing

From the very beginning, starting with the observation of colours from cholesteryl benzoate by Friedrich Reinitzer in the 1880s, the brightly coloured appearance of certain liquid crystals had been a source of interest and excitement. It turned out that these dazzling optical effects only occurred in one particular liquid crystal phase. This was the phase which Friedel had referred to as *liquides à plans* (see Chapter 4) and which he finally named the *cholesteric phase*.

The optical textures of these *liquides à plans*, we may recall, exhib-ited under the microscope some features of the smectic phases and other features of the nematic phase. The cholesteric textures exhibited the thin lines and brushes characteristic of nematics, which identified the defect lines along which the optical axis could not be defined. Yet at the same time one could find focal conics, indicating that there were sets of layers colliding at particular types of geometric surfaces. Friedel explained:[1]

> *They [the* optical elements of cholesteric material*] assemble themselves into a structure the details of which are not yet understood. However, it does involve strong twists around a direction normal to the positive optical axis [what we now call the director]... It is this complete complex structure which displays, at equal distances, structural discontinuities... parallel to the elementary positive axis. These are Grandjean surfaces. It is completely different in smectic materials. But nevertheless, these equidistant surfaces, whose nature is so different in the two cases, give rise to the same focal conic shapes.*
>
> *By this means it is possible to understand the appearance of one and the same structure in both cholesteric and smectic materials.*

Friedel's resolution of this apparent paradox had involved a twisted or *chiral* phase. The layers had formed spontaneously as the optical axis had twisted around on itself. Modern terminology refers to the cholesteric liquid crystal phase in just those terms: as the *chiral nematic* phase. The word *chiral* comes from the Greek word *kheir*, meaning hand. A chiral object is like a hand—it is not the same as its mirror image.

Some molecules are not chiral and are identical to their mirror image, but others possess a chiral structure. These molecules can exist in two separate and distinct forms: a left-handed and a right-handed form, each of which is the mirror image of the other. Cholesterol—and compounds formed from cholesterol—consist of chiral molecules. Chemists have found that the two distinct chiral forms of the same structure do not readily convert from one form to the other. Furthermore, different chiral forms of compounds, left-handed or right-handed, can behave chemically in different ways. In biology there are many chiral molecules whose biological function depends crucially on their chirality. Left-handed and right-handed versions of the same chemical drug can have very different effects on the human body.*

The structure of the cholesteric phase is twisted with periodic surfaces at right angles to the twist direction. However, Friedel notes that the separation of the surfaces is very different in the two liquid crystal phases. In smectic phases, the distances are of the order of a molecular length or two

---

* A tragic demonstration of this was with the drug thalidomide. A convention of chemists is to label the two chemical structures that are mirror images of each other as *R* (*rectus*) and *S* (*sinister*). The *R* form of the drug thalidomide was found to be a tranquiliser, pain-killer, and anti-emetic, and was given to pregnant women to alleviate morning sickness. After widespread incidence of deformed foetuses and live babies, further research revealed that the *S* form was teratogenic, i.e. it caused birth defects. Furthermore (and peculiarly for chiral molecules) the *R* form could be converted by body chemistry into the *S* form, so the use of thalidomide for pregnant women was an unexpected disaster.

(20–30 ångströms), and this gives rise to scattering or reflection of X-rays having comparable wavelengths. By contrast, in cholesteric materials the separation of the equidistant surfaces is anything from 100 to 1000 times larger. The structure still reflects electromagnetic radiation, but this time the wavelengths of the radiation reflected are of the order of thousands of ångströms, which correspond to visible light.

In the context of liquid crystals, chiral molecules interact in a special way with polarized light. When such light passes through any material (even a gas) containing chiral molecules, then the plane of polarization of the light is rotated. It is for this reason that chiral molecules and phases containing them are described as being as *optically active*. The amount of rotation depends on the nature of the molecules and their concentration. Two materials containing equal concentrations of either left-handed or right-handed versions of the same molecules will rotate the plane of polarization by exactly the same amount but in *opposite directions*, i.e. to the left or to the right. If a mixture is prepared containing equal amounts of left-and right-handed molecules (a so-called *racemic* mixture), however, then the plane of polarization of the transmitted light does not change. Indeed, the mixture behaves like an ordinary material.

The early work of Charles Mauguin on the optics of twisted liquid crystals, described in Chapter 4, provided a basis for understanding some of the behaviour of cholesteric liquid crystals, but it was Friedel who clearly explained their structure, as we have also described previously. In his monumental thesis on the mesomorphic state Friedel deduced the structure of the cholesteric phase from the experimental observations of Mauguin and Grandjean.[1] The argument is subtle and detailed, but contains the key elements of the unique structure of perhaps the most remarkable type of liquid crystal.

Reinitzer's original observations of the colour changes of cholesteric liquid crystals were thus finally explained. The important feature of those materials was that they consisted of chiral molecules. In the liquid crystal phase, the molecules assemble to form twisted structures, the twist being at right angles to the axis of the organizing molecules. It is this twist that is responsible for coloured light reflected by cholesteric liquid crystals. The twist results from the particular handed internal structure of the liquid crystal molecules and is absent in non-chiral liquid crystals such as nematics. The remarkable optical effects observed in cholesteric liquid crystals have their origins in the special way in which chiral molecules interact with light. However, the effects are magnified by the twisted structure of the cholesteric phase to give the observed dramatic optical effects.

There were still many details to be sorted out, some of which Friedel was aware of. For example, the reflected light was at its best and most recognizable if the cholesteric liquid crystal was aligned. In unaligned samples, such as those viewed initially by Reinitzer, the flashes of light were multi-coloured, but in a suitably aligned sample, the reflection of light was of a single colour, except that it varied with viewing angle and temperature.

In 1932 the Swedish theoretical physicist C.W. Oseen had considered the propagation of light through Friedel's twisted structure.[2] His calculations showed that there should be a particular wavelength for which this medium *completely reflects* circularly polarized light of the *same* spatial handedness as the structure. On the other hand, a circularly polarized wave of *opposite* handedness should pass right through the sample (see Technical box 12.1). It was not until 1951 that H. de Vries formulated a complete theory of the phenomenon.[3]

The conclusion of the theorists' work was that the wavelength of the maximally reflected light depends on the pitch of the helical structure. However, the cholesteric pitch is temperature-dependent. It is not such

TECHNICAL BOX 12.1 Selective reflection from cholesteric liquid crystals

The essence of a cholesteric liquid crystal is that it has a twisted structure. The rod-like molecules do not align in a parallel fashion, as in nematics and smectics, but they twist with respect to each other. The resultant structure is similar to those created by Mauguin and others (see Figure 9.1) in the early years of the twentieth century when investigating the optics of liquid crystals. In twisted nematic cells, the molecules are non-chiral (i.e. neither left-nor right-handed) and the twist is determined by the external alignment of the liquid crystal at the bounding surfaces of the film. Thus the *pitch* of the twist (the distance over which the rotation of the director completes a full turn) is determined by the external twist of the cell. In cholesteric liquid crystals, the twist is caused internally by the molecular structure, and the pitch of the consequent helix is normally much shorter than is achieved in twisted nematic cells. The formation of twist in liquid crystals is explored in a little more detail in a later section in this chapter on "structure, symmetry and shape".

Of interest to us in the context of cholesteric temperature-sensing devices is the optical behaviour of films of short-pitch cholesteric liquid crystals. Most liquid crystal displays use Polaroid film as polarizers. The liquid crystal responds to polarized light and the polarizers cause incident ordinary light to become plane-polarized, that is the electric field associated with the light wave oscillates in a direction perpendicular to the direction of travel and is restricted to the plane defined by the direction of travel and the oscillation direction (see Technical boxes 1.3 and 1.4). Light having these characteristics is sometimes referred to as linearly polarized light (an alternative name for plane-polarized light). Mauguin showed that a twisted nematic structure could rotate the orientation of the plane of polarization of light passing through a suitable cell.

There are other forms of polarized light that can be generated optically if plane-polarized waves are combined in particular ways. This is best explained by an analogy. Imagine a simple pendulum, swinging to and fro. This is an example of a linear oscillation, or plane-polarized since the string of the pendulum and the direction of the oscillation define a fixed plane. Now imagine that the point of attachment of the pendulum string is not fixed, but is itself swinging in a plane at right angles to the first plane. What is the motion of the bob at the end of the first pendulum? Actually it depends on the amplitude of the swings of the first pendulum and its attachment point. If these are exactly the same, it does not require much thought to realize that the pendulum bob will execute a circular motion. It is circularly polarized.

Exactly the same happens if two plane-polarized light waves, travelling in the same direction, but having perpendicular oscillations of the same amplitude, are combined. If the waves are travelling at the same velocity, but displaced with respect to each other by a quarter of a wave, then the resultant oscillation is a circular motion of the electric field about the direction of travel of the light. The resultant light is circularly polarized, and it can be circularly polarized to the left or to the right. In practice, it is not as difficult as it sounds to generate circularly polarized light. In the context of cholesteric liquid crystals, circularly polarized light figures strongly.

The phenomenon of refraction is caused by the speed of light being slowed by the medium through which it passes. We showed in Chapter 1 that the extent to which the light is slowed depends on its polarization. The larger the refractive index, the more the light is slowed. The mechanism of molecules slowing light is complicated, but perfectly understood. The process depends critically on the electronic structure of the molecules, and how the molecules are organized in the medium through which the light is passing. Cholesteric phases are very special: they are twisted, and the twist may have a pitch equal to the wavelength of the light that is illuminating the phase. In white light there will be a range of wavelengths, so for one particular wavelength (i.e. colour) there can be a light wave that is exactly the same as the pitch of the twist in the cholesteric liquid crystal. Under normal circumstances the light is unpolarized and there is no special direction of oscillation for the electric field associated with the light wave. Put another way, unpolarized light contains all possible polarizations equally. If a light wave passes through a medium that not only has a periodic structure equal to the wavelength of the light but also a twisted structure, then there will be a very strong interaction, exactly like fitting a round peg in a round hole. The interaction is so strong that the medium totally reflects back this light, but only the light of the appropriate wavelength. What is more, the reflected light is of exactly the same circular polarization as the twisted structure that it interacts with. The cholesteric liquid crystal is acting as a perfect mirror, but only for light of exactly the right wavelength and the correct circular polarization. Light of other colours and polarizations passes through the material essentially unaffected.* This is the selective reflection of cholesteric liquid crystals and is illustrated below. The effectiveness of the process depends on the alignment of the liquid crystal film and selective reflection works best if the twisted structure is uniformly aligned, with the twist axis perpendicular to the plane of the film. This is just the alignment to observe Grandjean planes (see Figure 4.3). Clearly for such a twisted structure the whole sample cannot be uniform—in the figure below there are lots of twists in a sample film. The regions between the uniform twists appear as special hexagonal defects (see Figure 4.5) and the optical texture is known as the Grandjean planar texture. It is called planar because to view it the molecules should be aligned parallel to the bounding surfaces.

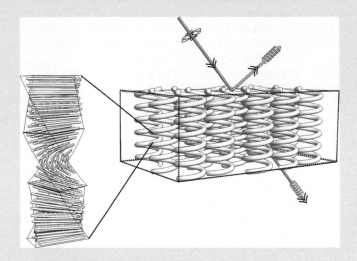

This figure is a representation of the selective reflection of circularly polarized light by a film of planar-aligned cholesteric liquid crystal. The twisted structure of the liquid crystal is represented by springs, all having the same handedness; the springs trace out the orientation of the director or optic axis through the film. The arrangement of molecules within the twisted structure is shown by the exploded view, where the rods represent groups of molecules. The incident light is unpolarized, and the reflected light is circularly polarized with the same handedness as the twist of the liquid crystal. The transmitted light is also circularly polarized, but with the opposite sense.

---

* This statement is an over-simplification, and there will be a narrow band of wavelengths (or colours) that will be selectively reflected. Also the colour depends on the angle of illumination and viewing angle of the reflected light.

a jump, therefore, to the idea that, at least in principle, the change in colour of the reflected light as the temperature changes can be used as a simple temperature indicator. It turns out that the apparent colour of the cholesteric depends very sensitively on the temperature just above the transition from the cholesteric phase to a smectic A phase, so materials that matched this condition would be ideal for temperature sensors.

The American liquid crystal inventor James Fergason, also known for his pioneering work on the twisted nematic display device, was the first to build a practical liquid-crystal-based thermometer. In fact, his work on thermal imaging devices using cholesteric liquid crystals pre-dated his research on display devices and was described in Fergason's first patent in 1963.[4] The principle of operation of liquid crystal temperature sensors is quite simple. By mixing different cholesteric liquid crystals it is possible to produce mixtures that change colour over different ranges of temperature. If a suitable mixture is painted onto a surface, ideally over a black background, then a colour map of temperature of the surface is produced. Such a system has found many applications to detect hot-spots in electronic circuits or faults in mechanical structures. The idea has been widely applied in medicine, for example tumours beneath the skin can be detected, as well as deep vein thrombosis.[5]

Since the first applications were identified in the 1960s, cholesteric liquid crystal temperature sensors have developed considerably. The materials have changed (it is no longer necessary to use compounds of cholesterol) and custom-synthesised liquid crystals having suitable optical activity work even better. Methods of application have also changed. Encapsulating the liquid crystal in polymer bubbles removes some of the messier consequences of application, especially to bare skin. Flexible films containing the polymer-encapsulated liquid crystals can be made, which are then reusable.[6]

While temperature maps are useful in some circumstances, often it is desirable to know the specific temperature of an item. This can be done by preparing different mixtures that change colour over a very small range of temperature, say 1° or less. A series of these can then be assembled to give something like a thermometer, with coloured strips or elements becoming visible at different specific temperatures, as in liquid crystal wine or fish-tank thermometers. One of the best applications is in fever-strip indicators (Figure 12.1).[7]

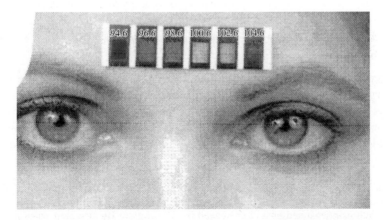

Figure 12.1 A liquid crystal body temperature indicator. LC Hallcrest Inc.

It would have been satisfying to Reinitzer and Lehmann to know that their visual observations of flashes of colour from liquid crystals have had such a direct application and benefit. To very large extent visual stimulation is what excites human animals of both sexes, and cholesteric liquid crystals have found application in decorative jewellery, cosmetics, and art and design. Finally, a possible application with great potential, although one that is beyond explanation here, is the use of cholesteric or chiral liquid crystals as media in which laser action can be induced.[8] Complicated as this is, the laser light generated relies on the very particular structural and optical properties of the liquid crystal.

Liquid crystals from disc-like molecules

In 1923 Daniel Vorländer published his comprehensive review of liquid crystal materials.[9] The focus of his review as expressed by its title was 'The investigation of molecular shape with the aid of liquid crystals'. In the review Vorländer asked the question 'Do star-, cross-, and flake-shaped molecules also form liquid crystals?' Despite his strongly held belief, stated many times before, that to form liquid crystals molecules had to be extended and rod-like, Vorländer was not so prejudiced as to totally discount other possibilities. What is more, he had an army of research assistants to investigate other ideas, however crazy. In the review referred to Vorländer acknowledges the 'indefatigable work of my assistants and disciples'. *Fifty two* names are listed, a co-authorship which in the present day would only be rivalled by papers on fundamental particle physics.

The results of an extended study of many different shapes of molecules were reported by Vorländer, and he recognized that:

> *It is true that one could think that flake-like molecules can be so arranged, superimposed with each molecule's large face laying on each others', in a packing resembling a Volta's pile-like structure*, so that anisotropic structures are generated.*
>
> *The results of our investigations, however, are against this hypothesis.*

Vorländer's conclusions remained the conventional wisdom in the liquid crystal community for the next 50 years.

As we described in Chapter 8, the years following World War II were a boom time for science in general. Research on liquid crystals began to spread around the world, not only throughout the West, but even as far as

---

* In 1800 the Italian scientist Alessandro Volta (1745–1827) demonstrated the first electrical battery, consisting of a pile or stack of alternating metal discs (copper and zinc), each pair separated by a cloth soaked in salt solution. By attaching wires to the each end of the stack an electric current could be generated, as evidenced by sparks between the connections.

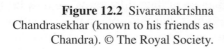

Figure 12.2 Sivaramakrishna Chandrasekhar (known to his friends as Chandra). © The Royal Society.

India. There had always been a tradition of scholarship in the old East India Company Administrative Capital of India, Calcutta,[†] and one of its sons, Sivaramakrishna Chandrasekhar (1930–2004), was to carry this tradition into the field of liquid crystals. It was to be Chandrasekhar who showed that Vorländer's investigations were incomplete.

Chandrasekhar (Figure 12.2) was born into a privileged and accomplished family. His father became Accountant General when India achieved independence, while his uncle, Sir Chandrasekhar Venkata Raman, won the Nobel Prize for physics in 1930. The young Chandra completed his early training in Madras and Nagpur, then moved to the newly opened Raman Research Institute in Bangalore, of which his uncle was the founding Director.

It was his years in England, between 1954 and 1961, that were to prepare Chandrasekhar for a highly successful research career in liquid crystals. A postgraduate scholarship at Cambridge was followed by research fellowships at the University of London and the Royal Institution. While in London Chandrasekhar came into contact with early pioneers in X-ray studies of materials, most notably Sir Lawrence Bragg as Director of the Royal Institution and the Nobel Laureate Dorothy Hodgkin.

On his return to India, Chandrasekhar joined the University of Mysore to establish a new Department of Physics. Looking around for a new field for research, he was inspired by his earlier contacts in England to investigate liquid crystals. He became a leading figure in this rapidly expanding new area of science, and in due course returned to the Raman Research Institute (RRI) in Bangalore to set up a group of international renown in liquid crystal studies.

[†] Calcutta (Kolkata in modern spelling) has nurtured three Nobel prize winners: Rabindranath Tagore (Literature 1913), Venkata Raman (Physics 1930), and Mother Theresa (Peace 1979).

Not put off by the unsuccessful researches of Vorländer, the newly formed group at the RRI looked again at flake-like, or as we now prefer to describe them *disc-like*, compounds, but this time they were successful.[10] Although not exactly identical, Chandrasekhar's liquid crystalline compounds bore an uncanny resemblance to those for which Vorländer had failed to find any sign of liquid crystalline properties, as we can see in Figure 12.3.

Why did Chandrasekhar find liquid crystalline behaviour in disc-like molecules, whereas success had eluded Vorländer? Part of the explanation, of course, is that Vorländer expected (and indeed hoped!) *not* to find liquid crystals, whereas by Chandrasekhar's time, ideas had changed and he *needed* to be successful. From a scientific viewpoint, the chemical key to Chandrasekhar's accomplishment lay in the greater regularity of the structure of the new materials, together with the longer attached carbon (alkyl) chains. In fact his triumph was two-fold. Not only were these new compounds liquid crystalline, that is they melted twice—once from crystal to liquid crystal and then again from liquid crystal to liquid—they were apparently a new type of liquid crystal, neither nematic nor smectic.

Chandrasekhar's group had discovered a new class of liquid crystal that is now referred to as *columnar* liquid crystals. Techniques to study the structures of liquid crystal phases had advanced considerably, and it was now standard practice to use X-ray scattering to reveal unequivocally the details of the arrangement of molecules. The results from Chandrasekhar's laboratory were surprising. The X-ray results showed that the molecules were indeed arranged in stacks, as foreseen by Vorländer, but the stacks

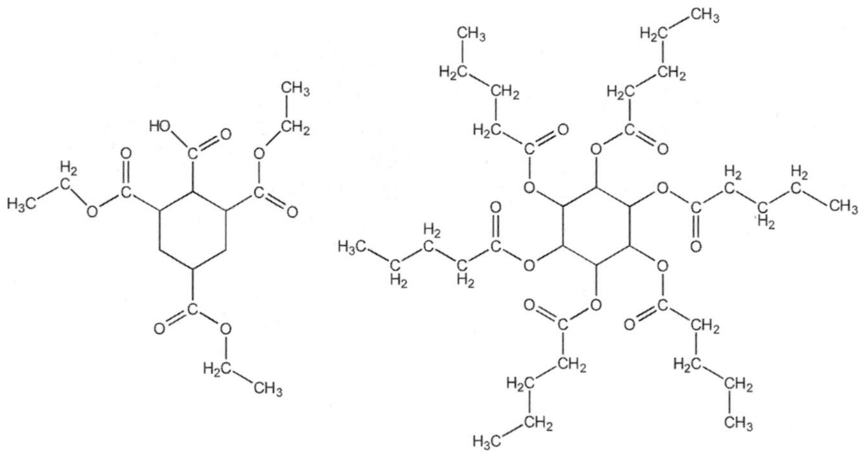

Figure 12.3 Left: Vorlander's flake-like molecule (not liquid crystalline). Right: Chandrasekhar's disc-like molecule forms a columnar liquid crystal.

themselves packed together in a semi-solid-like array, with each stack surrounded by six others (hexagonal close-packing).

What was the evidence that these materials were liquid crystalline? Firstly, the optical microscope revealed images of a soft, fluid, gooey texture, and sometimes the image was that of a hexagon, magically paralleling the molecular structure. Furthermore, the material was birefringent, the characteristic signature of liquid crystals. The determining result was that X-ray results showed that the columns were liquid.

We can now systematically describe nematic, smectic, and columnar liquid crystal phases in terms of their different types of molecular organisation. Nematic phases are true liquids with orientational order of molecules, but no positional order, and are thus three-dimensional liquids. Smectic phases also possess orientational order, but the molecules are arranged in layers, similar to solid crystals except that in smectic liquid crystals the material is liquid within the layers, leading to the conclusion that smectics are two-dimensional liquids and one-dimensional solids. Now there are also columnar liquid crystals. The columns of molecules pack together as though in a solid, but inside each column the molecules can flow just as they would in a liquid. A sketch of the structure of a columnar phase is illustrated in Figure 12.4, and an optical texture of a hexagonal columnar phase is shown in Figure 12.5.

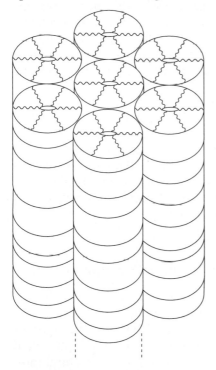

Figure 12.5 Optical texture of a columnar liquid crystal. A. Queguiner, A. Zann, J.C. Dubois, and J. Billard Liquid Crystals (ed. S. Chandrasekhar), Heyden, London, 1980.

Figure 12.4 Packing in columnar liquid crystals.

The set of liquid crystals was finally complete: columnar liquid crystals are one-dimensional liquids and two-dimensional solids.

It soon turned out that some materials forming the columnar liquid crystal phase could also exist in a nematic phase. It was clear that the phase had to be nematic because a schlieren optical texture was visible under the polarizing microscope. The obvious way to interpret these observations was to suppose that the disc-like molecules were aligned as a random fluid, but with the discs maintaining their orientational order. A *discotic* nematic is a phase in which molecules align so that their molecular planes are approximately parallel. This contrasts with the nematic phase of rod-like molecules in which the axes of the rods are approximately parallel.

Nowadays, many different varieties of liquid crystal phases are known which can be formed from disc-shaped molecules. They are distinguished by having different types of arrangement of the columns, and there are further variants formed when the molecules are tilted in the columns or indeed phases formed from chiral (optically active) disc-like molecules. The properties of columnar liquid crystal phases differ in many respects from those of nematics or smectics. Most importantly, their optical behaviour is different. In conventional nematics and smectics, the long axes of molecules are aligned. In columnar liquid crystals, by contrast, it is the *short* axes which align. This means that the refractive index along the alignment direction (the director) is smaller than that in a perpendicular direction: the birefringence is negative. This is an aspect of columnar liquid crystals that has been exploited in applications to make optical films that can be used to improve the appearance of some liquid crystal displays.

There are other possible applications for columnar liquid crystals in the arena of nanomaterials, such as molecular wires which conduct electricity down the columns of stacked molecules.[11] Another example concerns solar energy harvesters.[12] These devices trap light at the end of a column, then convert the light into electricity, and finally conduct the electricity away down the molecular stacks. Only time will tell if these ideas for columnar liquid crystals make an impact comparable to that made by the optical properties of nematics.

Structure, symmetry, and shape: ferroelectric liquid crystals

There was a time during the development of liquid crystal display technology when it seemed possible that nematic liquid crystal displays would be superseded by displays based on an exotic type of smectic liquid crystal. This smectic phase is a version of the tilted smectic C phase discussed in Chapter 8, but composed of chiral molecules. As we shall see, the combination of the layered structure of a smectic phase with tilt of molecules and chirality imposes some very special properties on the liquid crystal phase. These properties can be exploited in a variety of ferroelectric liquid crystal devices.* It is worth recalling that Vorländer had demonstrated experimentally that electric dipoles and ferroelectric organization of molecules played no part in stabilizing crystalline liquids, as he insisted on calling liquid crystals. However, that was in 1923, and the science has moved on considerably since then.

In order to understand the way in which ferroelectricity can develop in certain types of smectic liquid crystal, it is first necessary to explore in a little more detail the relationship between molecular shape and liquid crystal phase behaviour. This has been a recurring theme in the investigation of liquid crystals since Daniel Vorländer, but instead of thinking about molecules in terms of their structure and shape, it is helpful to introduce some simple ideas about symmetry.

Symmetry is about patterns and regularity. A carpet or wall hanging has images that repeat in two dimensions, and in three dimensions all are familiar with the symmetry experienced in traditional architecture. The human brain responds positively to visual symmetry in two or three dimensions. It is said that beauty is merely in the eye of the beholder. Would that it were true! Careful psychological experiments have shown that, in general, it is otherwise. Human perception of facial beauty is a function of symmetry. A beautiful face is one in which the features of one half exactly match those of the other half.

In physics, chemistry, and mathematics, symmetry has consequences more profound than just the aesthetic. It can exactly determine the properties of materials, such as which crystals are birefringent, that is visible in a polarizing microscope, and those that are not. In chemistry, symmetry decides which species of gas molecule in the atmosphere are harmful for the environment through global warming and which are not. In the mathematics of phase changes, the nature of the phase transition is influenced by the symmetry of the phases connected by the transition.

In terms of shapes that might be used to represent atoms or molecules, a sphere looks the same from all directions. When collected together, spheres will pack together to make solids in certain predictable ways. It was Daniel Vorländer who first realized that molecules which tended to form liquid crystals were extended in shape. Using the

---

* Ferroelectricity is the electric analogue of ferromagnetism (see Technical box 3.2). In a ferroelectric material, pairs of positive and negative charges (dipoles) line up in a regular array, just like the north and south poles of magnets. The electric dipoles form domains, and the domains are arranged so as to minimize the overall energy. It was just these arrangements that had been proposed by Max Born, eventually unsuccessfully, in 1916 to explain the nematic liquid crystal phase (see Chapter 3).

solid–object metaphor that has been so useful in the history of chemistry, he described liquid crystal-forming molecules (*mesogenic*, in the language of Georges Friedel), as *rod-like*. When forced together in a confined space, cylindrical rods will pack most efficiently if the cylinders are parallel. This gives us the picture we have for nematic phases, of parallel rod-like molecules, with a picture of smectic phases in which the parallel molecules also align in layers.

A cylindrical rod is symmetrical. If viewed from an end it looks like a circle in two dimensions, but viewed sideways on it is like a rectangle. This latter image disguises one intrinsic symmetry of the rod: if rotated about its axis, it always appears the same. In a liquid crystal phase of rod-like molecules the orientation of the molecule about its 'long axis' does not affect the way it packs in the phase, nor any of the properties measured for the phase. Furthermore, the mirror image of a rod is exactly the same as the original rod.

Now rods are not chiral and the phases they form are not chiral either, but we do know that chiral molecules can form liquid crystals. Friedel's cholesteric phase is *almost* nematic, as its modern name (the chiral nematic) attests. As we shall see, chiral molecules can also form smectic phases.

The cholesteric phases are made from molecules that are chiral and almost like rods, but how can a rod become chiral? The simplest way is to carve a screw-thread along it so resembles a bolt. The new object is still rod-like, but it is handed, depending on the cut of the thread as left-handed or right-handed. Let us now try to pack a set of threaded rods together, all of the same handedness. At first glance it may appear that nothing has changed, except that, if two threaded rods are forced together to be parallel, actually they will prefer to twist so that the threads engage (see Figure 12.6). Let us imagine a collection of chiral rod-shaped molecules forced to pack together in a parallel fashion. They also will twist, and this is the origin of the twisted structure we have already discussed in detail for cholesteric or twisted nematic liquid crystals. As with all analogies, we must be careful not to stretch its application too far. The real chirality of molecules arises from the way in which the atoms of the molecular structure are connected. There is no screw, but the chiral disposition of the atoms and groups along the molecular rod has the same effect and causes the long axes of adjacent molecules to twist a little.*

Compounds of cholesterol form smectic phases as well as nematic phases, so these also should be twisted. In the simplest smectic phase, designated A, the molecules arrange themselves in layers and the axes of the rod-like molecules are on average perpendicular to the layers. If

---

* The actual twist per molecule is extremely small, of the order of a fraction of a degree for each molecule.

Figure 12.6 Threaded rods fit together with a twist.

the molecules are chiral, then there will be a desire amongst the parallel molecules to twist, and the twist would be in the plane of the layers. However, the layers do not like to twist. In fact there is a huge energy barrier to twisting layers, so it just does not happen and the layers in the smectic A phase remain untwisted, except in bizarre and exotic phases known as twist grain boundary phases.[†13–15]

In our previous discussion of the tilted smectic C phase we did not attempt to explain why the molecules choose to tilt. Indeed there is no simple explanation applicable to all tilted phases. However, in what is sometimes disparagingly referred to as 'a hand-waving description', we will attempt to give an impression of why and how molecules in certain smectic liquid crystal phases tilt.

For a liquid crystal phase of rods, there is no reason at all for the molecules to tilt. As explained above, a rod is identical if rotated about its long axis, something called *axial symmetry*. Rods can pack just as well in a parallel fashion whether they are in a nematic (fluid) liquid crystal phase or in a layered (semi-solid) smectic phase. However, most molecules that form liquid crystal phases are not rod-like. They may have lumps and bumps that destroy the axial symmetry. Indeed we have

---

[†] In fact Nature has found a way of dealing with the twist in some layered liquid crystals in a rather curious fashion. For a few materials for which the properties are just right, the smectic layers break up into filaments of twisted blocks, forming another new smectic phase. This was predicted by de Gennes by analogy with a similar phenomenon in magnetic materials.[13] De Gennes was an expert in magnetic properties and he could make connections and predictions that would never have occurred to others. The theory of this unusual phase was developed by Renn and Lubensky,[14] and was discovered experimentally two years later.[15] The phase is known as the twist grain boundary phase.

already noted above that for chiral molecules these lumps and bumps cause the molecules to twist with respect to each other. The molecules are freely rotating about their long axes in a nematic phase. For a particular group of non-chiral or chiral molecules, the lumps and bumps are such that, on cooling to form a layered liquid crystal phase, the molecules fit together better by tilting.

The layers in smectic phases are defined in one way or another by the length of the molecule. The smectic layers we have described so far have been either monolayers, where the layer spacing is essentially the same as the molecular length, or bilayers, as in soap films and cell membranes, where the layer spacing is twice the molecular length. In addition to these simple structures there are many others, but the molecular length or some multiple of it is always the factor that determines the layer spacing.

Regardless of the details, in smectic phases it is the ends of the molecules that define the molecular length and hence the separation of the smectic layers. As molecules in a smectic layer tilt, the width of the layer is expected to decrease. This is indeed confirmed by experiment. If the tilt of molecules within smectic layers increased without limit until the molecules were lying along the layer planes, then the layers would effectively be destroyed, as illustrated in Figure 12.7. The struc-

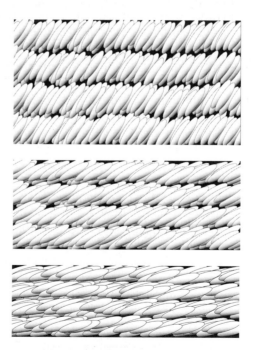

Figure 12.7 As the tilt angle of molecules in a tilted smectic increases, the layer thickness decreases. At the limit of 90° tilt, the layers are no longer defined and the structure resembles that of a nematic.

ture would be the same as a nematic phase. This is only a 'thought' experiment (it cannot actually be done), but it provides some insight into what happens with chiral molecules in tilted smectic phases.

If chiral molecules in a tilted smectic phase were subjected to the above thought-experiment and tilted through 90°, then the resultant phase would be a nematic. Since the molecules are chiral, there would be a twist developed in a direction perpendicular (at right angles) to the directions of the long axes of the molecules. Now reverse the thought experiment, that is allow the molecules to tilt back to form their layers. What happens to the twist? We know that the smectic layers cannot twist within the planes of the layers, but there is nothing to stop successive layers twisting with respect to each other, as shown in Figure 12.8. That is exactly what happens. As the molecules become less tilted with respect to the layer planes, the twist gets less, until it would be zero when the molecules are perpendicular to the layer planes. This is the liquid crystal phase we denote as the smectic A phase.

Our conclusion from these thought experiments is that if chiral molecules form a tilted smectic phase, then a twist is allowed, but it is a twist of layer planes rather than molecules and the twist is along a direction that is at right angles to the smectic layers. This is of course what happens. Thought experiments are of no use if they fail to give the observed result. It is now time to return to reality and give the background to one of the most exciting discoveries in post-war liquid crystal science.

What has been described above about a possible chiral smectic C phase would have been known to Bob Meyer, when in 1973 he took sabbatical leave from Harvard to spend some time working in the Solid State Physics Laboratory of the University of Paris South at Orsay (Figure 12.9). Meyer had been working for a few years on fundamental aspects of the physics of liquid crystals. In particular he had investigated

Figure 12.8 Successive layers of chiral molecules in a tilted phase are able to twist.

Figure 12.9 Bob Meyer (b. 1943).

what happens when electric fields are applied to nematic and smectic liquid crystals in various circumstances. During his sabbatical in France, Meyer was invited to lecture at the famous Physics Summer School of Les Houches in the French Alps. While preparing the lectures, he realized that there was a rather special consequence of symmetry for a tilted smectic phase consisting of chiral molecules. Certainly a twist between layers would be developed, but if one looked more closely (another thought experiment) at a single layer of chiral molecules, then there was something rather bizarre. This is best illustrated with a figure.

Imagine our exotic molecules that will form a tilted smectic phase as a shoal of fish (Figure 12.10). They pack together to give a compact monolayer of a tilted smectic C phase. What is rather remarkable is that we only see one side of the fish, the right side. If we wanted to see the left side, then the fish would have to tilt to the left instead of the right. Of course if both sides of the fish are the same, as is usually the case, then there is no physical difference between the two tilt directions. However, if one side of the fish was painted black and the other white, then indeed there would be a difference. One tilt direction would present a totally black image, while the other would be totally white. What is

Figure 12.10 Packing of fish tilted in a layer. Only the right-side is visible from the front: the fish are tilted to the right. Viewed from the back, only the left sides would be visible, and the fish would be tilted to the left. (Adapted from de Gennes[16] and Goodby[17].)

special about a fish with a black right side and a white left side? It is chiral! Change our fish into molecules, and the inescapable conclusion is that layers of a tilted smectic C phase will be distinguishable depending on whether the tilt is to the left or to the right.

Meyer was smart enough to jump to this conclusion without having to invoke the fish imagery. Moreover, if the sides of the molecule were distinguished not by colour, which is difficult, but by charge, which is easy, then he realized that there was a whole new regime of liquid crystals: *ferroelectric liquid crystals*.

The reasoning was faultless. In a smectic C phase of molecules which were chiral, and had a separated charge *across* the axis of the molecule, then individual layers would present to an observer a set of (say) positive charges for a tilt direction to (say) the right and a set of negative charges for a tilt direction to the left. A single layer would be positively charged if viewed from the front and negatively charged if viewed from the back, just like a simple battery. In technical terms the smectic layer is described as being polarized (nothing to do with polarized light): it has developed spontaneous electric polarization, like a magnet with the north and south poles replaced by positive and negative charges. There is a problem, however. In a chiral tilted smectic phase, the layers twist from one to the next, so the polarization predicted would spiral through the phase. Since we cannot resolve individual layers in the smectic phase, we would see nothing. The polarization would be averaged out to zero.

However, there was a hope that something dramatic might be observed if only the spiral twist of the layers could be unwound. Meyer's ideas were sufficiently novel and exciting to inspire interest from chemists, who set about trying to construct (synthesise) a molecule with all the right characteristics. They succeeded in producing a material known rather clumsily by its acronym, DOBAMBC.* The molecule was chiral, it had a charge separation across its axis (a molecular electric dipole), and most importantly it formed a smectic C phase.

All was set for the crucial experiment. A thin film of DOBAMBC was prepared so that smectic layers would form as far as possible parallel to the film and the surfaces of the containing microscope slides. The sample was placed in a polarizing microscope and an attempt was made to unwind the presumably twisted layers by applying a voltage across two wires that had been embedded in the liquid crystal film. Initially nothing happened, then, as the voltage was increased, the unmistakeable optical signature of the layers unwinding was seen. Eventually all the layers were unwound and a single domain of uniformly tilted ferroelectric smectic C liquid crystal was formed. Exactly the same happened if the voltage was reversed, but this time the uniform tilt direction was in the opposite direction. Meyer's inspired prediction had been

---

* *p*-decyloxybenzylidene-*p*'-amino-2-methylbutyl cinnamate.

confirmed experimentally[18] and a new area of liquid crystal research had been created.

Ferroelectric solid-state crystalline materials have been around for a long time and are used in a variety of devices. However, Meyer's ferroelectric liquid crystal was a revolution since it showed that under special circumstances it is possible to create ferroelectric domains in a fluid, something that had previously been deemed to be impossible. The trick that Meyer applied was to unwind the helix of layers in a smectic C liquid crystal with an electric field. Without the unwinding the liquid crystal would not be ferroelectric, and it was hard to see how the potential of ferroelectricity in liquid crystals could be exploited for applications.

There is nothing like a challenge, and two physicists, one in Sweden and one in the USA, rose to the challenge. Sven Lagerwall was another scientist inspired by the Orsay liquid crystal group and, following a sabbatical there in the early 1970s, Lagerwall established a dynamic research group at Chalmers University in his home town of Göteborg. Noel Clark at that time was a junior academic in the liquid crystal group at Harvard, which was Bob Meyer's home base. By 1977, Clark had moved to the University of Colorado at Boulder. He, of course, would have been well acquainted with the work of Meyer on ferroelectric smectic liquid crystals, and in due course he spent a sabbatical with Lagerwall's group in Sweden. There Clark and Lagerwall were able to prepare thin films of smectic C liquid crystals in which the helix was unwound. They prepared for the first time a genuine ferroelectric fluid, which was reported in *Applied Physics Letters* in 1980.[19]

The preparation was very delicate. The material was the same as that used by Meyer, DOBAMBC, and it was inserted into very thin cells of conducting glass that had been treated to give an alignment of molecules parallel to the surfaces. By cooling the film of DOBAMBC from its smectic A phase into the lower temperature smectic C phase, Clark and Lagerwall were able to produce a film in which the smectic layers were perpendicular to the surfaces of the glass plates. The twist of the layer planes of the chiral smectic C phase were now expected to be parallel to the glass plates, but there was no twist. The thinness of the sample and the surface treatment had forced the helix to unwind. What resulted was an array of ferroelectric domains in the film, with random polarizations up and down, corresponding to areas of tilt to the left and to the right.

There were now all manner of possibilities. By applying a voltage across the film it was possible to cause the domains having an electric polarization in the same direction as the applied electric field to grow at the expense of the others. Reversing the field allowed the other domains to be stabilized. Since the different domains corresponded to different tilt angles, the optic axes of the two sets of domains made an angle of twice the tilt angle with each other. The domains were therefore easily distinguishable in a polarizing microscope. Furthermore, by designing materials with an appropriate tilt angle, and adjusting the angles of the polarizers of the microscope, it was possible to make one set of domains appear black and the other set white.

If sufficient voltage was applied to the sample, then only the preferred set of domains was stabilized. On removing the voltage the sample stayed unchanged and one set of domains remained. If the polarity of the voltage was changed from positive to negative, then the domains switched from one set to the other, and this could be observed in the microscope as a change from black to white. The basis for a new type of liquid crystal display had been established. It has become known as the surface-stabilized ferroelectric liquid crystal (SSFLC) display.

TECHNICAL BOX 12.2 The surface-stabilized ferroelectric liquid crystal display

The twisted nematic display became the standard during the second half of the 1970s, but its limitations were all too apparent. The creation of complex images was made difficult by the restricted number of pixels that could be activated. If changing images were to be displayed, then there was also a problem concerning the speed with which the information could be transferred to the display. The faster the liquid crystal responded to a voltage, the better a moving image could be displayed. However, increasing the switching speed of a liquid crystal pixel also meant that it would lose its image faster, so multiplexing the display and refreshing activated pixels also had to be done faster. An ideal solution to this problem was a display based on a ferroelectric liquid crystal, which switched nearly a thousand times faster, in millionths of a second instead of the thousands of a second required by a twisted nematic display. Not only was the ferroelectric liquid crystal display faster, an activated pixel remained switched on when the voltage was removed. To change the pixel from one activated state to another required a voltage pulse of opposite sign. Such a display is known as a *bistable display*, and the ability to be switched from off to on and then on to off greatly simplified the representation of complex images.

The switching mechanism of a ferroelectric chiral tilted smectic C display is illustrated below. A thin film of liquid crystal is fabricated so that the layers appear as in (a). The molecules are chiral, so the front and back of the molecules are different. For the molecules used in the display, there is a separation of charge from front to back, that is an electric dipole perpendicular to the axis of the molecule. For the layers in (a), the optic axis is parallel to the molecular tilt (the axes of the ellipses in our model). If a positive voltage is applied to the front surface of the liquid crystal film, then the molecules will rotate so that their negative charge appears to the front. However, the constraints of the layers mean that the only way this realignment can occur is via a rotation on a cone, as illustrated in (b).

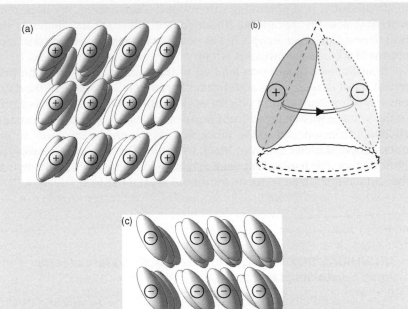

A single molecule rotates from left to right around the surface of an (imaginary) cone, and now presents a negative charge on the surface, but with a reversed tilt direction.

Now the front surface of the liquid crystal film presents a negative charge, as in (c), and the optic axis has changed its orientation by an amount equal to twice the initial tilt angle. By placing the film between a pair of crossed polarizers it is possible for the film to switch from black to white or vice versa.

The increased speed of switching of a ferroelectric film over a twisted nematic film is a direct consequence of the mechanism. For a twisted nematic film, the change in the direction of the optic axis is caused by the rotation of a rod-like molecule about an axis perpendicular to the length of the rod. This motion has a high moment of inertia associated with it. By contrast, the reorientation of the molecule (this time represented as an ellipse) on a cone has much less inertia, and so is faster.

This new type of display has several advantages over the traditional twisted nematic display. Firstly, switching between black and white is very much faster, perhaps a hundred times faster, than the switching time of a nematic display. This was very important as the difficulties in producing sharp fast-moving images on twisted nematic displays appeared almost insurmountable. There was another huge advantage of the SSFLC display over the twisted nematic display, which was that display pixels stayed on when the voltage was removed. This greatly simplified the problems of addressing. Technically, the behaviour was known as *bistability*; either set of domains of the smectic film could be switched and then would stay in the selected state after removal of a voltage. Perhaps the SSFLC display was an even better product in the search for the flat-screen television.

Not surprisingly, there were also problems with ferroelectric smectic liquid crystal displays. The separation of the glass plates that contained the liquid crystal had to be extremely small, about a micron (one thousandth of a millimetre) compared with a few microns for nematic display. Preparation of the liquid crystal alignment was even more complicated than for nematic displays, and even when achieved the alignment of the smectic layers was very fragile and could be disturbed by mechanical shock. Despite these difficulties, a few manufacturers took up the technology as a potential rival to nematic displays. The Japanese electronics company Canon produced a prototype large-area SSFLC television with some impressive viewing characteristics. However, the difficulties associated with high volume production of large-area SSFLC television displays were such that they never reached the market.

In other applications, however, SSFLC displays and other optical devices have been very successful. For displays of small size, of the order of a centimetre square, the problems of stability have been overcome, and the speed and addressing advantages make SSFLC displays extremely successful. One of the inventors of the SSFLC display, Noel Clark, set up a research and production company in 1985, Displaytech Inc., which has become the biggest manufacturer of small-area liquid crystal displays in the world.*

Ferroelectric liquid crystals are not only useful for displays. The characteristics of speed and bistability associated with SSFLC films make them ideal components in the fast-moving world of optical fibre communications. For example, spatial light modulators (SLMs) are highly sophisticated optical switches which control the information carried by optical fibres. They can be used to inject the information and

---

* Displaytech Inc. successfully developed microdisplays based on ferroelectric liquid crystal deposited onto a silicon substrate. These displays have found application as electronic viewfinders in digital cameras, camcorders, head-up displays, and other consumer electronic devices. In 2009, Displaytech was acquired by Micron Technology, which now provides a range of products derived from ferroelectric liquid crystal technology.

channel it along bunches of fibres. Switching information from one fibre to another is a bit like marshalling railway trains at a complex junction. SLMs can perform this task at lightning speed and can be made from ferroelectric smectic liquid crystals.[20]

Liquid crystals and disease

All living organisms are made from biological cells, from a single cell bacterium to the millions and millions of cells that make up a human being. Each cell is surrounded by a semi-permeable membrane. This membrane is a liquid crystal. It provides a support for many important proteins and allows the transport through the membrane of small bio-molecules, water, and ions. Within cells are yet smaller structures known as organelles, one example being the nucleus, which are also enclosed by membranes.

So important are cell membranes to the normal functioning of living systems that a breach of the membrane, or a change in its characteristics, can have a devastating effect on the operation of the cell and hence on the organism as a whole. This scientific statement prompts a medical question: can the converse be made to happen? Can some cell malfunctions be corrected by modifications or treatment of the cell membrane? We are straying now into the realm of speculation, but science fiction is often just a few decades in advance of reality. We have shown that liquid crystals are vital components of living systems. Now we explore the possibilities for moderating life processes, and even curing disease, by changes to the environment of biomolecules in cell membranes and inside cells.[21]

Modern medicine traditionally employs one of two alternative strategies when called upon to combat disease of the human organism. If the problem is localized, the surgeon will try to cut out the offending part of the body. If this approach is for some reason inappropriate, the offending cells must be killed, by drugs perhaps or radiation. If drugs are the chosen route, since almost all biological metabolism and function is via proteins the drugs will be directed to the relevant proteins.

An alternative to traditional medical approaches is the relatively new *gene therapy* strategy. The idea is that the genes which cause a particular protein disorder are targeted or modified, in the hope that they will re-programme the biochemistry. In this way the offending species will be exterminated or more of the right stuff (whatever it may be) produced.

In biology, the cell is king. It is the source of life and the origin of disease. The bilayer cell membrane is just like Lehmann's soap liquid crystal, except that it is composed of phospholipid molecules. However, unlike the liquid crystal phases formed in the laboratory from soap molecules or phospholipids, the membranes of cells contain over a

thousand different types of phospholipids. These molecules are synthesised in the cell by protein templates, and the composition of the membrane can change in response to external or internal circumstances. For
example, animal and plant cells change the membrane composition of
some of their cells to cope with changes in their environment, such as
extremes of temperature. One thing about Nature is its efficiency, and if
there are thousands of different lipid bio-molecules around, then there
must be a reason.

One important function of the cell membrane is to isolate the biochemistry of the cell, but it also provides a suitable environment for
bioactive molecules located in the membrane and acts as a suitable
attachment surface, inside or out, for other biomolecules. These functions could be performed by a relatively simple mix of membrane lipid
molecules and would not need the huge array of variants that are available in cells. The inevitable conclusion is forced on us that cell membranes are more than just flexible overcoats.

There is considerable evidence that many human diseases are accompanied by changes in cell membrane structure or composition. Some
examples are listed in Table 12.1, along with the observed membrane
abnormalities.

The big question is, are the membrane alterations the *cause* of the
cell malfunction or the *result* of the malfunction? Furthermore, can the
cell function be corrected by fixing the membrane? Is the appropriate
remedy a *hammer* in the form of a drug or will a spot of *grease* or membrane *massage* do the trick? These are big questions that are only just
beginning to be addressed by contemporary science. There are some
indications that targeting cell membranes can alleviate disease. For
example, the standard drug treatment for asthma is Intal® or disodium
cromoglycate, a well-known and much studied compound that readily
forms liquid crystal phases in the presence of water. The liquid crystal

Table 12.1 Diseases and membrane symptoms*

Disease	Membrane symptoms
High blood pressure and fatal heart attacks	Changes in phospholipids and cholesterol concentrations and changes in saturated and unsaturated fatty acid levels
Alzheimer's disease, ageing, and schizophrenia	Reduced levels of polyunsaturated fatty acids in phospholipid membranes
Respiratory diseases	Changes in the lipid composition of membranes
Triose phosphate isomerase deficiency (a rare debilitative disease)	Modification of membrane lipids and membrane fluidity

* Courtesy of E. Poetsch, formerly of E. Merck, Darmstadt, private communication (2008).

phases formed by disodium cromoglycate in water are similar to the lyotropic phases formed by soap or lipid species. However, the nature of the molecular organization is sufficiently different to warrant a new classification. These phases are now called chromonic liquid crystals,[22] a term that derives from the similarity in chemical structure between molecules that form chromonic liquid crystals and dye molecules which stain chromosomes.

TECHNICAL BOX 12.3 Chromosomes and chromonic liquid crystals[23,24]

Chromosomes carry some of life's vital ingredients and are so-called because it was discovered that they can be selectively dyed, or stained, by other chemicals so that they are more easily visible in the microscope. This technique of selectively staining cells, parts of cells, or other tissue has been a vital tool in animal and plant biology for more than 100 years.

Different stains highlight contrasting biological regions of cells or tissue, and so enable different active centres to be identified. The process of staining is a simple matter of chemical compatibility. If the molecules of the stain match to some extent those of the substrate, then staining will occur, otherwise it won't, just as non-permanent felt pens will write on some surfaces, but not others, and permanent felt pens will often write on yet other types of surfaces. Not surprisingly, the chemical compatibility of stains and biologically active surfaces sometimes results in the stain actually invading the bioactive surface and affecting its biological function. Thus, stains and drugs often have similar chemical structures. This is illustrated by the story of the birth of the synthetic dye industry.

In 1856 a young English chemist, William Henry Perkin (1838–1907), was working as an assistant in the Royal College of Chemistry (now part of Imperial College, London). He had been set to work by his boss, August Wilhelm von Hofmann, to synthesise chemicals that might have medicinal uses. One of these was quinine, an expensive drug extracted from natural products and much in demand for the treatment of malaria. Perkin was given the task of synthesising quinine, or something like it, from starting compounds derived from coal tar. Instead of quinine the product was a black mess, but careful treatment of the mess yielded a purple dye. This was the first synthetic dye and subsequently became known as mauveine. Perkin was sensible enough to abandon the search for drugs and his serendipitous discovery resulted in the founding of a huge chemical dyes industry. The similarities between some drugs and dyes goes beyond their chemical compositions, and it is now realized that certain drugs and dyes of a particular structural type form a new class of liquid crystals known as *chromonic* liquid crystals. This class of liquid crystals also includes other biomolecules, such as nucleic acids and proteins, which can form liquid crystal phases in water.

These liquid crystals are really a new type of lyotropic liquid crystal, but they show sufficient differences in behaviour to allow them to be categorized as a distinct variety. The molecules of chromonic liquid crystals have polar and non-polar parts which moderate their solubility in a polar solvent such as water. The molecules that form the lyotropic liquid crystals we have already encountered have polar and non-polar parts, and the polar parts are at the ends of the 'rod-like' molecules. To dissolve in water, these molecules aggregate into layers, cylinders, micelles, and vesicles to maximize the interactions between the polar parts of the molecules and the polar solvent.

The common structural feature that characterizes molecules which form chromonic liquid crystals is that they have a rigid non-polar core, disc-like or board-like in form, but with polar parts around the edges of the molecules. The only way these molecules can dissolve in water is by stacking in columns to present a polar outside to interact with the water, thereby minimizing the exposure of the non-polar molecular centres. In fact this is exactly the same mechanism adopted by biopolymers such as proteins and nucleic acids, which form liquid crystals in water solutions. Having formed stable cylinders, the chromonic dyes or drugs can then become even more organized to form a nematic or the more ordered hexagonal phase.

The chemical structure of disodium cromoglycate, the anti-asthmatic drug Intal, is illustrated below. Groups containing oxygen (O) atoms on the periphery of the molecule are able to interact with water and so solubilize the drug. At sufficient concentrations in water, columns of disodium cromoglycate form and then self-organize to form a nematic liquid crystal phase and then, at higher concentrations, a more ordered columnar hexagonal phase.

Left, nematic phase of Intal: the columns have no positional ordering. Right, hexagonal columnar phase of Intal: the columns form a hexagonal array.

The connection between the ability of certain drug and dye molecules to cure disease or stain cells and their propensity for forming liquid crystal phases is uncertain. What is clear, however, is that the interactions that drive the formation of chromonic stacks and then ordered liquid crystal phases in water are exactly the same as the interactions that permit the dye or drug to be absorbed by the cell membrane. For the moment, this is the limit of our knowledge. It is not known how the adsorbed species modify the cell membranes nor do we know how the behaviour of proteins in or attached to the membranes can be changed for the better to cure a disease.

The normal route to disease infection is by small organisms such as bacteria or viruses. Another source of disease can be through genetic modification or inherited genetic defects. Such malfunctions are transmitted to the cell proteins via the nucleic acids DNA and RNA. A further new category of disease has also recently been proposed in which infection has nothing at all to do with genes or their nucleotides.* The communication systems of the cell are by-passed and disease strikes the cell proteins directly. These diseases are categorized as 'protein-only' and embrace a range of conditions such as Alzheimer's, Parkinson's, and the so-called prion diseases, such as bovine spongiform encephalopathy (BSE or 'mad-cow disease') and Creutzfeld-Jakob disease (CJD, or 'human-mad-cow-disease').

What happens in these diseases is that normal folding of the polymer protein is disrupted. At the time of writing (2010) it is not clear why this happens. An alarming feature of the disease is that mis-folded protein molecules can induce other proteins to mis-fold, and infection can be transmitted between humans and animals directly through rogue proteins. One form of CJD, known as Kuru and found amongst the tribal people of New Guinea, is passed on by ritual cannibalism.[25] In all these protein-only diseases the affected biomolecules adopt a more rigid shape in the membranes, which apparently prevents normal cell functions. Perhaps an adjustment to the environment of the misshapen protein through an administered dose of lipid will induce a re-shaping of the protein and cure the condition. The mode of transmission of protein-only diseases is interesting because it seems to violate a fundamental tenet of biochemistry: that inherited biological information is always encoded by nucleic acids. Furthermore, protein-only diseases are also of interest in the context of liquid crystals, since the cell malfunction seems to be a consequence of a disorganization of the relevant cell proteins.

---

* Nucleotides are the chemical building blocks of the nucleic acids DNA (deoxyribonucleic acid) and RNA (ribonucleic acid).

At present, the pathological behaviour of proteins in membranes requires that the proteins themselves be targeted. Unfortunately, in order to do this it is often necessary to use clinical drugs that are highly poisonous to the organism as a whole. Such drugs work (when they work!) because they are *even more* toxic to the source of the illness than to the rest of the organism. Unsurprisingly, both doctors and patients are very wary of drugs of this kind, but it is possible that there is an alternative route. The activity of membranes could perhaps be moderated and adjusted by lipid or lyotropic liquid crystal treatments. If so, there is a chance that a number of diseases that are currently difficult to treat could be cured or controlled by the administration of membrane-active molecules with low toxicity and high efficiency.

Self-organization and molecular self-assembly are the hallmarks of liquid crystals. A pioneer in the study of connections between organization and function in living systems and liquid crystals is the veteran organic chemist Helmut Ringsdorf (b. 1929). In a long and active career (he is still working at the University of Mainz) Ringsdorf has brought together the fields of liquid crystals, biochemistry, and, as we shall see in the next section, polymers. It is appropriate to finish this section on the real significance of liquid crystals for living systems with a quotation from Ringsdorf:[26]

> Order and mobility are the two basic principles of Mother Nature... both are combined in liquid crystalline phases.

Liquid crystal polymers

Poly-liquid crystals: another poly-something to join the list of modern innovations. Poly substances are invariably good for you or do something for you which no other substance could do as well or as cheaply. Poly-cell is a great adhesive, polystyrene is the ultimate packing material, and the ultimate polluter, poly-filler is an aid for builders and decorators, poly-bags the ubiquitous containers, and poly-tunnels, a modern approach to the mass production of fruit and vegetables. The *polys* in these contexts refer of course to varieties of modern materials created from polymers. Their special properties derive from their chemical composition and molecular structure, and how the molecules of the polymer arrange themselves. It may be a surprise to discover that in some sense polymers emerged from the same laboratory as liquid crystals, the laboratory of Daniel Vorländer in the Martin Luther University of Halle.

The chemist's definition of a *polymer* is quite precise. It is a compound formed when small chemical units or small molecules of low molecular weight combine, that is react with each other, to form a larger chemical structure containing a few or many units, rather like creating a goods train by coupling a number of individual trucks together. The trucks may be all the same or they may be of two or more varieties coupled in a regular (A–B–A–B ...) or irregular sequence (A–A–B–A–B–B–B–A ...). There could be many varieties of trucks and these could be coupled in a completely random arrangement, but the result would still be a train.

If the chemical units are nucleotides then the resulting polymer is a nucleic acid. In the DNA of plants and cells, there are four different nucleotides.* Many thousands are linked together in the nucleic acid polymer molecules. It is the sequence of these nucleotides that defines the genetic make up of the cells of plants and animals, and it is the over-all structure of the nucleic acid, for example DNA or RNA, that determines its precise functions. These so-called *bio-polymers* are examples of a range of polymer molecules found in living systems. The number of nucleotides in a DNA or RNA molecule depends on its source. A nucleic acid from a virus may have just a few hundred, while human DNA contains millions of nucleotide units. The arrangement of individual units in these bio-polymers can vary even within a single type. For example, there are sections of regular sequences, often identified as genes, and other sections of random sequences. In terms of function, the latter seem to be useless.

Although lacking any understanding of their chemical make-up, early humans found many uses for naturally occurring polymers. Webs from spiders aided the treatment of flesh wounds, silk from silkworms could be woven, and resins from various sources found applications as glues and decorative lacquers. Later, in the eighteenth century, early explorers of South America were amazed to find local Indians using rubber latex for various purposes.

It was not until early in the twentieth century that it was possible for organic chemists to synthesise artificial polymers from existing chemical components. The consequence was that a whole set of new materials with hitherto unattainable properties became available. The process whereby small chemical units combine to produce polymers is called *polymerization*. A key figure in understanding the principles of polymerization was Hermann Staudinger (1881–1965), whose Ph.D. was completed in 1903 under the supervision of none other than Daniel Vorländer.

In due course Staudinger was to develop the basis for the new and important subject area of polymer chemistry. In recognition of this work he was awarded the Nobel Prize for chemistry in 1953. However, at the

---

* The units of DNA and RNA are slightly different. DNA is made up of the chemical units (known as bases) called adenine, guanine, cytosine, and thymine. These bases combine with sugars and phosphates to form the individual nucleotides that are the building blocks of DNA. RNA contains three of the bases of DNA, but the fourth is uracil, which replaces thymine.

time when he was a research student under Vorländer's supervision (1900–1903), Staudinger had been merely another young organic chemist. Vorländer's studies of liquid crystal chemistry had barely started; Staudinger would probably have had little inkling of the revolution in the understanding of 'states of matter' that was about to take place.

After Halle, Staudinger found himself in Karlsruhe—another seat of the liquid crystal revolution—and it was there that he began his work on polymerization. There is no evidence that at this stage Staudinger had any consciousness of liquid crystals or their potential importance. However, back in Halle, his former supervisor was certainly aware that polymers might have a role to play in liquid crystals. Vorländer, knowing that a rod-like structure was important for liquid crystalline properties, wanted to know what happened to the liquid crystals if the length of the constituent molecules was increased. He was able to lengthen his relatively short molecules by chemically joining them together into a chain. In his review of 1923[27] Vorländer reports some early findings:

> For higher members of the chain, the liquid crystalline characteristics increase...

> The data...shows very clearly the increase in liquid crystalline characteristics observed with increasing chain length... Furthermore they could be extended.

However, the limited facilities available at the time meant that liquid crystals formed by polymers could not really be purified and characterized, but as late as 1936, Vorländer was still optimistic about the prospects for polymer liquid crystals:

> ...one can study the change of molecular properties if one lengthens the molecules in one dimension. Will we perhaps arrive at a new state...(one which might be called)...supracrystalline-liquid.

The beginnings of the modern polymer industry can be traced to the 1930s, when such familiar materials as nylon, polythene (polyethylene), Teflon® (polytetrafluoroethylene), polystyrene, and polyvinyl chloride (PVC) were synthesised for the first time. Liquid crystal polymers took longer to be identified. In fact it took some time for one of the most remarkable materials of modern times to be recognized as a liquid crystal polymer.

In 1965, a young synthetic chemist, Stephanie Kwolek, working for Dupont in the USA, discovered a polymer of exceptional strength. The patent was filed in 1972,[28] and the material has since become known as Kevlar®. Fibres of Kevlar have five times the strength of comparable steel fibres and the material has found many applications, including body armour, helmets, and even suspension cables for bridges. The chemical constitution of Kevlar is relatively simple (see Technical box 12.4), but its great strength comes from its liquid crystallinity. The crude polymer can be dissolved in concentrated sulphuric acid, forming what we now recognize as a nematic liquid crystal solution. By spinning fibres from this solution, the long molecules of polymer are aligned and form a structure with extreme tensile strength.

TECHNICAL BOX 12.4 Polymer liquid crystals

The term 'polymer liquid crystals' carries a hint of what is special about liquid crystals. Polymers are a particular type of molecule, large with repeating units, which may in principle exist as a solid, liquid, or, less likely, gas. Scientists do not usually link the molecular or atomic species with a particular state of matter. However, the designation of polymer liquid crystals as something special reflects the fact that liquid crystals are defined by an orientational and sometimes spatial organization. For non-polymeric systems, molecular organization and the consequent state of matter is determined by *intermolecular* forces or *interatomic* forces, but in polymers macroscopic structures are determined much more by the internal chemical structure of the polymer molecule. The special properties of liquid crystals all derive from the particular interplay of the microscopic molecular structure and the molecular organization, which results in the formation of a liquid crystal phase. In polymer liquid crystals this structure–property relationship is especially strong.

Main chain liquid crystal polymers, as illustrated below, look like regular polymers—they are extended structures in which the units are connected end to end. However, they are made up of units, sub-molecules, which themselves have a tendency to form liquid crystal phases. Given sufficient flexibility of the links, polymer chains of rod-like sub-molecules can organize themselves to form nematic and smectic phases. If the molecular units are disc-like, then corresponding columnar phases are also possible.

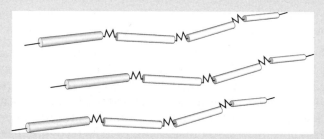

A main chain polymer liquid crystal made up of rods connected by flexible links.

For some polymers, there can be interactions between the chains, even chemical bonds, as in cross-linked polymers or elastomers, which modify the properties of the resultant materials. A particular instance of this is with the high-strength polymer Kevlar. The chemical structure of Kevlar is shown below, and its extraordinary high tensile strength comes from the bonds between the polymer chains. These are a special type of bond, called hydrogen bonds, which also provide the connectivity in biological polymer molecules.

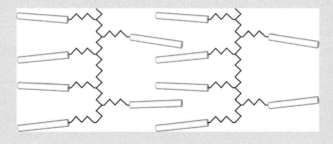

A schematic image of the liquid crystal main chain polymer Kevlar—the arrow indicates the alignment direction of the polymer fibres.

Side-chain liquid crystal polymers are more flexible than main-chain liquid crystals and share many of the properties associated with low molecular weight liquid crystals. An image of a side-chain liquid crystal polymer is shown below. Such polymers may form nematic and smectic phases, and can also be formed from disc-like molecular sub-units.

The first convincing proof for the existence of liquid crystal polymers came in 1975 with the publication of a brief paper on 'Mesophase structures in polymers'.[29] This gives all the required evidence of optical textures, X-ray structural data, and thermodynamic measurements to confirm the materials as genuine liquid crystals, although with hindsight other liquid crystal polymers such as Kevlar had already been reported.

These polymeric materials are examples of *main-chain* liquid crystal polymers. This means that the component units are connected to each other end to end, as in the classic polymer model. Such a structure has some limitations as far as possible applications are concerned, and so a number of researchers embarked on a search for *side-chain* liquid crystal polymers (see Technical box 12.4). In these structures the liquid crystal units are connected at only one end to a polymer chain. A consequence is that the materials are much more flexible, although they do not have the tensile strength of main-chain polymers.

The breakthrough in 1978 was due to German chemists, Heino Finkelmann and his senior colleague Helmut Ringsdorf,* from the University of Mainz, working with the polymer physicist Joachim Wendorff, who at the time was head of the Physics Department at the Deutsche Kunststoff-Institut (Plastics Institute) in Darmstadt. Their achievement was to attach liquid crystal monomer units to a polymer back-bone via a flexible link consisting of an alkyl chain.[30] The result was a polymer, but one which had many of the properties associated with low molecular weight liquid crystals. Indeed there are applications of side-chain liquid crystal polymers in optical displays and other devices for optical communications. The advantage of polymer liquid crystals in these situations is that they are structurally more stable. There is no need for containing surfaces and careful surface alignment. There are also disadvantages. For example, no liquid crystal polymer material could respond fast enough to act as a flat-screen TV, but for some applications physical robustness is more important than the speed of switching.

Polymers complete the catalogue of materials that can be formed in a liquid crystalline state. Some simple applications have been mentioned, but liquid crystalline biopolymers have great potential. Possible applications of biocompatible synthetic liquid crystal polymers are as muscle fibres, nerve fibres, and other elements of tissue.

There is no end in sight for the applications of liquid crystals, the fourth state of matter.

---

* Helmut Ringsdorf provides another link with the group at Halle. Ringsdorf was one of Hermann Staudinger's later students, completing his Ph.D. at the University of Freiburg in 1958.

Glossary

=====≍≍◇◇◇≍=====

Acid   A potentially corrosive liquid or water solution of a chemical that tends to react with some metals. The active chemical species in acids are positively charged hydrogen ions.

Active matrix   A term used to describe the arrangement of electronic elements in a semiconductor integrated circuit. For liquid crystal displays an *active matrix* is an array of *pixels* (picture elements) that can be separately activated by individual transistor switches (thin film transistors, TFTs) attached to each pixel.

Amphiphile   A molecule that contains different chemical groups that interact differently with water. One group may be hydrophilic (attracted to water), while the other group is hydrophobic (repelled by water). In many cases the hydrophobic part of the molecule is a flexible hydrocarbon chain. Fatty acids and lipids are examples of amphiphilic molecules encountered in liquid crystal phases.

Anisotropic   Adjective referring to objects (e.g. molecules or phases) that show unequal properties in different directions. Opposite of *isotropic*.

Birefringence   The property of a material which indicates that the speed of a light beam depends on its polarization. Sometimes called *double refraction*. When objects are viewed through a birefringent material, sometimes two distinct images can be seen. Birefringence is readily detected in a polarizing microscope.

Cathode-ray tube (CRT)   A cathode-ray tube is a vacuum tube having a fluorescent screen which lights up when activated by a beam of electrons. By arranging for the beam to scan across the screen inside the vacuum tube and deflecting the beam up and down, images can be created on the screen and viewed from outside the tube, hence acting as a display device. The electron beam acts as drawing device, activated by electrical signals applied to the electron beam source (the electron gun). Before 2000, most TV sets and desk-top computers had CRT screens. They have now been replaced by much less bulky liquid crystal displays.

Chiral   Adjective that can refer to a molecule, object, or material, from the Greek word for a hand. Chiral objects exhibit handedness. A chiral *molecule* or object is not equivalent to its mirror image. A chiral *material* will be able to distinguish left from right through physical and sometimes chemical interactions.

Chiral nematic   Modern description for the cholesteric liquid crystal phase identified by G. Friedel.

Cholesteric   Adjective referring to a liquid crystal *phase*.

Circularly polarized light   A kind of polarized light in which the associated electric and magnetic fields describe a spiral motion as the light propagates. Circularly polarized light can be left-handed or right-handed depending on the rotation sense of the spiral. It is often emitted from astronomical sources, and is regarded as strong evidence that in the region where the light is being emitted there are strong magnetic fields or very rapid rotations. In cholesteric liquid crystals, circularly polarized light is selectively propagated; depending on the handedness of the liquid crystal, left or right circularly polarized light will be transmitted or reflected to different extents.

Colloid A mixture of two phases, which normally do not mix. Two immiscible liquids shaken hard may form an apparently uniform fluid known as an emulsion colloid. Colloidal mixtures may be created by combining liquids and solids, or gases and liquids, or gases and solids.

Columnar phase A liquid crystalline phase consisting of disc-like molecules arranged in parallel columns. The columns pack together as in a two-dimensional crystal, but within each column the disc-like molecules can move freely.

Continuum theory A type of theory of a solid or fluid which regards the material as continuous and which follows flows of material and stresses through small elements of these continuous materials. Such a theory is usually used to discuss liquid and gas flows, solid deformations, and sound propagation. In the context of liquid crystals, the continuum theory is known as the *distortion theory*. A continuum theory is different from a *molecular* theory, which requires input of intermolecular forces. For liquid crystals, the original molecular theory was known as the *swarm theory*.

Crossed polarizers This is the usual arrangement of the two *polarizers* in a polarizing microscope. Because the directions of the polarizers which admit light are at right angles, then in the absence of a birefringent sample, no light passes through the polarizers. If the sample consists of an optically isotropic medium, or a birefringent sample viewed along one of its optic axes, then no light is transmitted. This set-up is particularly appropriate for investigating liquid crystals, since only optically anisotropic materials are visible.

Crystal A solid material in which the atoms or molecules are arranged in a precise repeating pattern (a crystal lattice). When viewed on a large scale, crystals often show sharp faces known as *facets*.

Defect Region (point, line or plane) in which the special directional properties of a liquid crystal change or become ambiguous. Analogue of a *grain boundary* between different crystal principal directions in a solid.

Dielectric properties These are properties of a material, whether a gas, liquid, liquid crystal or solid, that quantify or measure how that material behaves when an electric voltage is applied to the material. In general the dielectric properties of materials depend on many quantities, such as temperature, pressure, and, most importantly, the frequency of the electrical signal applied. Dielectric properties are always measured in comparison with the response of a vacuum. The dielectric response of a vacuum (measured as the *dielectric constant*) is conventionally set equal to 1. For liquid crystals, the most important dielectric property is the *dielectric anisotropy* (also known as the *permittivity anisotropy*). If the dielectric anisotropy is positive it means that the *dielectric constant* is greater *along* the director than *across* the director, and a liquid crystal will lower its energy by turning the director along the direction of the electric field. If the dielectric anisotropy is negative, then the liquid crystal will align its director perpendicular to the electric field.

Diode A diode is the basis of all semiconductor devices. It consists of a wafer of two different materials, the structures of which are such that electrons travel more easily from one of the materials to the other, rather than in reverse. It acts as an electrical rectifier. By particular choice of materials, it is possible to generate light from diodes by application of an electric voltage (hence, light emitting diode or LED). The light is only generated if the bias (positive or negative voltage) is connected the right way round to the different materials of the diode.

Director   A special direction in a *liquid crystal*. Often material properties which depend on direction (e.g. speed of light or sound) depend only on the angle between that direction and the director. Particularly used in discussion of the *nematic* phase.

Display device   An electronic device that selectively transmits, reflects, blocks, or produces light in such a way that information can be transmitted to the eye. Might be part of a watch, calculator, measuring device, computer screen, TV or similar.

Distortion theory   See *continuum theory*.

Electrode   This is an electrical conductor, usually a metal, or in the case of liquid crystal displays a conducting deposit of indium-tin oxide, which allows an electrical voltage to be applied to a material. The external source of voltage is applied to the electrode(s).

Electroluminescence   This is the generation of light from a material by applying a voltage to the material. The material is usually a solid crystal, but polymers can be made to exhibit electroluminescence. The materials may be pure compounds, but more usually are mixtures or composites. An LED is an example of an electroluminescent device.

Emulsion   A colloid in which two immiscible liquids are dispersed in each other. See the definition of *colloid*.

Fatty acid   A fatty acid is a compound containing a mildly acid group (carboxylic acid) attached to a hydrocarbon chain. The molecules are amphiphilic and so form lyotropic liquid crystals in water. Salts of fatty acids, for example with sodium, potassium, ammonium, etc., are classic soaps. As the name implies, fatty acids are traditionally derived from animal and vegetable fats.

Ferroelectricity   Ferroelectricity is the electric analogue of ferromagnetism (see Technical box 3.2). In a ferroelectric material, pairs of positive and negative charges (dipoles) line up in a regular array, just like the north and south poles of magnets. The electric dipoles form domains, and the domains are arranged so as to minimize the overall energy.

Focal conic liquid   Old terminology, introduced around 1910 by Georges Friedel for a material which exhibited the focal conic texture. Term abandoned by Friedel in 1922 when he introduced the term *smectic*.

Focal conic texture   A texture that looks like a set of overlapping fans or cones. The texture was identified by Georges Friedel in 1922 as being characteristic of a smectic phase. Layers in a non-uniform smectic film will collide with each other on particular types of so-called 'focal conic surfaces' if the layer thicknesses are to be maintained. The texture is also observed in cholesteric or chiral nematic phases.

Frederiks (or Fréedericksz) threshold/transition   The phenomenon in a liquid crystal (usually a nematic) whereby there is a critical minimum electric voltage or magnetic field necessary to cause the director in a thin liquid crystal film to begin to change direction.

Isotropic   Describes properties that do not change with orientation. In the context of *liquid crystals*, the higher temperature liquid phase is usually isotropic, but becomes *anisotropic* at lower temperatures. Thus *isotropic phase* or *isotropic liquid*.

LCD   *Liquid Crystal Display*. Display device using liquid crystals as the active element in the transmission or reflection of light. At the time of writing the LCD is the device most commonly used for the display of information on TV and computer screens. By contrast with the bulky CRT screens, LCDs are a type of flat screen.

Light polarization   Light is a type of electromagnetic wave. It is made up of rapidly vibrating electric and magnetic fields, with the direction of the electric and magnetic fields perpendicular to each other and perpendicular to the direction of propagation. The wavelength of visible yellow light is about 50-millionths of a centimetre. The *polarization* is the direction of the electric field.

Lipid   Lipids are a particular type of amphiphilic molecules which have a hydrophilic group and an attached hydrophobic hydrocarbon chain. In the presence of water, lipids will form micelles and bilayer membranes so that the hydrophilic groups are always in contact with the water.

Liquid crystal   A material *phase* intermediate between solid and liquid. More generally, a phase which exhibits some liquid-like and some anisotropic properties. Not all materials exhibit a liquid crystal phase. We have designated the liquid crystal phase as the *fourth state of matter*.

Lyotropic   Adjective describing materials which exhibit liquid crystal behaviour in solutions, usually, although not necessarily, with water as the solvent. The properties of such liquid crystals change with concentration. The changes observed often are comparable with the way the properties of liquid crystals formed by pure compounds change with temperature (*thermotropic liquid crystals*).

Mesogen   A molecule or compound with the propensity to become a *liquid crystal mesophase* under suitable circumstances (e.g. pressure, temperature, water concentration). Similarly *nematogen*, *smectogen*, etc.

Mesomorphic phase or mesophase   Another term for *liquid crystal*. The term was invented by Georges Friedel in 1922, from the Greek words for intermediate and form. Avoids the use of the word *crystal*, and hence is better terminology because liquid crystal might be thought to imply some *positional order*.

Micelle   A self-assembled collection of *amphiphilic* molecules, usually in an aqueous solution. The number of molecules in a micelle varies considerably, but is normally in the region of hundreds, although bigger (and smaller) micelles can be formed. Depending on the nature of the self-assembling molecules, their concentration, and the solvent, micelles can be spherical, cylindrical, or disc-shaped, and can themselves form a nematic phases in a *lyotropic* system.

Multiplexing   In the context of liquid crystal displays, this is an electronic process of switching a number of pixels (picture elements) at the same time or in rapid succession. This can be achieved by sending sequences of electrical pulses to arrays of pixels.

Nematic   Adjective (e.g. nematic *phase* or nematic material) describing circumstance in which there is one special director direction. Often now used as a noun, for example 'At room temperature this compound is a nematic'.

Nicol prism   Prism which diverts an undesired light *polarization* away from the desired direction, thereby acting as a polarizer. In the nineteenth century this was the main optical device used to obtain polarized light.

Nucleic acid   A polymer molecule (*biopolymer*) important in the coding and replication of genetic information in living cells. The building blocks of nucleic acids are nucleotides. DNA (deoxyribonucleic acid) and RNA (ribonucleic acid) are the nucleic acids normally encountered. Neither behave as acids in the usual chemical sense.

Nucleotide   These are small molecules that combine in particular sequences to form nucleic acids. It is the sequences of nucleotides that code the genetic information in biological cells.

Optic axis   An axis in an *anisotropic* medium, along which light polarized in any direction perpendicular to the propagation direction travels at equal speeds. There is no double refraction or optical interference for light travelling along this direction.

Order parameter Quantity, usually between 0 and 1, describing the degree of orientation of molecules in a liquid crystal (or, more generally, the degree of order in an ordered thermodynamic system).

Optically active Adjective describing a medium that will cause the plane of polarization of a plane-polarized light wave to rotate as it travels through the medium.

Orientational order Arrangement of the orientations of molecules in a *phase* in which some special directions are preferred. To be distinguished from *positional order*.

Passive matrix A passive matrix describes an array of pixels (picture elements) that do not have their own individual transistor switches. The pixels in a *passive matrix display* have to be activated by applying voltages to selected columns and rows of the matrix. This is achieved by a complex scheme of *multiplexing*, applying sequences of electrical pulses to the columns and rows. See *active matrix*.

Permittivity or electric permittivity See *dielectric properties*.

Phase In this book, this term is usually applied to a *thermodynamic* phase, which is a state of matter in which (traditionally) a material has the properties of a solid, liquid or a gas. Liquid crystals are a system of material phases. For the liquid crystalline materials discussed in this book, the term *phase* also applies to an *isotropic liquid*, a *nematic*, and a *smectic*. The change between thermodynamic phases as a consequence of a change in temperature or pressure is known as a *phase transition*.

The term is also used with an entirely different meaning in the context of wave physics to describe the relative positions of points on a wave: in-phase if the points are at the same relative position (0°, 360°, etc.) and out-of-phase if the relative positions of the points are exactly half a wavelength different (180°, etc.). Between these, the points are some number of specified number of degrees out of phase.

Phospholipid A particular type of lipid molecule (amphiphile) important in biological systems as the structural component of cell membranes. These molecules have a pair of hydrocarbon chains attached to the hydrophilic group, which itself contains a phosphate group.

Pixel Picture element. Depending on the resolution or picture quality, an LCD screen will have many thousands of pixels. For example, the standard visual graphics array (VGA) has 640 × 480 pixels, while the extended graphics array has 1024 × 768 pixels. Modern high definition screens have many more. Each pixel can itself be either on or off, but by combining them alphanumeric characters or pictures can be reproduced. Coloured images are created by having pixels with different coloured filters of red, green, and blue.

Plane polarization A particular type of polarized light in which the polarization of the light is confined to a plane defined by the oscillation direction of the electric field associated with the light wave and the direction of propagation of the light.

Polarized light Usually light that we see consists of a mixture of polarizations. In polarized light, a particular *polarization* has been selected, using a suitable *polarizer*. The light may be *plane-polarized*, *circularly polarized* (see above), or even *elliptically polarized* (not defined here).

Unpolarized light consists of all possible polarizations.

Polarizer An optical device which either absorbs or diverts light of a particular polarization, transmitting only light of a defined polarization. The device may be an assembly of birefringent elements or, more usually, a thin film of a suitable polymer. A polarizer illuminated by unpolarized light transmits only polarized light. Normal polarizers transmit *plane-polarized* light, but there are polarizers that will transmit *circularly polarized* light.

Polymer   A molecule consisting of many small units (*monomers*) bonded together in a chain. A *high* polymer may contain hundreds of thousands or even millions of monomers.

Positional order   Arrangement of atoms or molecules in a phase so that the atoms or molecules prefer certain regular positions or ranges of positions. Perfect (or near perfect) positional order occurs in a *crystal lattice*.

Schlieren texture   A texture commonly exhibited by thin films of a nematic phase and observed in a polarizing microscope very early in the history of liquid crystal studies. Dark thread-like areas are seen to permeate the material, ending at points anchored to the surface of the containing microscope slides.

Smectic   Liquid crystalline phase in which the molecules are organized more or less in layers, with only rather weak molecular interchange between the layers. There are many varieties of smectics, each a distinct phase in the thermodynamic sense, with different degrees of positional and orientational order. The two most common are smectic A and C.

Smectic A   Smectic phase in which the principal molecular direction is perpendicular to the smectic layers.

Smectic C   Smectic phase in which the principal molecular direction is tilted at some angle to the line perpendicular to the smectic layers.

Suspension   Similar to *colloid,* but may be more temporary.

Swarm theory   A form of molecular theory. See *continuum theory*.

Texture   Term used to describe the principal optical properties of a particular phase as seen under the microscope through crossed polarizers. The term is not precise and refers to the common features of the observed pattern: a qualitative rather than a quantitative term. In the early days the detailed origin of the texture was unknown. Nowadays the texture can usually be interpreted in terms of inhomogeneities in the optical properties of the material, which in turn are due to inhomogeneities in the director pattern. *Topological* methods, drawn from pure mathematics, enable the typical texture patterns for each phase to be matched to the kind of *defects* that occur in that phase, and hence (rather indirectly) to the details of the molecular order.

TFT   *T*hin *F*ilm *T*ransistor. Electronic component used in an LCD to provide individual pixels with an electrical switch. See active matrix.

Thermodynamics   The study of energy changes in materials enabling a classification of phases and phase transitions to be made. Particularly studied by physicists, physical chemists, and chemical engineers. Originally developed by nineteenth century engineers trying to understand the operation of machines and mechanical engines. Closely related to *statistical mechanics*, which attempts to explain thermodynamic processes at the molecular level.

Thermotropic   Adjective (but now sometimes noun) describing materials whose liquid crystalline properties change with temperature. To be contrasted with *lyotropic*.

Uniaxial liquid crystal   A liquid crystal with a single optic axis. The nematic and smectic A phases are examples of uniaxial liquid crystals.

Vesicle   A flexible droplet inside a lyotropic medium, with a flexible surface consisting of a bilayer of amphiphilic molecules. Conceptually related to *micelles* and important as a physical analogue of a biological cell.

X-ray Electromagnetic waves with wavelengths of the same order as molecular sizes (e.g. billionths of a metre). Wavelengths are much shorter (by a factor of about 1000) than the wavelength of visible light. Often used for studying materials on a molecular scale, although pictures as we normally understand them are not obtained, but rather patterns on a screen, which have to be analysed mathematically.

Also used to create images of structures of humans and other objects. X-rays, because of their short wavelength, are able to penetrate materials and so reveal their internal structures.

Timeline of events in the history of liquid crystals

1854	Rudolf Virchow first observes myelin fibres.
1888	Beginning of liquid crystal work as we know it. Friedrich Reinitzer (Prague) observes double melting in cholesteryl benzoate, with an intermediate cloudy phase. Also observes appearance of spectacular colours.
1888	Otto Lehmann (Aachen) sees crystallites under the polarizing microscope. Identifies flowing crystals and liquid crystals, and shows that on a microscopic scale they are birefringent.
1890	Ludwig Gattermann (Heidelberg) synthesizes *para*-azoxyanisole (PAA) and identifies 'stains' (*Schliere*), now identified with the Schlieren texture.
1902	Daniel Vorländer's students Meyer and Dahlem (Halle) observe double melting in Halle.
1904	Otto Lehmann publishes *Flüssige Kristalle.*
1905	Gustav Tammann (Göttingen) publicly challenges Lehmann (now at Karlsruhe) over the authenticity of liquid crystals. A Commission of Experts is set up by the German Physical Chemistry Society (*Deutsche Bunsen Gesellschaft*) to examine the claims.
1906	Daniel Vorländer begins serious chemical studies of crystalline liquids in Halle.
1907	Lehmann publishes his lectures *Flüssige Kristalle und Theorien des Lebens*.
1909	Lehmann visits Paris and Geneva and delivers lectures on liquid crystals.
1909	Georges Friedel (St Etienne) identifies the focal conic and thread-like textures.
1911	Charles Mauguin (Paris) observes that magnetic fields can orient liquid crystals and that the polarization plane of light will follow a rotating optic axis.
1914	World War I begins.
1916	François Grandjean (Paris) observes so-called Grandjean terraces.

1917	Grandjean proposes a molecular theory for liquid crystals. It is then forgotten.
1918	World War I ends.
1922	Lehmann dies aged 67.
1922	Friedel (now in Strasbourg) publishes *Les états mesomorphes de la matière.* Proposes the terms *nematic, smectic,* and *cholesteric* to describe different types of liquid crystal.
1923	Edmond Friedel and Maurice de Broglie (Paris) observe smectic layering using X-ray diffraction.
1923	Carl Wilhelm Oseen (Uppsala) first publishes a continuum theory of liquid crystals.
1924	Vsevolod Konstantinovich Frederiks begins work on liquid crystals in Leningrad.
1927	Frederiks effect is first reported in a Russian journal.
1928	Friedrich Reinitzer dies.
1931	Paul Ewald (Stuttgart) edits a definitive volume of papers on liquid crystals in German, known as the *Ewald Symposium.*
May 1933	Sir William Bragg and John Desmond Bernal organize a Faraday Discussion on 'Liquid crystals and anisotropic melts' at the Royal Institution in London. Proceedings are published as first set of liquid crystal papers in English. Water-soluble lyotropic liquid crystals are introduced by A. Stuart C. Lawrence.
December 1933	Georges Friedel dies.
1934	B. Levin and N. Levin (Marconi Company, London) apply for a patent on a liquid-crystal-based light valve.
1936	J.D. Bernal and collaborators observe liquid crystallinity in tobacco mosaic virus (TMV).
1936	V.K. Frederiks arrested (deported 1937).
1939	World War II begins.
1941	Death of Daniel Vorländer.
1944	Vsevolod Konstantinovich Frederiks dies while still under arrest.
1945	World War II ends.
1957	Review article on liquid crystals by G.H. Brown and W.G. Shaw from the USA.
April 1958	J.D. Bernal convenes a second Faraday Discussion on 'Macromolecules and liquid crystals'. F. Charles Frank presents a reformulation of Oseen's continuum theory, first proposed in 1923.
1958	Wilhelm Maier and Alfred Saupe publish a re-discovered version of Grandjean's 1917 molecular theory of liquid crystals. This time the theory endures.

1959	First identification of different smectic liquid crystal phases by H. Sackmann and H. Arnold, labelled A, B, and C.
1962	G.W. Gray publishes *Molecular Structure and Properties of Liquid Crystals.*
	Radio Corporation of America (RCA) begins research on liquid crystals. R. Williams at RCA reports electric-field induced light scattering from a liquid crystal.
1963	James L. Fergason and others file patent for thermal imaging devices using cholesteric liquid crystals.
1965	First official International Liquid Crystal Conference held at Kent State University, Kent, Ohio, USA.
	Liquid Crystal Institute opened at Kent State University.
1968	Pierre-Gilles de Gennes and others at Université de Paris-Sud begin research on liquid crystals.
May 1968	George Heilmeier's group at RCA announces a new display technology based on dynamic light scattering from liquid crystals.
1968	Frank M. Leslie publishes the definitive theory for the dynamics of liquid crystals. Now known as the Leslie–Ericksen theory, since it was jointly developed by Leslie and J.L. Ericksen in 1967 at the John Hopkins University, Baltimore, USA.
January 1970	J.L. Fergason publishes article in *Electrotechnology* on liquid crystals and their applications; he describes a twisted nematic effect.
August 1970	Leslie reports the theory of switching a twisted nematic with a magnetic field at the 3rd International Liquid Crystal Conference held in Berlin, Germany.
December 1970	Martin Schadt and Wolfgang Helfrich from Hoffmann-La Roche submit publication of details of a twisted nematic display and simultaneously patent the device.
1972	The high-strength liquid crystal polymer Kevlar is patented by S. Kwolek, Dupont, USA.
1973	G.W. Gray and colleagues publish details of new room-temperature nematic materials suitable for twisted nematic displays.
1975	Liquid crystal behaviour in a main-chain polymer is confirmed for the first time by A. Roviello and A. Sirigu.
1975	R.B. Meyer discovers ferroelectric behaviour in a chiral tilted smectic C phase.
1977	S. Chandrasekhar and colleagues from the Raman Research Institute, Bangalore, India, report the first discotic (disc-like) liquid crystals.

1978	H. Finkelmann, H. Ringsdorf, and J. Wendorff report for the first time liquid crystal behaviour in a side-chain liquid crystal polymer.
1980	N.A. Clark and S.T. Lagerwall report a liquid crystal display device using on a ferroelectric chiral smectic C material.
1982	C.W. Waters and E.P. Raynes patent a new supertwisted nematic display device.
1991	P.-G. de Gennes receives the Nobel Prize for Physics for research on liquid crystals and other soft matter.
1997	Sharp Japan introduce a 40-in diagonal colour liquid crystal television using active matrix thin-film transistor (TFT) technology.

Chapter notes

Science and history: the two cultures

1. Snow (1993).
2. Elkana (1974); Sambursky (1975).
3. Atkins (2007).
4. Snow (1959).
5. Kuhn (1957).
6. Watson (1968).
7. Hilary Rose, In the Shadow of the Men, *The Guardian* 15 June 2002, cited in http://www.spartacus.schoolnet.co.uk/SCfranklinR.htm (retrieved 30/12/09); review of Sayre (2002) by B.J. Martin at http://www.medscape.com/viewarticle/448302 (retrieved 30/12/09).
8. Sayre (1975); Maddox (2002).
9. Nasar (1998).

Chapter 1   It's all Greek to me: an introduction

1. Stegemeyer and Kelker (1991).
2. Poe (1838).
3. Kelker (1973).
4. Farrington (1944, 1947).
5. Dalton (1808).
6. Maxwell (1866).
7. Brush (1976).
8. Harman (1998).
9. Rowlinson (1988).
10. Rowlinson (2003).
11. Andrews (1869).
12. Cercignani (1998).
13. Brush (1983).
14. Pojman (2009).
15. Boas Hall (1970).
16. Sambursky (1975).
17. Farrington (1949).
18. Payne (1970).
19. Robinson (2006).

Chapter 2 Crystals that flow: fact or fiction

1. Knoll and Kelker (1988), p.50ff. Details of Lehmann's life, as well as the exchange of letters between F. Reinitzer and O. Lehmann, are well documented by Knoll and Kelker.
2. This exchange is recorded in Lehmann (1908b) (Chapter 3), Lehmann (1908c), and Reinitzer (1908).
3. Reinitzer (1888) (translated in Sluckin, Dunmur, and Stegemeyer (2004)).
4. Lehmann (1889) (translated in Sluckin, Dunmur, and Stegemeyer (2004)).
5. Gattermann and Ritschke (1890) (translated in Sluckin, Dunmur and Stegemeyer (2004)). Gattermann's letter to Lehmann is recorded in Lehmann's 1904 book. Ritschke was Gattermann's Ph.D. student.
6. See Lehmann (1890a–d, 1894, 1895, 1900, 1905).
7. Lehmann (1904).
8. For Schenck's reminiscences, see Schenck (1941).
9. Quincke (1894).
10. Tammann (1897).
11. Tammann (1901).
12. Lehmann (1901a).
13. Tammann (1902).
14. Lehmann (1902).
15. Schenck (1905b).
16. The presentation was written up as Schenck (1905a), which also includes Tammann's reply (both translated in Sluckin, Dunmur, and Stegemeyer (2004)).
17. Bredig and Schukowsky (1904).
18. Coehn (1904).
19. A summary of the exchange of letters is recorded in Knoll and Kelker (1988), translated by TJS (who accepts responsibility for any errors).
20. Lehmann (1905).
21. Tammann (1906).
22. Schenck (1909).
23. See, for example, the classic text *Ganot's Physics*, Ganot (1875).

Chapter 3 Liquid crystals, where do they come from?

1. Schenck (1905b).
2. Meyer and Dahlem (1903).
3. Vorländer (1906).
4. Vorländer (1907a).
5. Vorländer (1923).
6. Vorländer et al. (1910).
7. Vorländer (1908).
8. Lehmann (1908b).
9. Reinitzer, (1908).
10. Lehmann (1914).
11. Weiss (1907); see also the German version in Weiss (1908). Note that Bose cites the German version.
12. Bose (1907).
13. Bose (1908).
14. Brown and Shaw (1957).

15. Kramer (1911).
16. Born (1916).
17. Meyer *et al.* (1975), and see Chapter 12, p.14.
18. Szivessy (1925).
19. For a popular account of the development of quantum theory see Kumar (2008).
20. Lehmann (1901b), translated by TJS.
21. See Knoll and Kelker (1988), p.121ff.
22. Schleiermacher and Schachenmacher (1923).

Chapter 4   La Gloire Française

1. Mauro (1971).
2. von Helmholtz (1821–1894); von Helmholtz (1847).
3. Newton (1686).
4. Newton (1704).
5. Schenck (1905b).
6. Viola (1901).
7. Amerio (1901).
8. Hulett (1899).
9. Roozeboom (1904). For a brief discussion of Roozeboom's contribution, see Snelders (1997).
10. Rotarski (1903). In Russian Tadeusz became Fyodor, so sometimes he is credited with a forename beginning with F.
11. Vorländer (1908).
12. Wallerant (1905).
13. Wallerant (1906a, 1906b, 1906c).
14. Wallerant (1907).
15. Kelker and Knoll (1990), p.25.
16. Gaubert (1907).
17. Friedel (1907).
18. Reported in Knoll and Kelker (1988), pp.20–24, translated by TJS.
19. Lehmann (1909a).
20. Lehmann (1909b).
21. Mauguin (1911), translated by DAD.
22. Friedel (1995).
23. Wallerant (1909).
24. Friedel (personal communication).
25. See Friedel and Grandjean (1910a, b).
26. Friedel and Grandjean (1910c,d).
27. Friedel and Grandjean (1910b).
28. Grandjean (1916).
29. Grandjean (1917).
30. Friedel (1922). Partly translated in Sluckin, Dunmur, and Stegemeyer (2004), from where the quotations are drawn.
31. Reproduced in Kelker and Knoll (1990), p.37.

Chapter 5   The meeting that wasn't and the meeting that was

1. Vorländer (1923).
2. Friedrich, Knipping, and Laue (1912) (reprinted 1922).

3. See Wyart (1962a, b) for a discussion of this period of development of X-ray studies.
4. Friedel (1922).
5. Letter, Bragg Archive, Royal Institution, London.
6. Bragg Archive, Royal Institution, London.
7. Letter from Sir William Bragg to Georges Friedel dated 23 January 1925, Bragg Archive, Royal Institution, London.
8. Letter from Georges Friedel to Sir William Bragg dated 19 May 1924, Bragg Archive, Royal Institution, London.
9. Letter from Sir William Bragg to Georges Friedel dated 20 February 1925, Bragg Archive, Royal Institution, London.
10. Letter from P.P. Ewald to W.L. Bragg, Bragg Archive, Royal Institution, London.
11. Friedel and Friedel (1931).
12. Vorländer (1907b).
13. Vorländer, Daümer, Selke, and Zeh (1927); the remark cited is on p.468.
14. Weygand (1939).
15. Kast (1955).
16. Vorländer (1931).
17. Oseen (1929).
18. Oseen (1923).
19. For more information on Frederiks, see Sonin and Frenkel (1995) and Sonin (1988) (both in Russian).
20. Sonin (1988), p.27.
21. Repiova and Frederiks (1927).
22. Freedericksz and Repiewa (1927).
23. Frederiks and Zolina (1929).
24. Bragg Archive, Royal Institution, London.
25. Jenkin (2008), p.401.
26. Letter from Georges Friedel to Sir William Bragg dated 22 May 1933, Bragg Archive, Royal Institution, London.
27. Zocher and Birstein (1929).
28. Bernal and Crowfoot (1933).
29. Brown (2005), p.146.
30. Letter from Georges Friedel to Sir William Bragg dated 22 May 1933, Bragg Archive, Royal Institution, London.
31. Letter from Georges Friedel to Sir William Bragg dated 7 June 1933, Bragg Archive, Royal Institution, London.
32. Bragg (1934).

Chapter 6   The threads of life

1. Lehmann (1906).
2. Lehmann (1907).
3. Lehmann (1908a).
4. Shelley (1818).
5. Davies (1998), p.45.
6. Kauffman (1996).
7. Kauffman (2008).
8. Tanford (2004), p.83.

9. Franklin (1774).
10. Olby (1967), p.120.
11. Lehmann (1895).
12. Taken from Lehmann (1904), p.254, Fig. 468.
13. Redrawn from Lehmann (1904), p.255, Fig. 471.
14. Lehmann (1906).
15. Lehmann (1906), pp.789–793.
16. Haeckel (1917), when he was 83.
17. Haeckel (1917).
18. Zsigmondy and Bachmann (1912).
19. Wells (1921).
20. Lawrence (1929).
21. Bernal (1933).
22. Brown (2005).
23. Bawden, Pirie, Bernal, and Fankuchen (1936).
24. Stauff (1939).
25. Oken (1805).
26. Schwann and Schleyden (1847).
27. Gortner and Grendel (1925), p.439.
28. Copied with permission from Goodby (1998).
29. Guck (2009).

Chapter 7   The winds of war

1. Weygand (1943b).
2. See, for example, Kaplan (2001).
3. Yourgrau (1951).
4. For a biography of Weygand, see Remane (1994), p.183–191.
5. Weygand (1931, 1938, 1941).
6. Weygand (1942).
7. Remane (1994), p.188.
8. Kelker (1973).
9. Reitstötter (1963).
10. Donnan (1942).
11. Bodleian library file MS SPSL 227–8/456 (original in German, translated by TJS).
12. Bodleian library file /454 (translated by TJS).
13. See http://www.academic-refugees.org/history.asp (retrieved 5 May 2009) and Zimmerman (2006).
14. Bethe and Hildebrandt (1988).
15. Bodleian library file /451–3.
16. Reitstötter (1963).
17. Dr Aida Espinola, private communication.
18. See, for example, Bergen (2003), p.168ff. Jews married to Gentiles were lower down on the list for transportation.
19. Bodleian library file /465.
20. Bodleian library file /472.
21. Bodleian library file /470.
22. Bodleian library files/471, 473–480.
23. Espinola (1994).

24. Dreyer (1971).
25. Unna (2000).
26. Katz (2004).
27. See particularly the biography of Bernal by Brown (2005).
28. Bernal and Crowfoot (1933).
29. Bawden, Pirie, Bernal, and Fankuchen (1936).
30. Brown (2005), p.105ff.
31. Brown (2005), p.109.
32. Brown (2005), p.97.
33. Sonin and Frenkel (1995), p.75.
34. Brown (2005), p.115.
35. Brown (2005), p.134.
36. Brown (2005), p.169.
37. Brown (2005), p.266.
38. Sonin (1988) and Sonin and Frenkel (1995).
39. Sonin and Frenkel (1995), p.110.
40. Sonin and Frenkel (1995), p.113.
41. Saltykov-Shchedrin (1870), translated by Foote (1980).
42. Gorelik (1997) pp.72–77.
43. Sonin and Frenkel (1995), p.120. The authors are grateful to Professor I. Pynkevych (Kiev) for advice in translating this letter.
44. Sonin and Frenkel (1995), p.125.
45. Tsvetkov (1942).
46. Filippov (1999).

Chapter 8   Renaissance

1. Kelker (1973).
2. World War II History (www.104infdiv.org/saale).
3. *Newsweek*, April 30 1945.
4. Vorländer *et al.* (1937).
5. Weygand (1943a).
6. Gray, Jones, and Marson (1957).
7. Brown and Shaw (1957).
8. Arnold and Sackmann (1959).
9. Maier and Saupe (1958).
10. *The Times*, 4 October 1949; Brown (2005), p.311.
11. Bawden, Pirie, Bernal, and Fankuchen (1936).
12. Maddox (2002).
13. Bernal (1958).
14. Frank (1958).
15. Bills (1988).
16. Sackmann and Demus. (1966).
17. Chistyakov and Chaikowski (1969).
18. Chistyakov, Schabischev, Jarenov, and Gusakova (1969).
19. Bills (1988), p.215.
20. French Literature Companion, May 1968 (www.answers.com).
21. Ross (2002).
22. Edwards (1973).
23. de Gennes (1974).

24. Sluckin, Dunmur, and Stegemeyer (2004), p.357.
25. Landau (1937). For an assessment of Landau's career and impact, see, for example, Kojevnikov (2002).
26. Tsvetkov (1942).
27. Grandjean (1917).
28. de Gennes (1971).
29. Pieranski and Guyon (2007).
30. Orsay Group on Liquid Crystals (1969).
31. Mięsowicz (1946).
32. Oseen (1933).
33. Carlsson and Leslie (1999).
34. For full details of the contributions of Leslie, see Atkin and Sluckin (2003).
35. Fisher and Frederickson (1969).
36. Leslie (1969).
37. Leslie, Stewart, and Nakagawa (1991).
38. Parodi (1970).
39. Hansen (2007).
40. Interview between de Gennes, Herve Arribart, and Bernadette Bensaude-Vincent, 2 May 2002. www.sfc.fr/Material/hrst.mit.edu (17 September 2009).

Chapter 9   An unlikely story

1. Steele, and McConnell (2004); http://www.oxforddnb.com/view/article/343000.
2. Björnståhl (1918).
3. Levin and Levin (1936).
4. Lyons (1966), p.370.
5. Gray (1962).
6. Williams (1963).
7. Fergason (1964).
8. Castellano (2005), p.38.
9. Heilmeier, Zanoni, and Barton (1968).
10. *Chicago Tribune*, 30 May 1968.
11. See Castellano (2005), p.71.
12. W. Helfrich, private communication (2009).
13. Kelker and Scheurle (1969).
14. Kawamoto (2002).
15. Fergason, Taylor, and Harsch (1970).
16. Leslie (1970).
17. Fergason (1971).
18. Schadt and Helfrich (1971).
19. Helfrich and Schadt (1970).

Chapter 10   The light dawns in the West

1. Helfrich and Schneider (1965).
2. Williams and Schadt (1970, 1971).
3. Ambrose (1971).
4. Buntz (2005).
5. Schadt and Helfrich (1971).

6. Fergason, Taylor, and Harsch (1970).
7. Fergason (1971).
8. Castellano (2005), p.75.
9. Castellano (2005), p.77.
10. Buntz (2005).
11. Schumacher (1973).
12. Gray, Harrison, and Nash (1973).
13. Quoted in Kawamoto (2002).
14. Heilmeier (1973).
15. Hilsum (1977).
16. Kawamoto (2002).
17. Raynes (1974).
18. Gray, private communication, quoted in Sluckin, Dunmur, and Stegemeyer (2004), p.476.
19. Hilsum (1984). This gives an account of the involvement of RRE in developing liquid crystal devices and materials.
20. Merck Archives, Darmstadt, Germany.
21. Merck (2008), p.25.
22. Merck (2008), p.42.
23. Merck (2008), p.45.
24. Eidenschink, Erdmann, Krause, and Pohl (1977).
25. Belyaev, Litvak, and Samsonov (2001).
26. Blinov, private communication.

Chapter 11   The sun rises in the East

1. Kawamoto (2002).
2. Aftergut and Cole (1978).
3. Waters and Raynes (1982).
4. Waters, Brimmell, and Raynes (1983).
5. Amstutz, Heimgartner, Kaufmann, and Scheffer (1984).
6. Funada, Matsuura, and Wada (1980).
7. Wada, Wada, and Iijima (1986).
8. Fischer (1971).
9. Brody, Asars, and Dixon (1973).
10. Luo, Hester, and Brody (1978).
11. Snell, Mackenzie, Spear, LeComber, and Hughes (1981).
12. Castellano (2005), p.70.
13. Kawamoto (2002).
14. Kawamoto (2002), p.495.
15. Castellano (2005).
16. Castellano (2005), p.183.
17. Castellano (2005), p.285.
18. Baur, Kiefer, Klausmann, and Windscheid (1995). The effect was first reported at the 22nd German Liquid Crystal Workshop, Freiburg in 1993.

Chapter 12   The new world of liquid crystal materials

1. Friedel (1922).
2. Oseen (1933).

3. de Vries (1951).
4. Fergason, Vogl, and Garbung (1963).
5. Cameron, Sachdev, Gishen, and Martin (1991).
6. See *Hallcrest Handbook of Thermochromic Liquid Crystal Technology*, www.globalspec.com.
7. See information sheets from LCR Hallcrest at www.hallcrest.com.
8. Manabe, Sonoyama, Takanishi, Ishikawa, and Takezoe (2008).
9. Vorländer (1923).
10. Chandrasekhar, Sadashiva, and Suresh (1977).
11. Boden and Bushby (1999).
12. Li, Kang, Harden, Zhou, Dai, Jakli, Kumar, and Li (2008).
13. de Gennes (1972).
14. Renn and Lubensky (1988).
15. Srajer, Pindak, Waugh, and Goodby (1990).
16. de Gennes (1974).
17. Goodby (1991).
18. Meyer, Liébert, Strzelecki, and Keller (1975).
19. Clark and Lagerwall (1980).
20. O'Callaghan and Handschy (2001).
21. Escribá, González-Ros, Goñi, Kinnunen, Vigh, Sánchez-Magraner, Fernández, Busquets, Horváth, and Barceló-Coblijn (2008).
22. Lydon (1998b).
23. Lydon (1998a).
24. Tam-Chang and Huang (2008).
25. Perrett (1998).
26. Ringsdorf, Schlarb and Venzner (1988).
27. Vorländer (1923).
28. Kwolek (1972).
29. Roviello and Sirigu (1975).
30. Finkelmann, Ringsdorf, and Wendorff (1978).

Bibliography

General texts on the history of liquid crystals

Castellano, J.A. (2005) *Liquid Gold: the Story of Liquid Crystal Displays and the Creation of an Industry*. World Scientific, Singapore.

Friedel, J. (1995) *Graine de Mandarin*. Editions Odile Jacob, Paris (in French).

Hilsum, C. (1984) The anatomy of a discovery—biphenyl liquid crystals, in *Technology of Chemicals and Materials for Electronics*, Howells, E.R. (ed.). Ellis Horwood, Chichester, Chapter 3.

Kawamoto, H. (2002) The history of liquid-crystal displays, *Proceedings of the Institute of Electrical and Electronic Engineers* **90**, 460–500.

Kelker, H. (1973) History of liquid crystals. *Molecular Crystals and Liquid Crystals* **21**, 1–48.

Kelker, H. (1988) Survey of the early history of liquid crystals. *Molecular Crystals and Liquid Crystals* **165**, 1–42.

Kelker, H. and Knoll, P.M. (1989) Some pictures of the history of liquid crystals. *Liquid Crystals* **5**, 19–42.

Kelker, H. and Knoll, P. M. (1990) *Die Erforschung der flüssigen Kristalle zu Beginn des zwanzigsten Jahrhunderts: Ein Kapitel deutsch-französischer Wissenschaftsgeschichte* (Liquid crystal research at the beginning of the twentieth century: a chapter in Franco–German scientific history) (privately published in German).

Knoll, P.M. and Kelker, H. (1988) *Otto Lehmann: Erforscher der flüssigen Kristalle: Eine Biographie mit Briefen an Otto Lehmann* (Otto Lehmann, discoverer of liquid crystals: a biography of Otto Lehmann including letters) (privately published). Contains a lot of biographical detail on Lehmann, with reproductions of many letters, in German.

Korenic, I. (2000) A millennium of liquid crystals. *Optics and Photonics News* Feb, 16–22.

Sackmann, H. (1989) Smectic liquid crystals, a historical review. *Liquid Crystals* **5**, 43–55.

Schadt, M. (2009) Milestone in the history of field-effect liquid crystal displays and materials. *Japanese Journal of Applied Physics* **48**, 03B001.

Sluckin, T.J. (2007) *Fluidos fora da lei: a história dos cristais líquidos de curiosidade a technologia*. IST Press, Lisbon (in Portuguese).

Sluckin, T.J., Dunmur, D.A., and Stegemeyer, H. (2004) *Crystals that flow: classic papers from the history of liquid crystals*. Taylor and Francis, London and New York.

Sonin, A.S. (1988) *Doroga dlinuyu v vyek: iz istorii nauki o zhidkikh kristallov* (A road reaching back a century: on the history of liquid crystal science). Nauka Publishing House, Moscow (in Russian).

Sonin, A.S. and Frenkel, V.Ya. (1995) *Vsevolod Konstantinovich Freédericksz (1885–1944)*. Nauka Publishing House, Moscow (in Russian). Contains a lot of scientific detail, including equations.

Vill, V. (1992) Early history of liquid crystalline compounds. *Condensed Matter News* **1**(5), 25–28.

Alphabetical list of references consulted or containing more information on topics discussed in this history of liquid crystals

Aftergut, S. and Cole, H.S. (1978) Method for improving the response time of a display device utilizing a twisted nematic composition. US Patent 4,143,947.

Ambrose, E.J. (1971) Liquid crystalline phenomena at the cell surface. *Symposia of the Faraday Society 5 (Liquid Crystals)* 175–186.

Amerio, A. (1901) Sui cristalli liquidi del Lehmann *Nuovo Cimento* **2**, 281–297.

Amstutz, H., Heimgartner, D., Kaufmann, M., and Scheffer, T.J. (1984) Liquid crystal display, US Patent 4634229.

Andrews, T. (1869) On the gaseous and liquid states of matter. *Philosophical Transactions of the Royal Society* **16**, 575–590.

Arnold, H. and Sackmann, H. (1959) Isomorphiebeziehungen zwischen kristallinflüssigen Phasen 4. Mitteilung: Mischbarkeit in binären Systemen mit mehreren smektischen Phasen. *Zeitschrift für Elektrochemie.* **11**, 1171–1177 (translated in Sluckin, Dunmur, and Stegemeyer (2004)).

Atkin, R.J. and Sluckin, T.J. (2003) Frank Matthews Leslie. *Biographical Memoirs of Fellows of the Royal Society* **49**, 315–333.

Atkins, P. (2007) *Four Laws that drive the Universe*. Oxford University Press, Oxford.

Baur, G., Kiefer, R., Klausmann, H., and Windscheid, F. (1995) In-plane switching: a novel electro-optic effect. *Liquid Crystals Today* **5**, 13.

Bawden, F.C., Pirie, N.W., Bernal, J.D., and Fankuchen, I. (1936) Liquid crystalline substances from virus-infected plants. *Nature* **138**, 1051 (reprinted in Sluckin, Dunmur, and Stegemeyer (2004)).

Belyaev, V., Litvak, I., and Samsonov, V. (2001) Displays in USSR, Russian and Commonwealth of Independent States: lost and found priorities. *Proceedings of SPIE* **4511**, 1–12.

Bergen, D.L. (2003) War & genocide: a concise history of the holocaust. Rowman and Littlefield, Lanham, MD.

Bernal, J.D. (1933) Discussion contribution. *Transactions of the Faraday Society* **29**, 1082.

Bernal, J.D. (1954) *Science in History*. Watts, London.

Bernal, J.D. (1958) Introduction. *Discussions of the Faraday Society* **25**, 1.

Bernal, J.D. and Crowfoot, D. (1933) Crystalline phases of some substances studied as liquid crystals. *Transactions of the Faraday Society* **29**, 1032–1048.

Bethe, H.A. and Hildebrandt, G. (1988) Paul Peter Ewald 1888–1985. *Biographical Memoirs of the Fellows of the Royal Society* **34**, 134–176.

Bills, S.L. (ed.) (1988) Kent State/May 4. KSU Press, Kent, OH.

Björnståhl, Y. (1918) Untersuchungen über anisotrope Flüssigkeiten. *Annalen der Physik* **56**, 161–207.

Boas Hall, M. (1970) *Nature and Nature's Laws: Documents of the Scientific Revolution*. Harper and Row, New York.

Boden, N., Bushby, R.J., Clements, J., and Movaghar, B. (1999) Device applications of charge transport in discotic liquid crystals. *Journal of Materials Chemistry* **9**, 2081–2086.

Bodleian library file MS SPSL 227–8 (Zocher).

Born, M. (1916) Über anisotropen Flüssigkeiten. Versuch einer Theorie der flüssigen Kristalle und des elektrischen KERR–Effekts in Flüssigkeiten; Sitzungsber, Preuss. *Akademie der Wissenschaft* **30**, 614–645. This is the report of the maths and physics session of the Prussian Academy of Sciences dated 25 May 1916.

Bose, E. (1907) Für und wider die Emulsionsnatur der kristallinischen Flüssigkeiten. *Physikalische Zeitschrift* **8**, 513–517.

Bose, E. (1908) Zur Theorie der anisotropen Flüssigkeiten. *Physikalische Zeitschrift* **9**, 708–713 (translated in Sluckin, Dunmur, and Stegemeyer (2004)).

Bragg, Sir William (1934) Liquid Crystals, The Naft. *Association of Oil Producing Countries Magazine* **10**, 3–12.

Bredig, G. and Schukowsky, N. (1904) Prüfung der nature der flüssigen Kristalle mittels Kataphorese (Test of the nature of liquid crystals using cataphoresis). *Berichte der Deutschen Chemischen Gesellschaft* **37**, 3419–3425.

Brody, T.P., Asars, J.A., and Douglas Dixon, G. (1973) A 6 × 6 inch 20 lines-per-inch liquid-crystal display panel. *Institute of Electrical and Electronic Engineers Transactions on Electron Devices* **20**, 995.

Brown, A. (2005) *J.D. Bernal, The Sage of Science*. Oxford University Press, Oxford.

Brown, G.H. and Shaw, W.G. (1957) The mesomorphic state: liquid crystals. *Chemistry Reviews* **57**, 1049–1157.

Brush, S.G. (1976) *The Kind of Motion We Call Heat: A History of the Kinetic Theory of Gases in the 19th Century*. North Holland, Amsterdam (reprinted 1986, Elsevier).

Brush, S.G. (1983) *Statistical physics and the atomic theory of matter*. Princeton University Press, Princeton, NJ.

Buntz, G.H. (2005) *Twisted Nematic Liquid Crystal Displays*. Information No. 118, October 2005, Int. Treuhand AG, Basel, Geneva, Zurich.

Cameron, E.W., Sachdev, D., Gishen, P., and Martin, J.F. (1991) Liquid crystal thermography as a screening test for deep vein thrombosis in patients with cerebral infarction. *European Journal of Clinical Investigation* **21**, 548–550.

Carlsson, T. and Leslie, F.M. (1999) The development of theory for flow and dynamic effects for nematic liquid crystals. *Liquid Crystals* **26**, 1267–1280.

Castellano, J.A. (2005) *Liquid Gold: the Story of Liquid Crystal Displays and the Creation of an Industry*. World Scientific, Singapore.

Cercignani, C. (1998) *Ludwig Boltzmann: the Man who Trusted Atoms*. Oxford University Press, Oxford.

Chandrasekhar, S., Sadashiva, B.K., and Suresh, K.A. (1977) Liquid crystals of disc-like molecules. *Pramana* **9**, 471.

Chistyakov, I.G. and Chaikowski, W.M. (1969) The structure of p-azoxybenenes in magnetic fields. *Molecular Crystals and Liquid Crystals* **7**, 269–278.

Chistyakov, I.G., Schabischev, L., Jarenov, B.I., and Gusakova, L.A. (1969) The polymorphism of the smectic liquid crystal. *Molecular Crystals and Liquid Crystals* **7**, 279–284.

Clark, N.A. and Lagerwall, S.T. (1980) Submicrosecond bistable switching in liquid crystals. *Applied Physics Letters* **36**, 899–901 (reprinted in Sluckin, Dunmur, and Stegemeyer (2004)).

Coehn, G. (1904) Über Flüssige Kristalle. *Zeitschrift für Elektrochemie* **10**, 856–857.

Dalton, J. (1808) *A new system of chemical philosophy*. R. Bickerstaff, London. Available at http://www.archive.org/details/newsystemofchemi01 daltuoft.

Davies, P. (1998) *The Fifth Miracle*. Penguin Books, London.

de Gennes, P.-G. (1971) Short range order effects in the isotropic phase of nematics and cholesterics. *Molecular Crystals and Liquid Crystals* **12**, 193–214.

de Gennes, P.-G. (1972) An analogy between superconductors and smectics A. *Solid State Communications* **10**, 753–756.

de Gennes, P.-G. (1974) *The Physics of Liquid Crystals*. Clarendon Press, Oxford.

de Vries, H. (1951) Rotatory power and other optical properties of certain liquid crystals. *Acta Crystallographica* **4**, 219–226.

Donnan, F.G. (1942) Herbert Freundlich 1880–1941. *Obituary Notices of Fellows of the Royal Society* **4**, 27–50.

Dreyer, J.F. (1971) Citation: Prof. Dr. Hans Ernst Werner Zocher. *Molecular Crystals and Liquid Crystals* **12**, xxxix–xli.

Edwards, S. (1973) *Paris Commune 1871*. Quadrangle Press, New York.

Eidenschink, R., Erdmann, D., Krause, J., and Pohl, L. (1977) Substituted pheny-cyclohexanes – a new class of liquid crystalline compounds. *Angewandte Chemie, International Edition* **16**, 100.

Elkana, Y. (1974) *The discovery of the conservation of energy*. Hutchinson Educational, London.

Escribá, P.V., González-Ros, J.M., Goñi, F.M., Kinnunen, P.K.J., Vigh, L., Sánchez-Magraner, L., Fernández, A.M., Busquets, X., Horváth, I., and Barceló-Coblijn, G. (2008) Membranes: a meeting point for lipids, proteins and therapies. *Journal of Cellular and Molecular Medicine* **12**, 829–875.

Espinola, A. (1994) Fritz Feigl's School of Analytical Chemistry in Brazil. *Analytical Proceedings* **31**, 135–139.

Farrington, B. (1944) *Greek Science I*. Pelican Books, London.

Farrington, B. (1947) *Science in Antiquity*, 2nd edn. Oxford University Press, Oxford.

Farrington, B. (1949) *Francis Bacon, Philosopher of Industrial Science*. Henry Newman, New York.

Fergason, J.L. (1964) Liquid crystals. *Scientific American* **211**, 76–85.

Fergason, J.L. (1971) Display devices using liquid crystal light modulation. US Patent 3,731,986.

Fergason, J.L., Vogl, T.P., and Garbung, M. (1963) Thermal imaging devices utilizing a cholesteric liquid crystalline phase material. US Patent 3,114,836 (filed 4 March 1960, granted 17 December 1963).

Fergason, J.L., Taylor, T.R., and Harsch, T.B. (1970) Liquid crystals and their applications. *Electro-Technology* January, 41–50.

Filippov, A.P. (1999) Obituary Professor V. N. Tsvetkov (1910–1999). *Liquid Crystals* **26**, 943.

Finkelmann, H., Ringsdorf, H., and Wendorff, J. (1978) Model considerations and examples of enantiotropic liquid crystalline polymers. *Makromoleculare Chemie* **179**, 273.

Fischer, A.G. (1971) Design considerations for a future electro-luminescent TV panel. *Institute of Electrical and Electronic Engineers Transactions on Electron Devices* **18**, 802.

Fisher, J. and Frederickson, A.G. (1969) Interfacial effects on the viscosity of a nematic mesophase. *Molecular Crystals and Liquid Crystals* **8**, 267.

Frank, F.C. (1958) On the theory of liquid crystals. *Discussions of the Faraday Society* **25**, 19–28.

Franklin, B., Brownrigg, W., and Farish, M. (1774) Of the stilling of waves by means of oil. *Philosophical Transactions of the Royal Society of London*, **64**, 445–460.

Frederiks, V. and Zolina, V. (1929) On the use of a magnetic field in the measurement of the forces tending to orient an anisotropic liquid in a thin homogeneous layer. *Transactions of the American Electrochemical Society* **55**, 85–96 (reprinted in Sluckin, Dunmur, and Stegemeyer (2004)).

Freedericksz, V. and Repiewa, A. (1927) Theoretisches und Experimentelles zur Frage nach der Natur der anisotropen Flüssigkeiten. (Theory and experiment concerning the question of the nature of anisotropic fluids.) *Physikalische Zeitschrift* **42**, 532–546.

Friedel, G. (1907) (cited in Kelker and Knoll (1990)) *Bulletin de la Société Française de Minéralogie* **30**, 326.

Friedel, G. (1922) Les états mesomorphes de la matière. (The mesomorphic states of matter.) *Annales de Physique (Paris)* **18**, 273–474 (partly translated in Sluckin, Dunmur, and Stegemeyer (2004)).

Friedel, G. and Friedel, E. (1931) Les propriétés physiques des stases mesomorphes en general et leur importance comme principe de classification. (The physical properties of mesophases in general, and their importance in a scheme of classification.) *Zeitschrift für Kristallographie* **79**, 1–60.

Friedel, G. and Grandjean, F. (1910a) Les liquides anisotropes de Lehmann, 1re partie. *Comptes Rendus de l'Académie des Sciences* **151**, 327–329.

Friedel, G. and Grandjean, F. (1910b) Observations geometriques sur les liquides à coniques focales, 2ᵉ partie. (Geometric observations on focal conic liquids, 2nd part.) *Comptes Rendus de l'Académie des Sciences* **151**, 442–444.

Friedel, G. and Grandjean, F. (1910c) Les liquides à coniques focales. (Geometric observations on focal conic liquids.) *Comptes Rendus de l'Académie des Sciences* **151**, 762–765.

Friedel, G. and Grandjean, F. (1910d) Observations geometriques sur les liquides à coniques focales. (Geometric observations on focal conic liquids.) *Bulletin de la Société Francaise de Minéralogie* **33**, 409–465.

Friedel, J. (1995) *Graine de Mandarin*. Editions Odile Jacob, Paris.

Friedrich, W., Knipping, P., and Laue, M. (1912) Interferenz-Erscheinungen bei Röntgenstrahlen. (Interference phenomena in X-rays.) *Sitzungberichte der Bayerische Academie der Wissenschaften* **303**, 363. (Republished 1922 in *Naturwissenschaft* **10**, 361.)

Funada, F., Matsuura, M., and Wada, T. (1980) Interference colour compensated double layered twisted nematic display. US Patent 4,443,065.

Ganot, A. (1875) *Elementary treatise on physics, experimental and applied* (popularly known as *Ganot's Physics*), 7th English edn, translated from the French by E. Atkinson. Longmans Green & Co., London.

Gattermann, L. and Ritschke, A. (1890) Ueber Azoxyphenetoläther. (On azoxyphenetole ethers.) *Berichte der Deutschen Chemischen Gesellschaft* **23**, 1738–50 (partly translated in Sluckin, Dunmur, and Stegemeyer (2004)).

Gaubert, P. (1907) (cited in Kelker and Knoll (1990)) Sur les cristaux liquides de deux composés nouveauxde la cholésterine. *Comptes Rendus de l'Académie des Sciences* **145**, 722–725.

Goodby, J.W. (1991) *Ferroelectric Liquid Crystals: Principles, Properties and Applications*. Ferroelectricity and Related Phenomena, Volume 7. Gordon and Breach, Philadelphia.

Goodby, J.W. (1998) Liquid crystals and life. *Liquid Crystals* **24**, 25–38.

Gorelik, G. (1997) The top secret life of Lev Landau. *Scientific American* August, 72–77.

Gorter, E. and Grendel, F. (1925) On bimolecular layers of lipoids on the chromocytes of the blood. *Journal of Experimental Medicine* **41**, 439–443.

Grandjean, F. (1916) L'orientation des liquides anisotrope sur les cristaux. *Bulletin de la Société Française de Minéralogie* **39**, 164–213.

Grandjean, F. (1917) Sur l'application de la théorie du magnétisme aux liquides anisotropes. *Comptes Rendus de l'Académie des Sciences* **164**, 280–283.

Gratzer, W. (2000) *The Undergrowth of Science: Delusion, Self-deception, and Human Frailty.* Cambridge University Press, Cambirdge.

Gray, G.W. (1962) *Molecular Structure and Properties of Liquid Crystals.* Academic Press, London and New York.

Gray, G.W., Jones, B., and Marson, F. (1957) Mesomorphism and chemical constitution, part VIII: The effects of 3' substituents on the mesomorphism of the 4' n-alkoxydiphenyl-4-carboxylic acids and their alkyl esters. *Journal of the Chemical Society (London)* 393–401.

Gray, G.W., Harrison, K.J., and Nash, J.A. (1973) New family of nematic liquid crystals for displays. *Electronics Letters* **9**, 130–131.

Guck, J. (2009) Do cells care about physics? *Physics World* **22**, 31–34.

Haeckel, E. (1917) Foreword to *Crystal Souls–Studies of Inorganic Life. Forma* **14**, 1–204 (1999) (English translation by A. L. Mackay *Forma* **14**, 1–204 (1999).

Hansen, J.-P. (2009) Pierre-Gilles de Gennes, Obituary. *The Independent* May 29, 2007.

Harman, P.M. (1998) *The Natural Philosophy of James Clerk Maxwell.* Cambridge University Press, Cambridge.

Heilmeier, G.H. (1973) Displays–a Pentagon perspective, *Institute of Electrical and Electronic Engineers Transactions on Electron Devices* **20**, 921–925.

Heilmeier, G.H., Zanoni, L.A., and Barton, L.A. (1968) Dynamic scattering: a new electrooptic effect in certain classes of nematic liquid crystals. *Proceedings of the Institute of Electrical and Electronic Engineers* **56**, 1162–1171.

Helfrich, W. and Schadt, M. (1972) Optical device. Swiss Patent 532,261 (filed December 4, 1970, issued February 15, 1972).

Helfrich, W. and Schneider, W.G. (1965) Recombination radiation in anthracene crystals. *Physical Review Letters* **14**, 229–231.

Hilsum, C. (1977) A constructive philosophy on display research. *Institute of Electrical and Electronic Engineers Transactions on Electron Devices*, **24**, 791–794.

Hilsum, C. (1984) The anatomy of a discovery—biphenyl liquid crystals. In *Technology of Chemicals and Materials for Electronics*, Howells, E.R. (ed.). Ellis Horwood, Chichester, pp.43–58.

Hulett, G.A. (1899) Der stetige Übergang fest-flüssig. *Zeitschrift für Physikalische Chemie* **28**, 629.

Jenkin, J. (2008) *William and Lawrence Bragg, Father and Son.* Oxford University Press, Oxford.

Kaplan, D.R. (2001) The science of plant morphology: definition, history and role in modern biology. *American Journal of Botany* **88**, 1711–1741.

Kast, W. (1955) Die Molekul-Struktur der Verbindungen mit kristallin-flüssigen (mesomorphen) Schmelzen. (Molecular structure of compounds with crystalline-liquid [mesomorphic] melts.) *Angewandte Chemie* **67**, 592–601.

Katz, S. (2004) Berlin roots—Zionist incarnation: The ethos of pure mathematics and the beginnings of the Einstein Institute of Mathematics at the Hebrew University of Jerusalem. *Science in Context* **17**, 199–234.

Kauffman, S.A. (1996) *At Home in the Universe*. Penguin Books, London.

Kauffman, S.A. (2008) *Reinventing the Sacred*. Perseus Books, New York.

Kawamoto, H. (2002) The history of liquid crystal displays. *Proceedings of the Institute of Electrical and Electronic Engineers* **90**, 460–500.

Kelker, H. (1973) History of liquid crystals. *Molecular Crystals and Liquid Crystals* **21**, 1–48.

Kelker, H. and Knoll, P.M. (1990) Die Erforschung der flüssigen Kristalle zu Beginn des zwanzigsten Jahrhunderts: Ein Kapitel deutsch-französischer Wissenschafts-geshichte. (Liquid crystal research at the beginning of the twentieth century: a chapter in Franco–German scientific history.) Privately published.

Kelker, H. and Scheurle, B. (1969) A liquid-crystalline (nematic) phase with a particularly low solidification point. *Angewandte Chemie, International Edition* **8**, 884–885.

Knoll, P.M. and Kelker, H. (1988) Otto Lehmann: Erforscher der flüssigen Kristalle: Eine Biographie mit Briefen an Otto Lehmann. (Otto Lehmann, discoverer of liquid crystals: a biography of Otto Lehmann including letters. English translation see Knoll, P.M. and Kelker, H. (2010).

Knoll, P.M. and Kelker, H. (2010) *Otto Lehmann—Researcher of liquid crystals: a biography with letters to Otto Lehmann*. Books on Demand GmbH, Norderstedt.

Kojevnikov, A. (2002) Lev Landau: physicist and revolutionary. *Physics World* **15**(6), 5–9.

Kramer, F. (1911) Emil Boses Werken. *Physikalische Zeitschrift* **12**, 1244–1247.

Kuhn, T.S. (1957) *The Copernican Revolution*. Harvard University Press, Cambridge Mass.

Kumar, M. (2008) *Quantum*. Icon Books, London.

Kwolek, S.L. (1972) *Optically Anisotropic Aromatic Polyamide Dopes*. US Patent 3,671,542 (filed 23 May 1969, granted 20 June 1972).

Landau, L.D. and Landau, L.D. (1937) *К меории фазобых переходов*. (Towards a theory of phase transitions.) *Ж.Е.Т.Ф.* (paper 1) **7**, 19 (1937); (paper 2) **7**, 627 (1937).

English translation in *Collected works of L.D. Landau*. Pergamon, Oxford, 1965; pp. 193–216.

Lawrence, A.S.C. (1929) *Soap films: a study of molecular individuality*. G. Bell and Son, London. http://ia331338.us.archive.org/3/items/soapfilms030939mbp/soapfilms030939mbp.pdf).

Lehmann, O. (1889) Über fliessende Krystalle. (On flowing crystals.) *Zeitschrift für Physikalische Chemie* **4**, 462–472 (translated in Sluckin, Dunmur, and Stegemeyer (2004)).

Lehmann, O. (1890a) Einige Fälle von Allotropie. (Some cases of allotropy.) *Zeitschrift für Kristallographie und Mineralogie* **18**, 464–467.

Lehmann, O. (1890b) Ueber tropfbarflüssige Krystalle. (On drop-forming liquid crystals.) *Wiedemann's Annalen für Physik und Chemie* **40**, 401–423.

Lehmann, O. (1890c) Ueber krystallinischer Flüssigkeiten. (On crystalline liquids.) *Wiedemann's Annalen für Physik und Chemie* **41**, 525–537.

Lehmann, O. (1890d) Die Struktur krystallinischer Flüssigkeiten. (The structure of crystalline fluids.) *Zeitschrift für Physikalische Chemie* **5**, 427–435.

Lehmann, O. (1894) Ueber künstliche Färbung von Krystallen und amorphen Körpern. (On artistic colours in crystals and amorphous bodies.) *Wiedemann's Annalen für Physik und Chemie* **51**, 47–76.

Lehmann, O. (1895) Ueber Contactbewegung und Myelinformer. (On contact motion and myelin formation.) *Wiedemann's Annalen für Physik und Chemie* **56**, 771–788.

Lehmann, O. (1900) Struktur, System und magnetisches Verhalten flüssiger Krystalle und deren Mischbarkeit mit festen. (Structure, system and magnetic behaviour of liquid crystals and their miscibility in solid crystals.) *Annalen der Physik (Series 4)* **2**, 649–705.

Lehmann, O. (1901a) Flüssige Krystalle, Entgegnung auf die Bemerkungen des Hrn G. Tammann. (Liquid crystals: reply to Tammann's remarks.) *Annalen der Physik* **5**, 236–239.

Lehmann, O. (1901b) Physik und Politik, Festrede bei dem feierlichen Akte des Rektoratswechsels. (Braunsche Hofbuchdruckerei, Karlsruhe).

Lehmann, O. (1902) Ueber künstlichen Dichroismus bei flüssigen Krystallen und Hrn. Tammann's Ansicht. (On man-made dichroism in liquid crystals and Tammann's viewpoint.) *Annalen der Physik* **8**, 908–923.

Lehmann, O. (1904) *Flüssige Kristalle sowie Plastizität von Kristallen im allgemeinen, molekulare Umlagerungen und Aggregatzustandsändlerungen.* Wilhelm Engelmann, Leipzig.

Lehmann, O. (1905) Näherungsweise Bestimmung der Doppelbrechung fester und flüssiger Kristalle. *Annalen der Physik* **18**, 796–807.

Lehmann, O. (1906) *Physikalische Zeitschrift* **7**, 722–9, 789–96.

Lehmann, O. (1907) *Die scheinbar lebenden Kristalle.* (*Apparently Living Crystals.*) Verlag Schreiber, Esslingen und München.

Lehmann, O. (1908a) *Flüssige Kristalle und die Theorien des Lebens.* J.A. Barth, Leipzig.

Lehmann, O. (1908b) Zur Geschichte der flüssigen kristalle. (On the history of liquid crystals.) *Annalen der Physik* **25**, 852–860.

Lehmann, O. (1908c) Bemerkungen zu Hr. Reinitzers Mitteilung über Geschichte der flüssigen kristalle. (Remarks on Mr Reinitzer's contribution to the history of liquid crystals.) *Annalen der Physik* **27**, 109–1102.

Lehmann, O. (1909a) Les cristaux liquides. (Liquid crystals.) *Journal de Physique (Paris)* **7**, 713–735.

Lehmann, O. (1909b) Cristaux liquides et modèles moléculaires. (Liquid crystals and molecular models.) *Archives des Sciences Physiques et Naturelles (Geneva)* **28**, 205–226.

Lehmann, O. (1914) Die optische Anisotropie der flüssigen Kristalle. *Physikalische Zeitschrift* **15**.

Leslie, F.M. (1969) Continuum theory of cholesteric liquid crystals. *Molecular Crystals and Liquid Crystals* **7**, 407–420.

Leslie, F.M. (1970) Distortion of twisted orientation patterns in liquid crystals by magnetic fields. *Molecular Crystals and Liquid Crystals* **12**, 57–72.

Leslie, F.M., Stewart, I.W., and Nakagawa, M. (1991) A continuum theory for smectic C liquid crystals. *Molecular Crystals and Liquid Crystals* **198**, 443–454.

Levin, B. and Levin, N. (1936) Improvements in or relating to light valves. *British Patent* **441**, 274.

Li, L., Kang, S.-W., Harden, J., Zhou, X., Dai, L., Jakli, A., Kumar, S., and Li, Q. (2008) Nature-inspired light-harvesting liquid crystal porphyrins for organic voltaics. *Liquid Crystals* **35**, 233–239.

Luo, F.C., Hester, W.A., and Brody, T.P. (1978) Alphanumeric video performance of a 6in x 6in 30 lines per inch thin film transistor liquid crystal display panel. *Proceedings of the Society of Information Display* No.11, p. 94.

Lydon, J.E. (1998a) *Chromonics, Handbook of Liquid Crystals*, Vol. 2B. Demus, D., Goodby, J., Gray, G.W., Speiss, H.-W., and Vill, V. (eds), Wiley–VCH, Weinheim, pp. 981–1007.

Lydon, J.E. (1998b) Chromonic liquid crystal phases. *Current Opinion in Colloid & Interface Science* **3**, 458–466.

Lyons, E. (1966) *David Sarnoff*. Pyramid Books, New York.

Maddox, B. (2002) *Rosalind Franklin: The Dark Lady of DNA*. Harper Collins.

Maier, W. and Saupe, A. (1958) Eine einfache molekulare Theorie des nematischen kristallinflüssigen Zustandes. (A simple molecular theory of the nematic liquid-crystalline state.) *Zeitschrift Naturforschung* **13a**, 564–566 (translated in Sluckin, Dunmur, and Stegemeyer (2004)).

Manabe, T., Sonoyama, K., Takanishi, Y., Ishikawa, K., and Takezoe, H. (2008) Towards practical application of cholesteric liquid crystals to tunable lasers. *Journal of Materials Chemistry* **18**, 3040.

Mauguin, C. (1911) Sur les cristaux liquides de Lehmann. *Bulletin de la Société Française de Minéralogie* **34**, 71–117 (translated in Sluckin, Dunmur, and Stegemeyer (2004)).

Mauro, F. (1971) *Histoire de l'économie mondiale*. Sirey, Paris.

Maxwell, J.C. (1866) On the dynamical theory of gases. *Philosophical Magazine of the Royal Society* **32**, 390–393.

Merck (2008) *Coincidence and Courage, 40 years of liquid crystal research at Merck*. Trademark Publishing, Frankfurt.

Meyer, F. and Dahlem, K. (1903) Azo- und Azoxybenzoësäureester. *Annalen der Chemie* **326**, 331–346.

Meyer, R.B., Liébert, L., Strzelecki, L., and Keller, P. (1975) Ferroelectric liquid crystals. *Journal de Physique* **26**, L69–71 (reprinted in Sluckin, Dunmur, and Stegemeyer (2004)).

Mięsowicz, M. (1946) The three coefficients of viscosity of anisotropic liquids. *Nature* **158**, 27 (reprinted in Sluckin, Dunmur, and Stegemeyer (2004)).

Nasar, S. (1998) *A beautiful mind*. Simon and Schuster, New York (reprinted Faber and Faber, 2002).

Nehring, J. and Saupe, A. (1971) Elastic theory of uniaxial liquid crystals. *Journal of Chemical Physics* **54**, 337–343.

Nehring, J. and Saupe, A. (1972) Calculation of elastic constants of nematic liquid crystals. *Journal of Chemical Physics* **56**, 5527–5528.

Newton, I. (1686) *Mathematical Principles of Natural Philosophy*. First published 1686, translated into English by Andrew Motte, 1729. Modern version in two parts, revised by F. Cajori, University of California Press, Berkeley, 1962.

Newton, I. (1704) *Opticks or a Treatise on Light*. S. Smith and B. Walford, London. Available on the web at http://www.rarebookroom.org.

O'Callaghan, M.J. and Handschy, M.A. (2001) Ferroelectric liquid crystal SLMs – from prototypes to products. *Proceedings of SPIE* **4457**, 31.

Oken, L. (1805) *Die Zeugung*. Goebhardt, Bamberg, Würzburg.

Olby, R.C. (1967) *The origins of Mendelism*. Schocken Books, New York.

Orsay Group on Liquid Crystals (1969) Dynamics of nematic fluctuations in nematic liquid crystals. *Journal of Chemical Physics* **51**, 816–822.

Oseen, C.W. (1923) Versuch einer kinetischen Theorie der kristallinischen Flüssigkeiten, III Abhandlung. (Essay on a kinetic theory of crystalline fluids, part 3.) *Kungligar Svenska Vetenskapakademiens Handligar* **63**(12).

Oseen, C.W. (1929) Die anisotropen Flüssigkeiten. Tatsachen und Theorien. (Anisotropic liquids: experiments and theories.) *Fortschritte der Chemie, Physik und Physikalischer Chemie* **20**(2) Berlin.

Oseen, C.W. (1933) The theory of liquid crystals. *Transactions of the Faraday Society* **29**, 883–900 (reprinted in Sluckin, Dunmur, and Stegemeyer (2004)).

Parodi, O. (1970) Stress tensor for a nematic liquid crystal. *Journal de Physique* **31**, 581.

Payne, A.S. (1970) *The Cleere Observer: A biography of Antoni van Leeuwenhoek*. Macmillan, London.

Perrett, S. (1998) Misshapes and misfits: protein misfolding and disease. *Chemistry & Industry* 18 May, 389.

Perrin, J. (1918) La stratification des lames liquids. (Stratification of liquid layers.) *Annales de Physique (Paris)* **10**, 160.

Pieranski, P. and Guyon, E. (2007) Pierre-Gilles de Gennes, personal recollections. *Liquid Crystals* **34**, 995–1000.

Poe, E.A. (1838) *The narrative of Arthur Gordon Pym of Nantucket*. Harper & Bros., New York.

Pojman, P. (2009) Ernst Mach, in *The Stanford Encyclopedia of Philosophy*, winter 2009 edition, Zalta, E.N. (ed.). Available at http://plato.stanford.edu/archives/win2009/entries/ernst-mach/.

Queguiner, A., Zann, A., Dubois, J.C. and Billard, J. (1980) Mesogenic disc-like molecules with two-fold symmetry axis, in *Liquid Crystals*, Conference Proceedings, S. Chandrasekhar (ed.), Heyden, London, 35–40.

Quincke, G. (1894) Ueber freiwillige Bildung von hohlen Blasen, Schaum und Myelinformen durch ölsaure Alkalien und verwandte Erscheinungen, besonders des Protoplasmas. *Wiedemann's Annalen der Physik und Chemie* **53**, 593–631.

Raynes, E.P. (1974) Improved contrast uniformity in twisted nematic liquid crystal electro-optic display devices. *Electronics Letters* **10**, 141–142.

Reinitzer, F. (1888) Beiträge zur Kenntnis des Cholseterins. (Contributions to the understanding of cholesteryls.) *Monatshefte für Chemie (Wien)* **9**, 421–441 (translated in Sluckin, Dunmur, and Stegemeyer (2004)).

Reinitzer, F. (1908) Zur Geschichte der flüssigen Kristalle. Annalen der Physik **27**, 213–224.

Reitstötter, J. (1963) Hans Zocher. *Kolloid-Zeitschrift und Zeitschrift für Polymere* **190**, 97–98.

Remane, H. (1994) Conrad Weygand und die Deutsche Chemie, in *Medizin, Naturwissenschaft, Technik und Nationalsozialismus. Kontinuitäten und Diskontinuitäten*, Meinel, C. and Voswinckel, P. (eds). GNT–Verlag, Stuttgart, p. 183–191.

Renn, S.R. and Lubensky, T.C. (1988) Abrikosov dislocation lattice in a model of the cholesteric-to-smectic-A transition. *Physical Review A* **38**, 2132–2147.

Repiova, A. and Frederiks, V. (1927) К вопросу о природу анизотропно-жидкогососояния вещества. (On the problem of the nature of the anisotropic state of matter.) *Journal of the Russian Physical Chemistry Society* **59**, 183–200.

Ringsdorf, H., Schlarb, B., and Venzner, J. (1988) Molecular architecture and function of oriented systems: models for the organisation, surface recognition and dynamics of biomembranes. *Angewandte Chemie International Edition (English)* **27**, 113–158.

Robinson, A. (2006) *The last man who knew everything: Thomas Young, the anonymous polymath who proved Newton wrong, explained how we see, cured the sick, and deciphered the Rosetta Stone, among other feats of genius*. One World Publications, Oxford.

Roozeboom, H.W.B. (1904) *Die heterogenen Gleichgewichte vom Standpunkte der Phasenlehre, Erstes heft* (Heterogeneous equilibrium from the standpoint of the Phase Rule, Part I) Vieweg, Braunschweig.

Ross, K. (2002) *May '68 and its afterlives*. Chicago University Press, Chicago.

Rotarski, T. (1903) Ueber die sogenannten flüssigen Kristalle. *Berichte der Deutschen Chemischen Gesellschaft* **36**, 3158.

Roviello, A. and Sirigu, A. (1975) Mesophase structures in polymers. A preliminary account on the mesophases of some poly-alkanoates of p,p'-di-hydroxy-α,α'-di-methyl benzalazine. *Journal of Polymer Science: Polymer Letters* **13**, 455 (reprinted in Sluckin, Dunmur, and Stegemeyer (2004)).

Rowlinson, J.S. (1988) *On the continuity of the gaseous and liquid states*. (Translation of J.D. van der Waals, *Over de continueit van den gas- en vloeistof-toestand*, 1873). North Holland, Amsterdam.

Rowlinson, J.S. (2003) The work of Thomas Andrews and James Thomson on the liquefaction of gases. *Notes and Records of the Royal Society of London* **57**, 143–159.

Sackmann, H. and Demus, D. (1966) The polymorphism of liquid crystals. *Molecular Crystals* **2**, 81–102.

Saltykov-Shchedrin, M. (1870) История сдного города (*Istoriia odnogo goroda*) (1869–1870). (Translation by I.P. Foote, *The History of a Town*. Meeuws, Oxford, 1980).

Sambursky, S. (1975) *Physical Thought from the Presocratics to the Quantum Physicists*. Pica Press, New York.

Sayre, A. (1975) *Rosalind Franklin and DNA*. W.W. Norton (reprinted 2000).

Schadt, M. and Helfrich, W. (1971) Voltage-dependent optical activity of a twisted nematic liquid crystal. Applied Physics Letters **18**, 127–128 (reprinted in Sluckin, Dunmur, and Stegemeyer (2004)).

Schenck, R. (1905a) Über die Natur der kristallinischen Flüssigkeiten und der flüssigen Kristalle. (On the nature of crystalline fluids and liquid crystals.) *Zeitschrift für Elektrochemie* **11**, 951–955 (translated in Sluckin, Dunmur, and Stegemeyer (2004)).

Schenck, R. (1905b) *Kristallinische Flüssigkeiten und flüssige Kristalle*. Wilhelm Engelmann, Leipzig.

Schenck, R. (1909) Bericht über die neueren Untersuchungen der kristallinischen Flüssigkeiten. (Report on recent studies of crystalline liquids.) *Jahrbuch der Radioaktivität und Elektronik*, **6**, 572–639.

Schenck, R. (1941) Aus der Entwicklungszeit der Chemie des festen Zustandes. (Memories of the development of solid state chemistry.) *Zeitschrift für Elektrochemie und Angewandte Physikalische Chemie* **47**, 1.

Schleiermacher, A. and Schachenmacher, R. (1923) Otto Lehmann. *Physikalische Zeitschrift* **24**, 288–291.

Schumacher, E.F. (1973) *Small is beautiful*. Blonde and Briggs, London.

Schwann, T. and Schleyden, M.J. (1847) *Microscopical researches into the accordance in the structure and growth of animals and plants* (translated from German by H. Smith). Sydenham Society, London. Available on the web from http://vlp. mpiwg-berlin.mpg.de/library/data/lit28715/index_html?pn=7&ws=1.5.

Sheehan, H. (1993) *Marxism and the Philosophy of Science*. Humanity Books, Amherst, N. Y.

Shelley, M. (1818) *Frankenstein; or the Modern Prometheus*. Lackington, Hughes, Harding, Mavor and Jones, London. The first edition was published anonymously, only in later editions did Mary Shelley claim authorship.

Sluckin, T.J., Dunmur, D.A., and Stegemeyer, H. (2004) *Crystals that flow*. Taylor and Francis, London & New York.

Snelders, H.A.M. (1997) Zijn vloeibare kristallen levende organismen? (Are liquid crystals living organisms?) *Gewina* **20**, 129–142.

Snell, A.J., Mackenzie, K.D., Spear, W.E., LeComber, P.G., and Hughes, A.J. (1981) Application of amorphous-silicon field-effect transistors in addressable liquid-crystal display panels. *Applied Physics* **24**, 357–362.

Snow, C.P. (1993) *The Two Cultures*. Cambridge University Press, Cambridge.

Snow, C.P. (1959) *The Affair (Strangers and Brothers, book 8*. Penguin Books, London (reprinted by House of Stratus, Kelly Bray, 2000).

Sonin, A.S. (1988) *Doroga dlinuyu v vyek: iz istorii nauki o zhidkikh kristallov*. (*A road reaching back a century: on the history of liquid crystal science*.) Nauka Publishing House, Moscow (in Russian).

Sonin, A.S. and Frenkel, V.Ya. (1995) *Vsevolod Konstantinovich Freédericksz*. Nauka Publishing House, Moscow (in Russian).

Srajer, G., Pindak, R., Waugh, M.A., Goodby, J.W., and Patel, J.S. (1990) Structural measurements on the liquid-crystal analog of the Abrikosov phase. *Physical Review Letters* **64**, 1545–1548.

Stauff, J. (1939) Die Mizellarten wässeriger Seifenlösungen. (Micellar structure in dilute soap solutions.) *Kolloid-Zeitschrift* **89**, 224–233 (translated in Sluckin, Dunmur, and Stegemeyer (2004)).

Steele, R. and McConnell, A. (2004) Kerr, John, in *Oxford Dictionary of National Biography*. Oxford University Press, Oxford (retrieved from http://www. oxforddnb.com/view/article/343000).

Stegemeyer, H. and Kelker, H. (1991) Liquid Crystals already described in 1838! *Liquid Crystals Today* **1**(2), 3.

Szivessy, G. (1925) Zur Bornschen Dipolstheorie der anisotropen Flüssigkeiten. (On the Born dipole theory of anisotropic fluids.) *Zeitschrift für Physik* **34**, 474–484.

Tam-Chang, S.-W. and Huang, L. (2008) Chromonic liquid crystals: properties and applications as functional materials. *Chemical Communications* 1957–1967.

Tammann, G. (1897) Ueber die Grenzen des festen Zustandes. (On the boundaries of solid phases.) *Wiedemann's Annalen der Physik und Chemie* **62**, 280–299.

Tammann, G. (1901) Ueber die sogennanten flüssigen Krystalle. (On the so-called liquid cystal phases.) *Annalen der Physik* **4**, 524–530.

Tammann, G. (1902) Ueber die sogennanten flüssigen Krystalle II. (On the so-called liquid cystal phases II.) *Annalen der Physik* **8**, 103–108.

Tammann, G. (1906) Über die Natur der flüssigen Kristalle III (On the nature of liquid crystals III.) *Annalen der Physik* **19**, 421–425.

Tanford, C. (2004) *Ben Franklin Stilled the Waves*. Oxford University Press, New York.

Trebin, H.R. (1998) Defects in liquid crystals and cosmology. *Liquid Crystals* **24**, 127–130.

Tzvetkov, V.N. (1942) Über die Molekülanordnung in der anisotrop-flüssigen Phase. (On molecular order in the anisotropic fluid phase.) *Acta Physicochimica (URSS)* **16**, 132–147 (translated in Sluckin, Dunmur, and Stegemeyer (2004)).

Unna, I. (2000) The Genesis of Physics at the Hebrew University of Jerusalem. *Physics in Perspective* **2**, 336–380.

Viola, C.M. (1901) La legge degli indici razionali semplici e i cristalli liquidi. (The laws of simple rational indices and liquid crystals.) in *Processi verbali di Società Toscana di Scienze naturali* (read at meeting of 17 March 1901), pp.178–193.

von Helmholtz, H. (1847) *Über die Erhaltung der Kraft (On the conservation of force.)* G. Reimer, Berlin. Available on the web from Google books.

Vorländer, D. (1906) Über krystallinisch-flüssige Substanzen. (On crystalline-fluid substances.) *Berichte der Deutschen Chemischen Gesellschaft* **39**, 803–810 (translated in Sluckin, Dunmur, and Stegemeyer (2004)).

Vorländer, D. (1907a) Einfluß der molekularen Gestalt auf den krystallinisch-flüssigen Zustand. (Influence of the molecular structure on the crystalline-liquid state.) *Berichte der Deutschen Chemischen Gesellschaft* **40**, 1970–1972 (translated in Sluckin, Dunmur, and Stegemeyer (2004)).

Vorländer, D. (1907b) Neue Erscheinungen beim Schmelzen und Kristallisieren. (New observations of melting and crystallisation.) *Zeitschrift für Physikalische Chemie* **57**, 357–364.

Vorländer, D. (1908) *Kristallinisch-flüssige Substanzen*. Ferdinand Enke, Stuttgart.

Vorländer, D. (1923) Die Erforschung der molecularen Gestalt mit Hilfe der krystallinischen Flüssigkeiten. (The discovery of molecular structure with the help of crystalline liquids.) *Zeitschrift für Physikalische Chemie* **105**, 211.

Vorländer, D. (1931) Über die krystallinen Flüssigkeiten. (On crystalline liquids.) *Zeitschrift für Physikalische Chemie* **79**, 269–289.

Vorländer, D., Huth, M.E., and Wilke, R. (1910) Verhalten der Salze organischer Saüren beim Schmelzen. (The behaviour of organic acid salt melts.) *Berichte der Deutschen Chemischen Gesellschaft* **43**, 3120.

Vorländer, D., Däumer, E., Selke, W., and Zeh, W. (1927) Die einachsige Aufrichtung von festen and weichen Kristalmassen und von kristallinen Flüssigkeiten. (Uniaxial orientation of hard and soft crystal bodies and of liquid crystals.) *Zeitschrift für Physikalische Chemie* **129**, 435–474.

Vorländer, D., Wilke, R., Hempel, H., Haberland, U., and Fischer, J. (1937) Die kristallin-flüssigen und festen Formen des Anisal-p-amino-zimtsaureäthyl-esters $C_{19}H_{19}O_3N$. *Zeitschrift für Kristallographie* **97**, 485.

Wada, H., Wada, S., and Iijima, C. (1986) Liquid crystal display device. *Showa* **64**, 519.

Wallerant, F. (1905) Sur la constitution des corps cristallisés. *Comptes Rendus de l'Académie des Sciences* **141**, 768–770.

Wallerant, F. (1906a) Sur les cristaux liquids d'oléates d'ammonium. *Comptes Rendus de l'Académie des Sciences* **143**, 555–557.

Wallerant, F. (1906b) Sur l'origine des enroulements hélicoïdaux dans les corps cristallisés. *Comptes Rendus de l'Académie des Sciences* **143**, 605–607.

Wallerant, F. (1906c) Sur les cristaux liquides de proprionate de cholésteryl. *Comptes Rendus de l'Académie des Sciences* **143**, 694–695.

Wallerant, F. (1907) Les liquides crystallisés. (Crystallised liquids.) *Rivista di Scienza (Bologna)* **1**, 221–236.

Wallerant, F. (1909) *Cristallographie, déformation des corps cristallisés, groupements, polymorphisme-isomorphisme.* Librairie polytechnique Ch. Béranger, Paris.

Waters, C.M. and Raynes, E.P. (1982) Liquid crystal devices. UK Patent 2,123,163.

Waters, C.M., Brimmell, V., and Raynes, E.P. (1983) Highly multiplexable dyed liquid crystal displays, in *Proceedings of the Third International Display Conference, Kobe, Japan,* p.396–399.

Watson, J.D. (1968) *The Double Helix: A Personal Account of the Discovery of the Structure of DNA.* Readers Union (reprinted by Touchstone Books, New York 2001).

Weiss, P. (1907) L'hypothèse du champ moléculaire et la propriété ferromagnétique. *Journal de Physique et le Radium* **6**, 661–690.

Weiss, P. (1908) Molekulares Feld und Ferromagnetismus. *Physikalische Zeitschrift* **9**, 358–367.

Wells, P.V. (1921) Sur l'épaisseur des lames stratifiées. *Annales de Physique* (Paris) **16**, 69.

Weygand, C. (1931) *Quantitative analytische Mikromethoden der organischen Chemie in vergleichender Darstellung.* Akademische Verlagsgesellschaft, Leipzig.

Weygand, C. (1938) *Organisch-chemische Experimentierkunst.* J.A. Barth, Leipzig.

Weygand, C. (1939) Discussion remark on an article by W. Kast on anisotropic fluids. *Zeitschrift für Elektrochemie* **45**, 200–202.

Weygand, C. (1941) *Chemische Morphologie der Flüssigkeiten und Kristalle.* Akademische Verlagsgesellschaft, Leipzig.

Weygand, C. (1942) *Deutsche Chemie als Lehre vom Stoff.* Max Niemeyer Verlag, Halle.

Weygand, C. (1943a) Über formbeständige, isolierte kristallin-flüssige Binldungen. 4. Beitrag zur chemischen Morphologie der Flüssigkeiten. *Zeitschrift für Physikalische Chemie* B **53**, 75–84.

Weygand, C. (1943b) Daniel Vorländer 11.6.1867–8.6.1941. *Berichte der Deutschen Chemischen Gesellschaft* **76**, 41–46.

Williams, R. (1963) Domains in liquid crystals. *Journal of Chemical Physics* **39**, 384–388.

Williams, D.F. and Schadt, M. (1970) DC and pulsed electroluminescence in anthracene and doped anthracene crystals. *Journal of Chemical Physics* **53**, 3480–3487.

Williams, D.F. and Schadt, M. (1971) Electroluminscent device with light. US Patent 3,621,321 (filed October 1969, granted November 1971).

Wyart, J. (1962a) The new crystallography in France, in *Fifty Years of X-ray Diffraction*, Ewald, P.P. (ed.). Oosthoek, Utrecht, pp.446–455.

Wyart, J. (1962b) Personal reminiscences, in *Fifty Years of X-ray Diffraction*, Ewald, P.P. (ed.). Oosthoek, Utrecht, pp.685–690.

Yourgrau, W. (1951) Reflections on the natural philosophy of Goethe. *Philosophy* **26**, 69–84.

Zimmerman, D. (2006) The Society for the Protection of Science and Learning and the politicization of British Science in the 1930s. *Minerva* **44**, 25–45.

Zocher, Z.H. and Birstein, V. (1929) Beiträge zur Kenntnis der Mesophasen (Zwischenaggregatzustände) V: Über die Beeinflussung durch das elektrische und magnetische Feld. *Zeitschrift für Physikalische Chemie* **142**, 186–194.

Zsigmondy, R. and Bachmann, W. (1912) Über Gallerten. Ultramikroskopische Studien an Seifen-lösungen und -gallerten. (On gels: high resolution microscopic studies of soap solutions and soap-based gels.) *Kolloid-Zeitschrift* **11**, 145–157 (translated in Sluckin, Dunmur, and Stegemeyer (2004)).

Acknowledgements for figures and photographs

——⊃◦◦⊂——

We are grateful to the following for providing figures and photographs, and / or allowing their reproduction.

Figure	Source
Figure 1.1 Cholesteric drop	Courtesy of Professor H Stegemeyer
Figure 2.1 F Reinitzer	Courtesy of Deutsche Bunsen Gesellschaft (Kelker collection)
Figure 2.3 O Lehmann	Courtesy of Deutsche Bunsen Gesellschaft (Kelker collection)
Figure 2.7 G Tammann	Courtesy of the University of Tartu
Figure 3.1 D Vorländer	Courtesy of Deutsche Bunsen Gesellschaft (Kelker collection)
Figure 3.3 Vorländer's cigar boxes	Courtesy of Professor W Weissflog and Professor G Pelzl, University of Halle
Figure 3.4 E Bose	Courtesy of Deutsche Bunsen Gesellschaft (Kelker collection)
Figure 3.6(a) and (b) Lehmann's hammer mill	Courtesy of Professor Peter Knoll
Figure 4.1 G Friedel	Courtesy of Deutsche Bunsen Gesellschaft (Kelker collection)
Figure 4.2 F Grandjean	Courtesy of Acarologia©
Figure 4.4 Smectic fan texture	Courtesy of Dr. Ingo Dierking©, University of Manchester
Technical Box 4.3 Computer simulation images	Courtesy of Dr Martin Bates © University of York
STM Image of liquid crystal	Courtesy of Dr Jane Frommer, IBM Almaden Research Center
Figure 5.1 Sir William Bragg	Courtesy of State Library of South Australia. SLSA: B 3991.
Figure 5.2(a) Contents of Ewald symposium	Courtesy of Oldenbourg Wissenschaft

Figure	Source
Figure 5.2(b) Contents of London Symposia	Courtesy of Royal Society of Chemistry
Figure 6.1 R Virchow	Courtesy of Bildarchiv Preussicher Kulturbesitz BPK, Berlin
Figure 6.6 Tobacco Mosaic Virus	Courtesy of Professor Seth Fraden, Brandeis University
Figure 6.8 Cell membrane	Courtesy of Professor J W Goodby © University of York
Figure 7.3 V Frederiks	Courtesy of Professor A S Sonin
Figure 8.1 Institute of Chemistry, University of Halle	Courtesy of Professor W Weissflog and University of Halle
Figure 8.2 H Sackmann	Courtesy of Professor Stegemeyer and the Sackmann family
Figure 8.4 A Saupe	Courtesy of Peter Palffy Muhoray © Liquid Crystal Institute Kent State University
Figure 8.5 G H Brown	Courtesy of the Liquid Crystal Institute, Kent State University
Figure 8.8 F M Leslie	Courtesy of the Royal Society ©
Figure 10.1 W Helfrich	Courtesy of Professor W Helfrich
Figure 10.2 Watch cartoon	Courtesy of Professor H Stegemeyer and Dr M Schadt
Figure 10.3 M Schadt	Courtesy of Dr M Schadt
Figure 10.4 First TN display	Courtesy of Dr M Schadt
Figure 10.5 Hoechst advertisement for on the wall TV	Courtesy of Professor H Stegemeyer
Figure 10.6 G W Gray	Courtesy of Professor G W Gray
Figure 10.7 Merck catalogue 1907	Courtesy of E Merck, Darmstadt, Germany
Figure 10.8 V Titov	Courtesy of Professor Blinov and Mrs V Titov
Figure 11.2 Sharp flat screen	Courtesy of Sharp Europe
Figure 12.2 S Chandrasekhar	Courtesy of the Royal Society ©
Figure 12.9 R Meyer	Courtesy of Professor R B Meyer, Brandeis University
Tech box 12.3 Chromonic phases	Courtesy of Dr J E Lydon, University of Leeds

Index